Geography, Environment and Development in the Mediterranean

Edited by

Russell King, Paolo De Mas and
Jan Mansvelt Beck

sussex
ACADEMIC
PRESS

BRIGHTON • PORTLAND

The right of Russell King, Paolo De Mas and Jan Mansvelt Beck to be identified as editors of this work has been asserted in accordance with the Copyright, Designs and Patents Act 1988.

2 4 6 8 10 9 7 5 3 1

First published 2001 in Great Britain by
SUSSEX ACADEMIC PRESS
PO Box 2950
Brighton BN2 5SP

and in the United States of America by
SUSSEX ACADEMIC PRESS
5804 N.E. Hassalo St.
Portland, Oregon 97213-3644

British Library Cataloguing in Publication Data
A CIP catalogue record for this book is available from the British Library.

Library of Congress Cataloging-in-Publication Data

Geography, environment and development in the Mediterranean / edited by Russell King, Paolo De Mas and Jan Mansvelt Beck.

p. cm.
Based on papers originally presented at the 28th International Geographical Congress, held in The Hague, Aug. 5–10, 1996, most of the papers were presented at the Symposium on the Mediterranean Basin.
Includes bibliographical references and index.
ISBN 1–898723–89–3 (alk. paper) — ISBN 1–898723–90–7 (pbk.)
1. Mediterranean Region—Geography—Congresses. I. King, Russell, 1945–
II. De Mas, Paolo. III. Mansvelt-Beck, J. IV. International Geographical Congress
(28th: Hague, Netherlands: 1996)

D973/G42 2000
910'.91822—dc21

00 041300

Printed by Bookcraft, Midsomer Norton, Bath
This book is printed on acid-free paper

GEOGRAPHY, ENVIRONMENT AND DEVELOPMENT
IN THE MEDITERRANEAN

Contents

List of Tables

List of Figures

Preface and Acknowledgements

This book is based on papers originally presented at the 28th International Geographical Congress, held in The Hague during 5–10 August 1996. Within the ambit of this major gathering, most of the papers published here were presented at the Symposium on the Mediterranean Basin which was convened by the three editors, although a few of the contributions were presented to other sessions at the Congress. In the four years since the meeting, the authors have revised and updated their papers in conformance with the aim of this volume to be an integrated text on the geography, environment and development of an important world region. Given the European origin of the book's contributors and the availability of existing literature, the volume tends to convey a predominantly European perspective on the Mediterranean, although several chapters explicitly embrace other parts of the Basin such as North Africa and the eastern Mediterranean states.

We extend our thanks to several people and institutions. First, we thank the authors of the chapters for their patience and for their readiness to make the changes to their manuscripts requested by the editors. Second, we gratefully acknowledge the financial help of the Royal Dutch Geographical Society and of the Amsterdam Research Institute for Global Issues and Development Studies. Third, we thank Susan Rowland of the Cartographic Unit at the University of Sussex for her splendid work on the maps and diagrams which enrich the text. Our greatest practical debt, however, is to Jenny Money of the School of European Studies, University of Sussex, who skilfully managed the logistical process of bringing the diverse contributions together in readiness for the publisher and who dealt with a myriad of editorial details with meticulous care.

RUSSELL KING, PAOLO DE MAS, JAN MANSVELT BECK

The Contributors

Ewan Anderson is Professor of Geopolitics in the Centre for Middle Eastern and Islamic Studies, University of Durham.

Berardo Cori is Professor of Geography at the University of Pisa.

Paolo De Mas is Assistant Dean in the Faculty of Social and Behavioural Sciences, University of Amsterdam.

Michael Dunford is Professor of Economic Geography in the School of European Studies, University of Sussex.

Maria Lucinda Fonseca is Associate Professor of Geography and Pro-Rector of the University of Lisbon.

Don Funnell is Lecturer in Geography in the School of African and Asian Studies, University of Sussex.

Russell King is Professor of Geography and Dean of the School of European Studies at the University of Sussex.

Emile Kolodny is a retired geographer, former Directeur de Recherche at the Centre National de la Recherche Scientifique (CNRS), Aix-en-Provence.

Lila Leontidou is Professor of Geography at the University of the Aegean.

Elio Manzi is Professor of Geography at the University of Pavia.

Jan Mansvelt Beck is Senior Lecturer in Geography at the University of Amsterdam.

Armando Montanari is Associate Professor of Geography at the Università 'Gabriele D'Annunzio', Chieti, Italy.

Romola Parish is a former Lecturer in Geography at the University of St. Andrews; currently she is doing postgraduate work in Environmental Management at the University of Aberdeen.

Vicente Rodríguez Rodríguez is Senior Research Fellow in Geography at the Consejo Superior de Investigaciones Científicas in Madrid.

Pere Salvà Tomàs is Professor of Geography at the University of the Balearic Islands, Palma de Mallorca.

John Thornes is Professor of Geography at King's College London.

Adalberto Vallega is Professor of Geography at the University of Genoa and a Vice-President of the International Geographical Union.

W. Jan van den Bremen is a former Professor of Economic Geography, Faculty of Spatial Sciences, Groningen University; he is now retired.

Allan M. Williams is Professor of Geography and European Studies at the University of Exeter.

—————————— *1* ——————————

Unity, diversity and the challenge of sustainable development: an introduction to the Mediterranean

Russell King, Berardo Cori and Adalberto Vallega

What do we mean by 'the Mediterranean'? Is it valid to focus on the countries and regions of the Mediterranean Basin as a single, coherent geographical unit? If so, what are the essential characteristics of this unity? Are these unifying features stronger than the forces which diversify and fragment the Mediterranean? What are the major challenges facing the development of this complex region where three continents meet? These are the key questions posed – and, to an extent, answered – in this introductory chapter, which also functions as a series of signposts to the more detailed analyses of individual themes and topics in the chapters which follow.

Definition of the Mediterranean

It should be stressed at the outset that there is no commonly-agreed boundary which defines the Mediterranean region. The Mediterranean has been used as a flexible concept whose spatial extent varies according to the perspective being used – environmental, cultural, economic, geopolitical – as well as to the views of individual authors, bodies or commentators (Cori and Vallega 1996).

A first and common definition, which tends to be employed by the standard (but now dated) geography textbooks on the region, is the countries bordering the Mediterranean (Branigan and Jarrett 1975; Robinson 1970; Walker 1965): see figure 1.1a. This 'riparian state' definition is convenient, but simplistic and problematic. It recognises that statistics on the region are compiled largely from national datasets and it indicates an appreciation of the fact that nation states are the principal decision-takers in matters of politics, economics and environmental management, despite recent trends to globalisation and an awareness of shared responsibility in environmental issues. But the interior boundaries of these countries

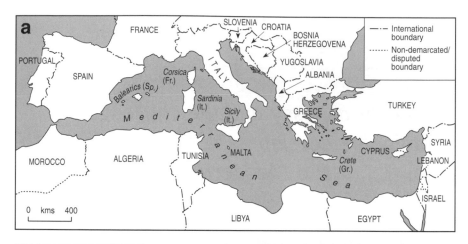

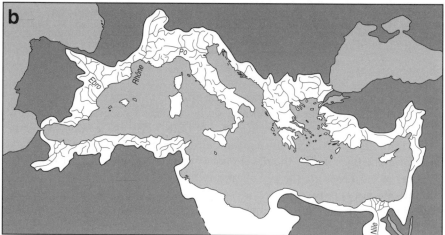

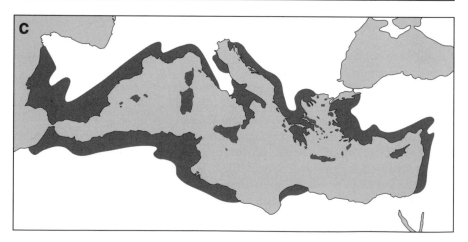

Figure 1.1   Delimiting the Mediterranean: (a) Mediterranean countries, (b) the Mediterranean watershed, (c) limit of cultivation of the olive

introduce problems: France extends to northern Europe, Algeria to the Sahara, Egypt to the Upper Nile Valley, and so on. Even countries which are unequivocally Mediterranean such as Spain and Italy extend to regions which are in many respects un-Mediterranean: Spain to the humid Atlantic coasts of Galicia, Italy to the snow-clad Alps. The break-up of Yugoslavia has produced successor-states with variable and tenuous links to the Mediterranean: only Croatia has a major portion of Mediterranean coastline. And then there is Portugal, which has no Mediterranean coastline but many environmental and cultural features which would tend to define it as a Mediterranean country.

Second, there is a less commonly used wider definition which extends the Mediterranean to the Black Sea and to the Middle East as far as the Gulf.[1] The former status of the Black Sea as a kind of lake of the Soviet Union and its client states has tended to blank out this area from western consciousness; nevertheless there are those who would argue for Bulgaria, Romania, Georgia etc. to be included within the broad definition of the Mediterranean (Montanari and Cortese 1993). An eastward extension of the Mediterranean to the Gulf recognises the ecological, cultural and economic similarity of countries such as Syria, Lebanon, Jordan and Iraq, as well as specific links through geopolitics and oil. In fact geopolitics – the Gulf War, the Israel/Arab conflict, the Balkans – provides many examples of the difficulty of drawing a 'containing' boundary around the Mediterranean and stresses how any such notional boundary must be 'open' to powerful links just beyond the region (Gillespie 1994, p. 2). These issues are further developed in **Chapter 2**.

Third, there are more spatially restrictive definitions. These usually hinge on notions of Mediterranean geography, ecology, landscape and culture. Several variants may be mentioned. The Blue Plan' definition invokes the concept of the 'Mediterranean Basin' defined by the Mediterranean Sea (excluding the Black Sea) and the adjacent coastal zone, where socio-economic activities are largely governed by their relations with the seaboard (Grenon and Batisse 1989, pp. 15–16). This 'proximity to the sea' criterion is itself a nebulous concept, and the Blue Plan authors in fact go on to remark that 'these coastal zones might vary in territorial depth from one area to another, depending on the problems to be considered and the nature of the disciplines involved' (Grenon and Batisse 1989, p. 16). This is a sensible argument for a flexible, pragmatic approach to defining the region. But when it comes down to the operational need to draw lines on maps, Grenon and Batisse opt for two 'solutions', neither totally satisfactory. For all matters to do with water resources and requirements the boundary is the Mediterranean watershed (figure 1.1b). This produces two major anomalies, the Rhône and the Nile, and at least one minor one (the Po), which originate from non-Mediterranean environments. The second Blue Plan boundary is statistical, made up of all the sub-national administrative units which border the sea – regions, provinces, departments etc. – for which statistical data on social, economic and demographic trends are available. This forms a variable coastal strip of around 50–100 km, plus of course the islands. Whilst this is clearly a solution dictated by practical considerations, Grenon and Batisse maintain (1989, p. 18) that 'it does not diverge too much from a frequently used biogeographical definition, that of the area in which the olive-tree grows'.

This leads naturally into definitions of the Mediterranean based on ecological and physical-geographic criteria. Ecologists recognise the Mediterranean as a distinct region made up, in particular, of certain diagnostic species and associations of species of plants, crops and wildlife. Geographers would be the first to stress the common climatic regime which is so characteristic that it has been taken as a type in meteorological studies. Climate in turn is largely responsible for the distinctive associations of crops and plants which typify the Mediterranean landscape. Pride of place is usually given to the olive, not only because of its widespread diffusion throughout the zones possessing a Mediterranean climate (the olive being a kind of expression of this climate because the summer drought is necessary for the build-up of oil in the fruit), but also because of its cultural and landscape symbolism, especially its role in Mediterranean diets and mythologies. Hence the limit of the olive is a common definition of the Mediterranean realm (figure 1.1c). But other Mediterranean plant indicators are also worth noting, if only because they often reinforce the broad pattern portrayed by the olive. Here we may mention the holm and kermes oaks, classic plants of the 'natural' Mediterranean woodlands, and the Aleppo pine. And of cultivars there are the fig, the pistachio, the carob and others. Standing at the interface between physical geography and culture, these species embody a more synthetic definition of 'Mediterraneanism' with links to climate, diet, history, landscape, farming systems and ways of life (King 1997).

The unity of the Mediterranean

Writings on the Mediterranean reveal a tension between centripetal and centrifugal forces: unity versus diversity. Even the unity is expressed in different ways: by recourse to history, geography, ecology, functionality or homogeneity. In drawing the contours of this debate we will examine first the factors and interpretations of the Mediterranean as a *unity*. In the next section we will look at the Mediterranean as an expression of diversity and fragmentation, and at its role as a 'frontier'.

Several *geographical* and *ecological* elements have been mentioned already. In synthesis these are climate, vegetation and landscape. At the risk of sounding over-deterministic, climate dictates the vegetation, controls the land-use regimes of farming, influences the seasonal and daily patterns of life, and lends colour and light to the landscape. McNeill (1992) stresses that the Mediterranean is also a *mountain* landscape, a theme analysed in depth in Chapter 12. With the exception of Libya and Egypt, one is almost never out of sight of mountains. Geologically the Mediterranean mountains exhibit recurrent themes. Almost every range runs parallel to the coast, pushed up by the collision of the African and European tectonic plates. The dominance of young sedimentary rocks and the presence of unstable fold mountain systems explain the frequency of earthquakes in the region. The juxtaposition of steep mountain chains with narrow coastal plains and inter-montane basins fashions the amalgam of land-use patterns and farming systems: irrigated vegetable gardens and orchards; dry-farmed wheat on the low hills; terraced vineyards and olive-groves; goats on the scrubland; transhumance of sheep between the winter pastures on the lowlands and summer grazing on the mountains (Houston 1964).

The geographical and ecological integrity of the Mediterranean Basin leads to a second important unifying element: the Mediterranean as an area of *shared environmental responsibility*. Much has been written of the Mediterranean as an arena for environmental conflict and fragility.[3] Particularly since the 1960s, the exploitation and transport of oil, rapid population growth around the coastal strip and the development of mass tourism have given the Mediterranean economic dynamism but precipitated an environmental crisis: coastal waters are polluted, water resources are depleted and once-fine landscapes spoiled by chaotic and ugly urbanisation. These issues reappear in several chapters in this book. Issues of environmental management and sustainable development are picked up towards the end of this chapter; sustainable development is examined in the context of population growth in Chapter 6; the challenges posed by tourism for planning and resource management are touched on in Chapters 9 and 10; Chapter 11 looks at the diverse landscape expressions of some of these pressures; and Chapter 14 addresses the notion of the Mediterranean 'environmental crisis'.

The third element of Mediterranean unity is the *historical* dimension. There is no doubt that the Mediterranean had greater unity in the past than it does at present. From antiquity to the late Middle Ages and the Renaissance there was a strong pattern of intra-Mediterranean relations which emphasised the unity of the Basin (Amin 1989, p. 2). Some of these historical patterns and features have survived to become part of the region's present character. The greatest unity was achieved in ancient times, when the Mediterranean exhibited considerable political, economic and cultural coherence and homogeneity. Under the Romans, for whom the Mediterranean was simply '*Mare Nostrum*', the whole of the Basin (and beyond, to north-west Europe and the Danubian Plains) was imperial territory. Centralised political control, massive urban development, rural transformation and a transcontinental transport system were some of the centripetal forces binding all parts of the region to the imperial capital located near the centre of the Mediterranean. In particular, the long tradition of urban life fostered by the Romans is one of the enduring features of Mediterranean geography: the city has long been, and continues to be, a 'magnet' for Mediterranean peoples (Chapter 5). The picture of the Mediterranean as a coherent cultural and economic unit is the most prevalent motif in Fernand Braudel's famous treatise (Braudel 1972, 1973). For him the essence of the Mediterranean was the product of intellectual and commercial intercourse (Braudel 1972, p. 14) – a clearly functional definition.

Fourthly, we have the notion of a distinct *Mediterranean economy* (Tovias 1994). Part of this is built around geographical and historical features already noted: the strength of towns as centres of trade, services and culture (less so of industry); and rural landscapes dominated by olives, vines, and grazing sheep and goats. Agriculture retains a key role in the economy of all Mediterranean countries (with the exception of micro-states like Malta and Gibraltar), despite a steady exodus from agricultural employment in recent decades (Popovic 1992). To give a few typical examples, agriculture's share of total employment in 1990 was 40 per cent in Egypt, 24 per cent in Greece and 47 per cent in Turkey. Italy (9 per cent), Spain

5

(10 per cent) and Israel (4 per cent) had lower figures but even in these countries high percentages of the total population live in rural areas, from which they pursue non-farming activities. Hence, despite the long tradition of urbanism, a rural way of life is maintained more or less throughout the region. Agricultural production has seen a reorientation from peasant farming to specialised products geared towards North European (and, for some, Middle Eastern) markets (Pratt and Funnell 1997). Other important features of Mediterranean economies are the lack of a nineteenth- or early twentieth-century industrial revolution (though modern small-scale industry may be widespread); reliance on the service sector and especially mass tourism; the strength of the informal economy (which is reckoned to account for about one-third of Greek GDP, for instance); and the relevance for many countries of revenues remitted by their emigrants living and working abroad. Divisions obviously exist between the more prosperous EU economies and most of the countries outside the EU, and between oil and non-oil producers. Economic patterns and contrasts are analysed in depth in **Chapter 3**, whilst in **Chapter 4** van den Bremen argues that trading patterns are a powerful integrating force for the Mediterranean Basin countries.

Finally, there is the notion that the Mediterranean is a relatively *homogeneous cultural region* in which Mediterranean society (or societies) can be meaningfully identified as a distinct type. There has been a lively debate in social anthropology about this question (Goddard 1994). Few would dispute the fact that between the 1950s and the 1970s the Mediterranean, and especially the remoter parts of the European Mediterranean like the Andalusian mountains and the far south of Italy, became a classic field location for studies of 'traditional' rural communities (some of the pioneering studies were Banfield 1958; Davis 1973; Pitt-Rivers 1954). Many of these investigations focused on the unique rather than the general: subsequently Davis (1977) made a powerful plea to 'Mediterraneanist' anthropologists to sacrifice some of the myopic detail of their particular neck of the Mediterranean backwoods and try to say something meaningful about Mediterranean cultures, family structures, rural economies, politics, migration processes etc. There *are* important generalisations to be made – about the role of land tenure, for instance; or about honour and shame, gender relations and the concept of *machismo*; or about the curious fact that belief in the 'evil eye' exists in so many different parts of the Mediterranean. Nowadays, however, the idea of the 'Mediterranean anthropological region' is receding, partly because anthropology itself is changing and partly because of the realisation that the construct was basically a European one – the key features that were posited as 'Mediterranean' were only typical of those areas of southern Europe where field research has been carried out, namely the southern parts of Spain and Italy plus Greece, Cyprus and Malta (Goddard *et al.* 1994).

Nevertheless, despite the self-doubts of anthropologists, it is possible to suggest that, in the Mediterranean, geography, history, economics, culture and much else besides, share many common patterns and rhythms which are unique to the region and distinct from the rest of Europe, Africa or Asia.

Mediterranean diversity and fragmentation

Against these unifying features can be set other elements which work to fragment the Mediterranean, particularly in the modern era. First, a fairly obvious point. Because the Mediterranean countries are split amongst three continents, they are more commonly viewed as parts of these continents than as a specifically Mediterranean group. Geographers have tended to write about them as parts of their textbooks on Europe, or Africa, or Asia or the Middle East. Similarly the United Nations, the World Bank and other international bodies publish national statistics in continental groupings – virtually no organisation treats the Mediterranean as a region for statistical purposes (Tovias 1994).

In historical and cultural terms, too, the Mediterranean is often portrayed as a series of distinct if overlapping units or spheres of influence: southern Europe, the Maghreb, the Mashreq, the Middle East, the Arab states, the Ottoman Empire etc. Religious and ethnic heterogeneity is a current theme of the region. There are four major religious spheres – the Catholic and Orthodox churches, Judaism, Islam – plus a political fragmentation into more than 20 states and a multitude of languages and dialects. Many states have important religious and linguistic minorities. The differential experience and impact of colonialism is a further dimension of historical, cultural and economic complexity. Spain's colonial empire was largely outside the Mediterranean, except in the later Middle Ages. The French colonial stamp was particularly intense on the Maghreb states, especially Algeria, but it was also felt in the eastern Mediterranean (Egypt, Lebanon, Syria). Italy planted colonies in Libya in the interwar years. Britain's colonies were Gibraltar, Malta and Cyprus, strategic points for a complete control of the 'sea-space' of the Mediterranean.

In 1981 Nurit Kliot divided the region's then 19 states into four groups: countries belonging to the European 'core' (France, Italy), countries of the European 'periphery' (Greece, Turkey etc.), socialist states (Albania, former Yugoslavia etc.) and the Arab states. Twenty years later the number of Mediterranean littoral states has increased to 21 with the break-up of Yugoslavia, and the geopolitical situation has become, if anything, more complex as Europe's old East/West ideological divide of the Iron Curtain has been replaced by a multifaceted North/South division running across the width of the Mediterranean. This Mediterranean fault-line or 'Rio Grande' has become a powerful construct for the north/south dichotomisation of the Mediterranean because it represents a boundary between what are perceived (especially in Europe) as two different 'worlds' – the First World and the Third World or, more insidiously, 'us' and 'them' (King 1998). However much one may condemn the contemporary European geopolitical discourse which presents all non-Europeans as 'others', the complex realities of intra-Mediterranean divisions and contrasts do need to be appreciated. The basic parameters of this trans-Mediterranean divide are geopolitical, economic and demographic, and these themes are the respective subjects of **Chapters 2, 3 and 6**.

As Ewan Anderson demonstrates in **Chapter 2**, the geopolitical dimensions of the Mediterranean are extremely complex and generally work to divide or fragment the Basin. With its various examples of parliamentary democracy, military dictatorship and quasi-feudal monarchy, the Mediterranean is a veritable laboratory for the political scientist. More important have been the repercussions of global political change on the region. Before 1990 many countries within the region were exposed to East/West superpower competition: both NATO and the USSR (and the British) had military bases in the Mediterranean and the overall level of tension was high. The Mediterranean was 'paracolonial' territory, coveted not just for itself but as a facet of imperial global strategy (Anderson and Fenech 1994, p. 11). The function of the Mediterranean Sea as a corridor for the transport of much of Europe's and the world's oil supply was a further factor which heightened the geopolitical stakes. In the last decade the disintegration of state socialism, the weakening of Russia as a world power and the emergence of the USA as a single superpower, have had diverse impacts on the Mediterranean. On the one hand, the level of East–West tension in the Mediterranean has sharply diminished; on the other hand, this has led to the exacerbation of smaller-scale regional conflicts which had somehow been stabilised by the regime of superpower competition (Kliot 1997). The obvious example was the long-running conflict in ex-Yugoslavia which, in the more open geopolitical environment of the 1990s, took so long to settle (if it has been settled), and in fact was only concluded by the legitimisation of ethnic and nationalistic hegemonies. Other potential causes and symptoms of conflicts can easily be listed: boundary disputes, water, arms proliferation, religious fundamentalism, demographic trends, terrorism (Gillespie 1994, p. 3). The Mediterranean has also felt the effects of other major conflicts located around its margins such as the Gulf War and the Arab/Israeli saga, whilst some aspects of 'North–South' conflict are being played out *within* Mediterranean states, such as Cyprus, Lebanon and the Israeli-occupied territories.

As Dunford and King show in **Chapter 3**, relative levels of economic well-being within the Mediterranean have changed and in some respects the trans-Mediterranean contrast has become sharper. Whereas before the 1950s virtually all Mediterranean countries could be regarded as semi-developed or as a periphery or semi-periphery of Europe, now the gap has significantly widened between the EU Mediterranean states on the one hand and the countries of the southern and eastern seaboards on the other. Italy and, to a lesser extent, Spain, Portugal and Greece have moved close to the economic levels of northern Europe, whilst most of the North African states, in particular, have languished.

No less marked are the demographic contrasts between the northern and southern Mediterranean countries (**Chapter 6**). As with the changing economic situation, the key has been the rapid switch in the profile of the southern European countries: from a medium-high fertility regime which operated up to the 1950s (when fertility was already low in northern Europe) to an unprecedently-low fertility regime today. By the early 1990s the fertility rates of Spain and Italy were the lowest ever recorded in the world. Total fertility rates (TFR) of 1.2–1.4 children per woman (Italy, Spain, Portugal, Greece) are now lower than those of northern Europe (for example Germany 1.5, France 1.8, Sweden 2.1). More to the point, TFR values around the southern Mediterranean flank are 3–4 times those of southern Europe and, at least for some decades to come, future population growth is assured by the young age profile of southern Mediterranean populations compared to the rapidly ageing population structures of the northern shore countries. Thus whereas Italy

(17 per cent), Spain (20 per cent), Greece (20 per cent) and Portugal (21 per cent) have small proportions of their population under 15 years of age, the percentages are at least twice as high in Morocco (42 per cent), Algeria (46 per cent), Tunisia (39 per cent), Egypt (40 per cent) and Syria (49 per cent). Using UN medium-scenario projections, Montanari and Cortese (1993) calculate an increase in working-age population in the Asian and African Mediterranean countries (Turkey round to Morocco) of 133 million between 1985 and 2025. This is a quantity of new workers which local, national labour markets cannot possibly absorb: hence the growing trans-Mediterranean migration flows of the last two decades (King 1996). As Montanari shows in Chapter 6, migration is now one of the major elements in Mediterranean international relations.

Subdivisions and conceptualisations of the Mediterranean

Unity versus diversity is undoubtedly a reductive dichotomisation of how the Mediterranean can be viewed. Two further elaborations can be made which imply a plurality of ways of thinking about the geography of the region: subdivisions of the Mediterranean, and different conceptualisations of the Basin.

The channel between Sicily and Tunisia divides the Mediterranean Sea into two basins, each with limited outlets   the Strait of Gibraltar to the west, the Dardanelles to the north-east and the Suez Canal to the south east. The Western Mediterranean is relatively self-contained and spatially coherent, with three large island regions – Corsica, Sardinia and the Balearics. The larger Eastern Mediterranean is much more spatially fragmented, both in terms of the shape of its coasts (especially the northern one which alternates large peninsulas with the northward-extending Adriatic and Aegean Seas), and in terms of the multitude and variety of its islands (Chapter 13 explores the geography of the Mediterranean islands). This twofold division also sets the scene for a pragmatic division of the Mediterranean into land-based quadrants (Joannon and Tirone 1990, p. 70):

- to the north-west, a group of advanced European countries, members of the EU, with high levels of development (but also some less developed regions such as Andalusia and the Italian Mezzogiorno) and very low levels of demographic fertility;
- to the south-west, the Maghreb states, with low levels of development and economic growth, high demographic pressure, and strong economic and migratory links to the countries of the north-western quadrant;
- to the south-east, a heterogeneous group of countries, most with low levels of development, dominated by their relationships to the oil-producing countries of the Gulf and by their involvement in the Arab-Israeli conflict;
- to the north-east, a group of countries dominated by Turkey and the successor-states of Yugoslavia, which generally aspire to the European model of development and eventual EU membership.

Other divisions, many of them dualistic, can also be readily identified: the sharp juxtaposition between mountain and plain which is such a recurrent feature of Mediterranean landscapes; the contrast between the coast and the interior; or the social divisions between urban and rural life; or between rural societies based on settled agriculture and those devoted to pastoralism, which may be nomadic or transhumant. The Mediterranean is constantly changing and offering new models of settlement, economy and spatial organisation. Amongst these is the 'tourist city' – the linear conurbation which is the product of the commodification of the holiday and leisure industries; or the more flexible and balanced uses of urban/rural space associated with mixed industrial/service/rural development in the so-called Third Italy (and elsewhere); or the strength and dynamism of the informal economy. None of these examples of the 'originality' of the Mediterranean is unproblematic, either conceptually or in terms of their use and exploitation of people and space. Above all, they focus our attention on the future development and environmental sustainability of the region.

On a more macro scale, there are several conceptualisations of the Mediterranean which attempt to distil the unique character of the region either in terms of its internal essence or in terms of the special position of the region within the global system (or that part of the global system that derives from the Old World). Some of these conceptualisations overlap; others are antithetical. Many hinge around the paradox of the Mediterranean being a 'mid-land' sea that separates, but at the same time connects, Europe, Africa and Asia: the ambiguous role of the Mediterranean as both a barrier and a bridge. Without doubt the spatial integrity of the Mediterranean, with its ecological unity, largely defined by climate, and its historical role as a melting-pot not only of its rimland populations but also as a rendezvous for people beyond the mountain ranges and deserts of the Old World, is compromised by the fact that, uniquely of the world's subcontinental macro-regions, it draws three continents to its shores, as well as juxtaposing, under the Old World Order, the First, Second and Third Worlds.

Reflecting this historical and geographical complexity, Brunet (1995) posits seven models of the Mediterranean:

- as a closed basin with few external relations – the schott model;[4]
- as a maritime space between two continents – the straits model;
- as a space of maritime connections – the lake model;
- as a bridge between continents – the isthmus model;
- as a hostile barrier – the frontier model;
- as a region organised around poles of development – the focal model;
- as a zone structured according to relationships between regions of different levels of development – a model based on the tension between growth and uneven development.

Space does not allow a full evaluation of these Mediterranean models, but each has some validity and some correspondence to reality, either now or in the past, as subsequent chapters will show. Some aspects of the frontier model are brought out in Chapter 2 on Mediterranean geopolitics. The last two on the list of seven are explored in more detail in **Chapter 3** on the economic geography of the Mediterranean. The straits, lake and isthmus models are implicit in **Chapter 4** on intra-Mediterranean trading patterns. The function of the Mediterranean as a *connecting space*, especially in the past when transport by ship was easier and faster than by land, is brought out in this extract from the writer Jean Giono (quoted in Bort 1998, p. 27):

> Exchanges did not take place across, but with the help of this sea. If it were replaced by a continent, nothing from Greece would have passed on to Arabia, nothing from the East would have reached Provence, nothing from Rome would have journeyed to Tunis. But on this water, murder and love have been exchanged for thousands of years, and a uniquely Mediterranean order has emerged.

Another powerful interpretation of the Mediterranean, one which responds to the contemporary reality of global political and economic relations but which draws on an ancient concept, is that of a *limes*. This view has emerged strongly from the writings of French theorists such as Rufin (1991) and Foucher (1998) who see the Mediterranean as a newly emerging *limes* dividing Europe from its 'near South' of poor, underdeveloped 'barbarians'. Of course the re-use of the Roman term 'barbarian' is deeply insulting to the populations of the southern and eastern Mediterranean, many of whom are descendants of civilisations which are much older than that of Rome; but the term reflects one version of the new view of the Mediterranean as a border region which is 'closed to migrations and citizenship but open to trade, ideas and languages, and in an asymmetrical relationship which remains a permanent source of tension' (Foucher 1998, p. 236).

So the Mediterranean is a plural space of overlapping and hazy boundaries and complex economic, political, socio-demographic, cultural and environmental relations. In the next and final section of this chapter we focus on the last of these, examining how growing environmental problems reflect the pressures of economic and population growth concentrated especially along the coastal strip, and tracing the evolution of a common awareness of the need to put in place an environmental action plan to restore the purity of the sea, rescue the environmental quality of the coastal regions, and build an environmentally sustainable development model for the future.

The challenge of sustainable development

The ancient Mediterranean has traditionally been viewed as a resilient ecosystem, experiencing little change until the wholesale cutting of the forests and the development of intensive agriculture. Even irrigation was viewed as part of the 'hydraulic civilisation' where nature was tamed to bring about a kind of resourceful harmony. As Thornes shows in **Chapter 14**, in reality this argument is not substantiated. Already by the Bronze Age the land was badly ravaged; the forest has been cleared and has regrown many times over succeeding millennia; rampant floods have been a persistent occurrence down the centuries; and the unpredictable recurrence of severe drought challenges the image of the traditional peasant farmer in harmony with the environment. Moreover, reforestation and the great works of hydraulic engineering over the past century have done little to assuage the repeated failures of the past. Indeed, forest fires, siltation of artificial lakes and accelerated erosion of over-grazed, over-ploughed land show that many lessons in Mediterranean environmental management have still to be learned.

Over the past 25 years environmental policy in the Mediterranean has proceeded along a twin track of environmental protection (initially pollution control) and planning for sustainable development. The Mediterranean Basin has been recognised as a crucial area both to essay a regional interpretation of global change and to implement a path of resource-efficient development. The challenges are enormous, for the physiographic and ecological unity of the Basin is overlain by so many dimensions of complexity: a history of dense settlement and exploitation of the natural environment stretching back over several millennia; a political fragmentation into twenty or so states, making collective decision-making over shared environmental resources extremely difficult; and an extensive spectrum of natural, semi-natural and human hazards ranging from frequent seismic and volcanic activity to human-induced atmospheric warming and heavy soil erosion, as well as social processes such as northward migrations, ethno-religious conflict and intense settlement and pollution of coastal areas.

The human landscape of the Mediterranean still bears the weight of its troubled past. The timeless rural societies beloved of a generation of postwar anthropologists often reflected, or resulted from, a ruthless exploitation of a poverty-stricken population by a greedy landowning class tied in turn to rampant political power. The contemporary environmental history of the Mediterranean is one of broken promises, of resource greed and of ceaselessly exploitative and ecologically damaging agriculture and tourism, of nations clutching at the economic straws blowing in the European wind and struggling to compete in a fiercely aggressive world of change (Thornes 1995).

The trigger for action was marine pollution. Rapid development, without much concern for the consequences, took place during the 1960s and by the 1970s the Mediterranean had become one of the most contaminated seas in the world, its beaches plastered with oil, its waters poisoned by chemicals and many of its wildlife species on the verge of extinction (Pastor 1991). Industrial waste, untreated domestic sewage, agricultural chemicals, unmonitored oil spills and relentless exploitation of fish stocks were the main causes of the abuse of the Mediterranean. Danger signs began to appear: in 1973 there was an outbreak of cholera in Naples, traced to contaminated shellfish harvested in the badly polluted bay. An article in *Nature* predicted that there would be deaths from mercury poisoning amongst the fish-eating people of the Mediterranean by the year 1995 (Dorozynski 1975). Other scientific evidence indicated that the Mediterranean was dying (Osterberg and Keckes 1977).

The result was the Barcelona Convention for the Protection of the Mediterranean Sea against Pollution (1975), a declaration set up under the aegis of the United Nations Environmental Programme's 'regional seas' initiative, and eventually signed by all the shoreline states. Protocols banned the dumping of certain substances from ships and aircraft, created an oil pollution monitoring system, and began to tackle land-based pollution. The Barcelona Convention was the foundation for the Mediterranean Action Plan, a framework for bringing together the very different political and ideological regimes of the countries which border the Basin, including traditional enemies such as Greece and Turkey, and Israel and the Arab states. Although Haas (1990) has argued that the 'Med Plan' was mainly achieved by cooperation amongst the various countries' scientific elites and that real reductions in pollution have been modest, the unification of the Mediterranean Basin member-states around the environmental issue is heartening in its geographical symbolism.

Reviewing progress after the first ten years of the Mediterranean Action Plan in the light of the 1985 Genoa meeting to monitor the Barcelona Convention, King (1990) concluded that the results were mixed but that on the whole the earlier pessimism about the future of the Mediterranean had been overstated. In other words, 'the Mediterranean was sick but not dying' (King 1990, p. 11). Extremely serious pollution had been thoroughly documented in certain coastal areas near to big cities and industrial ports, and oil pollution was always to be regarded as a serious threat, with one third of the world's oil transported across the Mediterranean, which has a very slow rate of water turnover.[5] Inshore cultivation of shellfish, especially mussels, was regarded as unsafe. Discharge of chemical fertilisers down the Po Valley was held to be responsible for the glutinous masses of algae which invaded the bathing waters of Adriatic seaside resorts during the late 1980s. On the other hand, open sea pollution was found to be still at a low level and, moreover, significant improvements had been achieved as a result of more careful control over oil discharges (cleaning equipment at terminals and the banning of tank-washing at sea) and over some industrial pollution (Clark 1989, pp. 169–71).

As stated above, environmental management in the Mediterranean region has evolved along two teleologically linked tracks: environmental protection and sustainable development (Vallega 1995). The first derived from the application of the Barcelona Convention on Pollution and the first phase of the Mediterranean Action Plan. The second track is aimed at the economic and social goals of sustainable development and this has emerged strongly in the light of the adoption of Agenda 21 by the United Nations Conference on Environment and Development held in Rio de Janeiro in 1992. Further key meetings in Tunis in 1994 (Conference on Sustainable Development of the Mediterranean) and in Barcelona (1995) resulted in important new initiatives, including the second stage of the Mediterranean Action Plan (1996–2005), the establishment of the Mediterranean Commission on Sustainable Development, and the Agenda 21 for the Mediterranean (Vallega 1998). Meanwhile, somewhat independently of these initiatives, the Blue Plan had been carrying out research and building up scenarios for the future economic and social development of the Mediterranean (Grenon and Batisse 1989).

According to the Convention on Environment and Development agreed by the contracting partners in Barcelona in 1995, two major goals for sustainability are the protection of biodiversity and integrated coastal management. The latter of these recognises that the Mediterranean coastal strip represents the major geographical site for development pressures and environmental conflicts, due above all to the intense process of 'littoralisation' of the population and of economic activities (industry, intensive agriculture, tourism, transport) in recent decades (Leontidou *et al.* 1998). If the other major goals, deriving from 'Med Agenda 21', are included, the final result is a spectrum of objectives which, as can be seen in table 1.1, give a coherent shape to an efficient sustainable development-oriented approach to the Mediterranean's future.

Although eminently sensible, even visionary, these objectives, and the whole discourse of sustainable development in the Mediterranean, are castles in the air. A major problem arises from the fact that Med Agenda 21 is not a legal tool; therefore, automatic coordination between the Convention system and this long-term programme is not possible. In fact, Med Agenda 21 was not adopted by the Conference of the Plenipotentiaries on the Barcelona Convention so its approach has perforce come to be linked to a set of statements and *mises-au-point* identifying areas of possible collaboration between Mediterranean states (Vallega 1998, p. 28). It is also self-evident that most states agree to implement sustainable development only as a vague act of principle and with regard to a very limited range of objectives. As a result conflict could arise between states, anxious to protect their own national interests, and with the wider ambitions of the Mediterranean Action Plan. Resolution of these conflicts will depend on the extent to which the Commission on Sustainable Development will be oriented toward a rigorous and highly-principled pursuit of the goals and guidelines specified in the Convention and Med Agenda 21, or will be sensitive to pragmatic compromise. Development versus ecosystem will be another major concern, since most countries consider economic development in conventional economic growth terms, not in terms of long-term economic efficiency, resource conservation and the pursuit of sustainable development.

On a wider scale, other developments during the 1990s have served to frustrate the prospects for an agreed policy of sustainable development and have led analysts to draft more cautious scenarios (Vallega 1998, p. 18). As revealed in more detail in Chapter 2, the Mediterranean has become one of the most troubled geopolitical regions of the world. Conflicts associated with the break-up of Yugoslavia, especially the terrible wars in Bosnia and Kosovo; the continued Arab-Israeli conflict; the perceived threat of Islamic fundamentalism in North Africa; tensions between Greece and Turkey; Russian imperialism in the Black Sea; instability in

Table 1.1 Mediterranean cooperation and development

Convention on Environment and Development	Med Agenda 21
Integrated coastal management	Protection of the atmosphere from pollution, with special reference to urban and industrial sites
Protection of biodiversity	Forest management Desertification Water management Technology transfer

Albania – all are evidence of this. Now the Mediterranean is criss-crossed by major trafficking routes for drugs, illegal labour and other illicit activities.

Hence, it seems that the question as to whether the Mediterranean might be the first multinational region to put sustainable development into practice must remain unanswered.

Conclusion

This chapter has laid the foundations for the rest of the book by considering some basic questions regarding the definition and character of the Mediterranean as a geographical macro-region. Definitions are many, and their utility varies according to purpose – economic analysis, explaining geopolitical conflicts, generalising socio-cultural patterns, modelling environmental processes, or evaluating the possibilities and limitations for sustainable development. For the general purposes of this book as an academic text to be read by students interested in the region, our working definition will be twofold. For issues involving international affairs or national statistics (for instance on economic development or tourism) we adopt the definition of 'the countries bordering the Mediterranean Sea plus Portugal'. For issues dependent on physical, ecological or cultural criteria, we adopt a more spatially restricted view of 'Mediterraneanism' which encompasses those regions which lie close to the sea and are somehow influenced by it. In this case, fixing a line on a map is more difficult; the realm of the olive (figure 1.1c) provides one approximation.

The tension between the conceptualisation of the Mediterranean as a unified region as opposed to one that is fragmented and diversified is even more difficult to resolve, and the chapters which follow evince different interpretations, leading to the plural notion of several 'Mediterraneans'. On the whole, we feel that there are sufficient arguments to present the Mediterranean as a meaningful geographical unit, but we would also stress that this unit is an extremely flexible concept, having many forms and expressions and, at a smaller scale, many sub-divisions and particularities.

In one of the most detailed geographical texts on the Mediterranean, Houston (1964) defines the essence of this region as the intense interaction, over a long period of time, between physical, cultural and economic (especially rural, agrarian) elements. Whilst traditional analyses of the Mediterranean have tended to project this interaction as a kind of harmonious symbiosis, it is doubtful if this has really, or ever, been the case: the fault has been geographers' traditional reluctance to confront social and political processes. Now all this has changed. Population increase and littoral concentration, the growing threat of major pollution events, global warming and the competition for scarce and diminishing water resources all indicate that the Mediterranean will be one of the world's crucial testing-grounds for finding a path to development which delivers real improvements in living standards whilst respecting the rights of individuals and preserving the sustainability of resources.

Notes

1   This is the geographical frame of reference adopted by Birot and Dresch (1953, 1956) in their classic two-volume geography of the Mediterranean and the Middle East. However, we are critical of the hypothesis of extending the boundary of the Mediterranean to the Persian Gulf. Of course Lebanon and Syria border the Mediterranean, so their status as Mediterranean countries is clear. Their similarity with Jordan and Iraq does tempt one to extend the border further east, but in the same way Egypt extends to Sudan, or Italy to Switzerland, and these are not considered as Mediterranean countries.

2   The Blue Plan for the Mediterranean is a coordinated plan for environmental improvement in the Mediterranean operated under the aegis of the United Nations Environment Programme (UNEP); see Grenon and Batisse (1989).

3   Much of this work recently has derived from the MEDALUS (Mediterranean Desertification and Land Use) Project coordinated by John Thornes at the Department of Geography, King's College London. For important outputs from this research see Brandt and Thornes (1996), Mairota et al. (1998), and also Thornes' contribution to the present volume (Chapter 14).

4   Named after the shallow, brackish ephemeral lakes which occupy desert depressions in North Africa and the Middle East.

5   It takes approximately 80 years for the water of the Mediterranean Basin to replace itself.

References

Amin, S. (1989) Conditions for autonomy in the Mediterranean region, in Yachir, F. (ed.) *The Mediterranean between Autonomy and Dependency.* Tokyo: UN University Press, pp. 1–24.

Anderson, E. and Fenech, D. (1994) New dimensions in Mediterranean security, in Gillespie, R. (ed.) *Mediterranean Politics, Vol. 1.* London: Pinter, pp. 9–21.

Banfield, E. C. (1958) *The Moral Basis of a Backward Society.* Glencoe: The Free Press.

Birot, P. and Dresch, J. (1953, 1956) *La Méditerranée et le Moyen-Orient.* Paris: Presses Universitaires de France, 2 vols.

Bort, E. (1998) Gulf or bridge: the Mediterranean frontier of the European Union, in Anderson, M. and Bort, E. (eds) *Schengen and the Southern Frontier of the Mediterranean Union.* Edinburgh: International Social Sciences Institute, University of Edinburgh, pp. 25–38.

Brandt, C. J. and Thornes, J. B., eds (1996) *Mediterranean Desertification and Land Use.* Chichester: Wiley.

Branigan, J. J. and Jarrett, H. R. (1975) *The Mediterranean Lands.* London: MacDonald and Evans.

Braudel, F. (1972, 1973) *The Mediterranean and the Mediterranean World in the Age of Philip II.* London: Collins, 2 vols.

Brunet, R. (1995) Modèles des Méditerranées, *L'Espace Géographique,* 24(3), pp. 200–2.

Clark, R. B. (1989) *Marine Pollution.* Oxford: Clarendon Press.

Cori, B. and Vallega, A., eds (1996) Human dimensions of regional changes: the case of the Mediterranean, *Bollettino della Società Geografica Italiana,* Ser. XII, 1(1), pp. 1–121.

Davis, J. (1973) *Land and Family in Pisticci.* London: Athlone Press.

Davis, J. (1977) *People of the Mediterranean.* London: Routledge and Kegan Paul.

Dorozynski, A. (1975) Mediterranean poison fish forecast, *Nature,* 254, pp. 549–51.

Foucher, M. (1998) The geopolitics of European frontiers, in Anderson, M. and Bort, E. (eds) *The Frontiers of Europe.* London: Pinter, pp. 235–50.

Gillespie, R., ed. (1994) *Mediterranean Politics, Vol. 1.* London: Pinter.

Goddard, V. A. (1994) From the Mediterranean to Europe: honour, kinship and gender, in Goddard, V. A., Llobera, J. R. and Shore, C. (eds) *The Anthropology of Europe.* Oxford: Berg, pp. 57–92.

Goddard, V. A., Llobera, J. R. and Shore, C. (1994) Introduction: the anthropology of Europe, in Goddard, V. A., Llobera, J. R. and Shore, C. (eds) *The Anthropology of Europe.* Oxford: Berg, pp. 1–40.

Grenon, M. and Batisse, M. (1989) *Futures for the Mediterranean Basin: the Blue Plan.* Oxford: Oxford University Press.

Haas, P. M. (1990) *Saving the Mediterranean.* New York: Columbia University Press.

Houston, J. M. (1964) *The Western Mediterranean World: an Introduction to its Regional Landscapes.* London: Longman.

Joannon, M. and Tirone, L. (1990) La Méditerranée dans ses états, *Méditerranée*, 70(1–2), pp. 1–70.

King, R. (1990) The Mediterranean: an environment at risk, *Geographical Viewpoint*, 18, pp. 5–31.

King, R. (1996) Migration and development in the Mediterranean region, *Geography*, 81(1), pp. 3–14.

King, R. (1997) An essay on Mediterraneanism, in King, R., Proudfoot, L. and Smith, B. (eds) *The Mediterranean: Environment and Society.* London: Arnold, pp. 1–11.

King, R. (1998) The Mediterranean: Europe's Rio Grande, in Anderson, M. and Bort, E. (eds) *The Frontiers of Europe.* London: Pinter, pp. 109–34.

Kliot, N. (1981) The unity of semi-landlocked seas, *Ekistics*, 290, pp. 345–58.

Kliot, N. (1997) Politics and society in the Mediterranean Basin, in King, R., Proudfoot, L. and Smith, B. (eds) *The Mediterranean: Environment and Society.* London: Arnold, pp. 108 25.

Leontidou, L., Gentileschi, M. L., Aru, A. and Pungetti, G. (1998) Urban expansion and littoralisation, in Mairota, P., Thornes, J. B. and Geeson, N. (eds) *Atlas of Mediterranean Environments in Europe.* Chichester: Wiley, pp. 92–7.

Mairota, P., Thornes, J. B. and Geeson, N., eds (1998) *Atlas of Mediterranean Environments in Europe.* Chichester: Wiley.

McNeill, J. R. (1992) *The Mountains of the Mediterranean World: an Environmental History.* Cambridge: Cambridge University Press.

Montanari, A. and Cortese, A. (1993) South to North migration in a Mediterranean perspective, in King, R. (ed.) *Mass Migration in Europe: the Legacy and the Future.* London: Belhaven Press, pp. 212–33.

Osterberg, C. and Keckes, S. (1977) The state of pollution in the Mediterranean Sea, *Ambio*, 6(6), pp. 321–6.

Pastor, X. (1991) *The Mediterranean.* London: Collins Brown for Greenpeace.

Pitt-Rivers, J. (1954) *People of the Sierra.* London: Weidenfeld and Nicholson.

Popovic, B. (1992) Employment growth and change in the Mediterranean basin during the 1980s, *International Labour Review*, 131, pp. 297–312.

Pratt, J. and Funnell, D. (1997) The modernisation of Mediterranean agriculture, in King, R., Proudfoot, L. and Smith, B. (eds) *The Mediterranean: Environment and Society.* London: Arnold, pp. 194–207.

Robinson, H. (1970) *The Mediterranean Lands.* London: University Tutorial Press.

Rufin, J.-C. (1991) *L'Empire et les Nouveaux Barbares.* Paris: Editions Jean-Claude Lattès.

Thornes, J. B. (1995) Global change and regional response: the case of the Old World Mediterranean, *Transactions of the Institute of British Geographers*, 20(3), pp. 357–67.

Tovias, A. (1994) The Mediterranean economy, in Ludlow, P. (ed.) *Europe and the Mediterranean.* London: Brassey's, pp. 1–46.

Vallega, A. (1995) Towards the sustainable development of the Mediterranean, *Marine Policy*, 19(1), pp. 47–64.

Vallega, A. (1998) The Mediterranean after Rio, in Conti, S. and Segre, A. (eds) *Mediterranean Geographies.* Rome: Società Geografica Italiana (Geo-Italy, vol. 3), pp. 17–42.

Walker, D. S. (1965) *The Mediterranean Lands.* London: Methuen.

2

The Mediterranean Basin: a geopolitical fracture zone

Ewan Anderson

The focus of this chapter is Mediterranean geopolitics; all the issues discussed involve the interplay of geography and politics. They are geopolitical in that the concern is with the geographical setting in which political decisions are taken or, at worst, conflict occurs. The emphasis is upon those geographical factors with potential to influence such political events (Anderson 1993). It will be shown that in the Mediterranean, a region in which so many vital geopolitical issues interdigitate, the policy focus of the West is fragmented. Moreover, the Mediterranean is a region in which the causes of instability are intensifying rather than attenuating. It is not unreasonable to characterise the Basin as a geopolitical core region surrounded by intimately interrelated and overlapping potential problem areas.

Diplomats and strategists from Mackinder (1904) to Walters (1974) have been concerned with geography and the exercise of power and their arguments have depended largely upon a consideration of the influence of physical geography on politics at the international level. Over the past 20 years or so, the concept of geopolitics has changed considerably. A broader view of the contribution of geography has been taken by, among others, Kissinger (1982), whilst Van der Wusten (1998) offers a recent view from within the discipline. The most obvious extension has been into the realm of economic geography, with the result that 'resource geopolitics' is now a commonly-accepted term. Social geography, broadly defined to subsume cultural, ethnic, linguistic and other such variables has also offered an important contribution to the development of the subject. Thus, geopolitics embraces all aspects of geography which could potentially influence political strategy. Furthermore, geopolitics is not scale-specific and may be significant at a variety of levels from global to the local. Within the Mediterranean Basin, the demise of the bipolar world and the development of geopolitical flashpoints both need to be considered.

Geopolitical change

Between 1989, the fall of the Berlin Wall, and 1991, the disintegration of the Soviet Union, the all-embracing global confrontation between NATO and the Warsaw Pact ended. In contrast to Central Europe, the effects of this change were not immediately obvious in the Mediterranean Basin. Although the Mediterranean had been a prominent theatre in East–West relations, it was, nonetheless, geographically peripheral to Europe. More importantly, the great diversity within the region and the long history of many of its problems both militated against any abrupt change. The Cold War had effectively overshadowed rather than eliminated local and regional differences.

As we saw in the previous chapter, the old division between East and West has been replaced by a fracture zone between North and South, a fracture zone which passes through the centre of the Mediterranean.[1] The new cleavage is deeper than that which it superseded; East–West differences were largely externally induced, but North–South contrasts represent real deeply ingrained differences. Thus, as was shown in the early 1990s, the East–West division evaporated as a result of decisions and circumstances which were quite external to the Mediterranean region. In sharp contrast, it is clear that the North–South division can only be narrowed by the wishes and actions of the indigenous actors themselves. The old balances have been upset, but new ones have yet to emerge (Evert 1991).

The demise of the bipolar world has undoubtedly diminished the possibility of global conflict and has provided greater opportunities for international cooperation. At the same time, however, the probability of regional and local crises has increased. The Mediterranean region is one in which there is a particularly concentrated combination of negative factors; indeed the range and complexity of potential problems rival those in any other region of comparable size. Therefore, while the effects of the changes from 1989 onwards were less obvious in the Mediterranean than in several other parts of the world, the aftermath, exemplified particularly by the case of the break-up of the former Yugoslavia, has been seismic. The Mediterranean region is littered with potential geopolitical flashpoints, most of which have a long history. Problems such as the Arab–Israeli conflict, the Cyprus question, the Yugoslav issue, the external policies of radical states, terrorism, militant fundamentalism in some southern Mediterranean states, or south–north migration, do not owe their existence to the Cold War. The Cold War served to complicate the issues through foreign intervention and patronage or to contain them or simply to distract attention away from them (Anderson and Fenech 1994). The Mediterranean is both a crossroads and a frontier; a point of contact between two dissimilar worlds (Rathbone 1992). The political, economic, social and strategic differences between the 'near North' and the South are more obviously accentuated in the Mediterranean region than elsewhere, the more so now that the role of the superpowers is no longer so overbearing. The southern littoral of the Mediterranean is, in broad terms, an area of burgeoning population, growing poverty and increasing radicalism, while the northern shore, dominated by the development of the European Union (EU) into one of the major economic world power blocs, is characterised by increasing prosperity and declining birth rates. Thus, the fracture zone through the centre of the Mediterranean separates what must be increasingly seen as the 'have' nations from the 'have not' nations: this is a central theme of the next chapter.

These massive changes have largely left the security policy of the Western alliance in disarray. As the Southern Flank of NATO, the Mediterranean was considered significant but definitely secondary in importance compared to the Central Region. However, the end of the Cold War has fundamentally changed the geo-strategic map of Europe so that there is no meaningful distinction between the centre and the southern periphery (Asmus *et al.* 1996). Indeed, the strategic significance of the Mediterranean itself is an open question.

As with Western Europe, the divisions between security regions have been eliminated; a similar development can be followed in many of the regions around the margins of the Mediterranean itself. The various theatres of North Africa, the Red Sea region, but more importantly the Persian-Arabian Gulf, the Black Sea periphery, the Caucasus and even Central Asia are all becoming increasingly implicated in the security of the Mediterranean. Indeed, in this sense, the Mediterranean has been transformed from an area of marginal importance to a focus of security interest.

This breakdown of traditional boundaries has been well illustrated by Asmus *et al.* (1996), using the example of energy geopolitics. The increasingly close relationship between Europe, the Middle East and Central Asia is very clear. As new oilfields come on stream in Central Asia and the Caucasus, and as Iraqi supplies return to the world market, so the pipelines are likely to converge upon the Mediterranean. Furthermore, the expanding network of gas pipelines between North Africa and Southern Europe will affect perceptions of vulnerability. At present, in a daily total of some 3,000 ships, 65 per cent of the oil and natural gas required in Western Europe transits the Mediterranean. Thus, Europe is bound to have an increasingly obvious economic interest in the stability of the Mediterranean region.

While at the theoretical level there would seem to be a marked coincidence of concerns, the reality is that geopolitical viewpoints vary widely. For the EU, particularly France, Spain and Italy, the emphasis is upon the western basin of the Mediterranean and relations with the Maghreb. In Central and Northern Europe interests are less visceral and more broadly based. For the US, the focus is upon the broader national security context (Dismukes and Hayes 1991). Four areas have been defined as being critical to the growth of the world economy: North America, Western Europe, North-East Asia and the Persian-Arabian Gulf. The Mediterranean abuts two of these regions. Furthermore, the Black Sea region is increasingly implicated in the global oil economy. It seems likely that to the complicated geopolitics of Middle Eastern oil will be added the Byzantine geopolitics of Central Asian oil and both will exercise a major influence upon Mediterranean security concerns.

The macro-political agenda

Accelerated by the end of the Cold War, the inter-connectedness of the international system has resulted in the recognition of what may be termed a 'macro-political agenda'. This highlights problems such as the massive migration of people, nuclear proliferation, terrorism and international debt, which are both sources and consequences of conflict. They are so characterised not merely because they have assumed global proportions, but because they constitute potential global disorder on a scale which exceeds the capacity of even the most powerful state to resist on its own (Camilleri 1994). Many of these issues are highly relevant in the Mediterranean.

The Mediterranean has long been acknowledged as the cradle of civilisation, but it is actually also the cradle of terrorism, precisely because so many flashpoints and grievances are concentrated in the region. In the period from 1970 to 1990, more than one third of all international terrorist incidents occurred within the Mediterranean region. While such violence has normally been associated with the southern littoral, it should be realised that all the states along the northern shore have also been involved. Mediterranean states have been implicated in government deals with terrorists, in terrorist training and in government terrorism itself. In October 1985, the Italian cruise ship *Achille Lauro* was hijacked in the Mediterranean (Clutterbuck 1990) and this act has remained as a symbol of terrorism at sea. Virtually every Mediterranean country has witnessed acts of terrorism, whether in southern Europe, the Balkans, the Middle East or North Africa, although the major concentration of active terrorist organisations is undoubtedly in the Levant. Apart from that area, terrorist activities are currently of major significance in Turkey, Egypt and Algeria. Furthermore, terrorism although usually locally or regionally generated, knows no boundaries and is effectively exported from Syria and, possibly, Libya, while in the case of Israel it can be considered to be state-sponsored.

By its nature, terrorism is often linked to international crime, particularly weapon smuggling and drug trafficking. As the actions of terrorists constitute a defiance of the established order, so in turn punitive action often entails violations of international sovereignty or human rights. Thus, the high incidence of terrorism in the Mediterranean illustrates not merely the range of problems of the region, but a rejection in many quarters of the international order.

Another example of such rejection, sometimes accompanied by violence, is Islamic fundamentalism; the development of this type of radicalism in North Africa has been responsible for the growing fears of Europeans over immigration from that region. Muslim radicals in Algeria, led by the Islamic Salvation Front, despite scoring a victory in the first round of the national elections (December 1991), were prevented from capitalising upon this by military action. The crisis was further escalated by the assassination of President Boudiaf on 29 June 1992 and the situation remains fraught. Furthermore, Tunisia and Morocco are aware that they could face similar problems. Apart from the Maghreb, civil wars in the former Yugoslavia have developed Christian/Muslim overtones. The situation has been further complicated by the fact that Turkey, the most powerful state in the region, has claimed responsibility for the security of Muslims in Bosnia and Macedonia.

While many observers consider the threat of Islamic radicalism to be greatly exaggerated, it should certainly not be underestimated. States with Muslim populations continue to be concerned with radicalisation and, as a result, instability. The hostility between the state and the fundamentalists in most countries has resulted in greater recourse to violence by the latter, increasing the repressive measures by the former. These manifestations of rejectionist tendencies further underline the division between the European and the non-European littorals of the Mediterranean.

One of the most deep-seated causes of insecurity lies in the obvious differences between the demographic structures on both sides of the Mediterranean, considered in more detail by Armando Montanari in **Chapter 6**. The population of the south, from Egypt to Mauritania, was estimated to number some 153 million by the year 2000 (de Vasconcelos 1991). At present, over half of this population is aged under 15. By 2025, the population could be in the range of 260 million (Asmus *et al.* 1996). Unemployment is rising fast and the possibilities of emigration to Europe are being constantly reduced. Indeed, given the relatively high unemployment in Western Europe and the movement of immigrants from Eastern Europe, the situation is likely to worsen. As *The Times* (5 February 1992) stated: 'the Mediterranean is to Europe what the Rio Grande is to the United States'.

The movement of labour across the Mediterranean has been a permanent feature of the economies of the region, but it has always involved the potential for social problems. When the migrants are refugees, possibly connected with Islamic radicalism, the situation can be greatly exacerbated. Even the guestworkers from Turkey, long welcome in the industrial areas of Central and Western Europe, are now perceived as an increasing burden on the welfare system. Numbers have already been reduced and there have been moves for expulsion, thus raising the spectre of greater animosity between Turkey and the EU or NATO.

For many reasons, amongst them the number of potential flashpoints in the region, the instability of the regimes, the role of the armed forces in government or simply national prestige, many states within the Mediterranean Basin maintain large defence expenditures. Indeed, Turkey, Egypt, Greece and Israel are four of the ten leading recipients of major conventional weapons (SIPRI 1996, pp. 463–86). It must be remembered also that the implementation of the Conventional Forces Europe (CFE) programme has resulted in a vast surplus of available conventional weapons within the region.

France and Israel are nuclear powers, while the Russian Federation maintains nuclear stockpiles within range of the Mediterranean. For the region as a whole, nuclear proliferation is seen as a threat, particularly in the case of Libya, Iran and Iraq. Chemical weapons, seen as a less expensive substitute for nuclear weapons, also pose a problem in that the Mediterranean region includes the greatest concentration of chemical weapons holders in the world. It is also assumed that there are stocks of biological weapons within the region, since they and chemical weapons are seen by many states as the only possible counter-weight to the known nuclear weapons stockpiles of Israel. To the dangers of nuclear, chemical and biological weapon proliferation must be added the spread of ballistic missile technology.

Within the Southern Tier, seven countries (Egypt, Iran, Iraq, Israel, Libya, Saudi Arabia and Syria) possess tactical ballistic missiles. The great danger lies in the coupling of ballistic missile capabilities with mass casualty-producing weapons (Johnsen 1993). Technology is therefore eliminating the barriers of security within the region and within a decade it is likely that every European capital will be within the range of ballistic missiles based in the Middle East or North Africa (Asmus *et al.* 1996).

These examples from the macro-political agenda illustrate clearly the potential for increased instability within the Mediterranean region. However, the full force of the argument is only perceived when it is viewed against the background of more local problems and flashpoints.

Indigenous problems

Among the many potential problems – physical, political and socio-economic – within the Mediterranean environment, two issues are of particular concern: boundaries and water.

Boundaries

The most intractable boundary disputes concern Israel and its neighbours, although the situation has been alleviated somewhat by settlements with Egypt and Jordan. However, the boundaries with Lebanon, Syria and Palestine are far from settled, although the election of the new Labour Government under Barak in Israel does appear to promise an increased spirit of compromise.

The boundary dividing Cyprus is also a source of tension and part of the continued confrontation between Turkey and Greece. The other obvious boundary problems involve the former Yugoslavia, where religious, linguistic and ethnic variables are mixed in such a complex pattern that allocation, let alone demarcation, appears almost impossible. These problems, the most obvious concerning onshore boundaries, are matched by a range of issues offshore.

Greece and Turkey, both members of NATO, have been locked in dispute over the allocation of the Aegean Sea for more than 20 years and on three occasions have been close to armed conflict. If Greece were to obtain its full entitlement, Greek sovereignty would be extended over 71 per cent of the Aegean. Turkey, on the other hand, claims a median line boundary, citing special circumstances which include possible presence of oil and the necessity to protect shipping transiting from the Black Sea. The issues of Cyprus and the Aegean, both seemingly intractable problems at the present time, represent the most obvious geopolitical outcomes of the unremitting hostility between Greece and Turkey. The psychological aspects of this run so deep that the achievement of any viable solutions must be considered as problematic as anything in the Eastern Mediterranean Basin, including the long-term settlement of the Arab-Israeli disputes. After World War I, Greece and Turkey had approximately the same populations, but while that of the former has remained virtually static, the population of Turkey has increased more than sixfold and this demographic picture takes no account of the Islamic dimension. In the case of Turkey, the perception from within the country is that the continuing prosperity of its major economic region, together with the developments linking it to Central Asia, remains at the mercy of another state. One glimmer of light is the movement towards agreement over Open Skies above the Aegean Sea and this may provide the impetus for further accords. On the other hand, negotiations over the future entry of the Greek part of Cyprus into the EU can only exacerbate tensions. Elsewhere in the Mediterranean, much of the median line between the northern and southern states has still to be settled, as have the maritime boundaries of Cyprus.

Hydropolitics

With dry summers and small drainage basins restricting supply while burgeoning population and tourism enhance demand, the Mediterranean is a region of increasing water shortage. Israel, Jordan, Egypt, Libya and Malta rank globally among the countries with the lowest per capita consumption of water in the world.

Two Mediterranean catchments, those of the Jordan and the Nile, have been a particular concern of hydropolitics (Agnew and Anderson 1992). Many models for potential conflict or cooperation in drainage basins have been advanced, but most agree upon the significance of three factors: the relative geographical positions of the states, their degree of interest in water problems and the power from both internal and external sources which can be brought to bear to influence decisions. In both basins there is a high degree of interest in hydrological issues, but in neither is the upstream state the most powerful and therefore the one able to control events. Egypt and Israel are clearly the principal instigators of developments within their own catchments.[2]

A power asymmetry can be the trigger for conflict in what is already a non-cooperative setting with an environmental imbalance. In both basins cooperation has been, at best, partial and there are records of hostility. Furthermore, in the Nile catchment there are already deficits while in the Jordan Basin there is a zero-sum situation. Thus, while it is difficult to isolate water as a specific originating cause of conflict, it can often be seen as a contributory factor in an already hostile environment. In the Nile Basin there have already been several disagreements between Egypt and Sudan, but the most likely scenario for discord occurs between Egypt and Ethiopia, the state which controls the headwaters of the Blue Nile.

The annual average flow of the Nile is 76 billion cubic metres, while that of the Jordan is 1.5 bcm. Thus, conflict in some form involving water is more likely in connection with the River Jordan. Several hydropolitical foci, all related to the advantageous position of Israel, can be identified. The Yarmuk Triangle gives Israel some control over the intake for the major irrigation system of the Jordan, the East Ghor Canal. Occupation of the Golan Heights guarantees control over the headwaters of the Jordan, while the occupation of southern Lebanon offers potential for using water from the Litani River. Most importantly and most significantly for the current peace process, the Israeli occupation of the West Bank allows water to be abstracted from aquifers previously inaccessible to Israel and raises fundamental questions about the future territorial extent of the new state of Palestine.

Geopolitical flashpoints

Many obvious geopolitical flashpoints have already been identified and there is a clear coincidence of view between the various writers on the subject. Indeed, in characterising the Southern Flank of NATO as progressing 'from southern region to southern front to southern flashpoint', Boorda (1993), by exaggeration, highlights the overall security problem of the Mediterranean. Johnsen (1993) identified the points of crisis as Bosnia, the Caucasus, eastern Turkey, Cyprus, Lebanon and Libya.[3] In North Africa, Vormann (1995) singled out Algeria and Egypt as the two states most subjected to increased violence over the recent past. Despite the historic Gaza–Jericho agreement between Israel and the Palestine Liberation Organisation of September 1993, the question of Israel and the Occupied Territories remains an obvious area of potential conflict. Also, of course, the relationship between Turkey and Greece continues to provide a cause for alarm.

This inventory of flashpoints concerns only the littoral states of the Mediterranean. If the interrelated regions in which there are geopolitical tensions are added, then flashpoints with a potential for affecting the security of the Mediterranean can also be discerned in Central Asia, Eastern Europe, the Middle East and parts of Africa. On a smaller geographical scale, Jerusalem, the Turkish Straits, the Suez Canal, the boundary zone of Cyprus and a number of Aegean islands can all be identified as possible flashpoints with widespread repercussions.

The future

In 1995, NATO launched a dialogue with Egypt, Tunisia, Morocco, Mauritania, Jordan and Israel. This dialogue continues, but while NATO is opening many of its training programmes to nationals from these countries, they have not, as yet, been invited to join the Partnership for Peace. However, the Supreme Allied Commander Europe (SACEUR) has stated that 'the NATO leadership welcome the participation of those countries which are not members of the Partnership for Peace programme' (Medvedenko 1999).

Furthermore, NATO is anxious to avoid any duplication of EU initiatives, particularly since the ambitious project launched by the EU for a Euro–Mediterranean Partnership at the Barcelona Conference in November 1995. The partnership concept is for complete bilateral cooperation between the EU and the states or entities of the southern shore of the Mediterranean; the overall objective is political stability in the region. Other cooperative bodies of interest are the Arab Maghreb Union, established in 1989, and the Mediterranean Forum, the only comprehensive body limited in membership to those countries which are actually littoral to the Mediterranean. Since most of the European Mediterranean states are already members of the Conference on Security and Cooperation in Europe (CSCE), the idea of a Conference on Security and Cooperation in the Mediterranean (CSCM) must be a welcome possibility. The concept was launched by Italy and Spain in 1990 at the CSCE Conference on the Protection of the Ecological System of the Mediterranean, held in Palma di Mallorca. Since the CSCM would provide a unifying forum for the entire region and would be concerned with security, it remains a particularly compelling suggestion.

The Mediterranean Dialogue was given added impetus at the Madrid Summit (July 1997) when the Mediterranean Cooperation Group (MCG) was created. Through the MCG, NATO is directly involved in political discussions with Mediterranean Dialogue countries and can therefore, with them, monitor the security situation in the Mediterranean. The creation of the MCG has increased the public profile of NATO's Mediterranean dimension (Bin 1998). However, the Barcelona process of the EU and the Middle East peace process are also concerned with the enhancement of stability and the improvement of security in the region. Therefore, the future of the Mediterranean Dialogue will be greatly influenced by the level of success achieved by these two processes.

Within the Mediterranean Basin there is a wide-ranging array of actual and potential geopolitical problems, many of which seem more acute since the breakdown of the bipolar international system. As a flashpoint region, only the Middle East takes precedence over the Mediterranean. Initiatives such as the Mediterranean Dialogue offer some hope for future stability but, for the Balkans at least, they appear to be too little, too late.

Notes

1   Although tensions across the fraction zone remain the expectation, the most recent examples of actual conflict have occurred immediately to the north of the line in the Balkans, notably in Bosnia and Kosovo.
2   While not strictly Mediterranean rivers, the Tigris and Euphrates need to be factored into Mediterranean hydropolitics as a result of the potential for conflict over water in their upper basin states: Turkey and Syria. The combined annual flow of the two rivers, approximately 84 billion cubic metres (bcm), is rather greater than that of the Nile but the overall water budget will be greatly affected when the Southeast Anatolian Project (GAP) is completed.
3   Johnsen (1993) might well have identified the Balkans as the point of crisis. Bosnia enjoys an uneasy peace but, since the NATO bombardment (March–June 1999), Kosovo and the neighbouring areas of Macedonia, Montenegro and Albania remain flashpoints.

References

Anderson, E. W. (1993) *An Atlas of World Political Flashpoints*. London: Pinter.
Agnew, C. T. and Anderson, E. W. (1992) *Water Resources in the Arid Realm*. London: Routledge.
Anderson, E. W. and Fenech, D. (1994) New dimensions in Mediterranean security, in Gillespie, R. (ed.) *Mediterranean Politics, Volume 1*. London: Pinter, pp. 9–21.
Asmus, R. D., Larrabee, F. S. and Lesser, J. O. (1996) Mediterranean security: new challenges, new tasks, *NATO Review*, May, pp. 25–31.
Bin, A. (1998) Strengthening cooperation in the Mediterranean: NATO's contribution, *NATO Review*, Winter, pp. 24–7.
Boorda, J. M. (1993) The Southern Region – NATO forces in action, *NATO's Sixteen Nations*, 2, pp. 5–9.
Camilleri, J. A. (1994) Security: old dilemmas and new challenges in the post-Cold War environment, *GeoJournal*, 32(4), pp. 135–46.
Clutterbuck, R. (1990) *Terrorism, Drugs and Crime in Europe after 1992*. London: Routledge.
de Vasconcelos, A. (1991) The New Europe and the Western Mediterranean, *NATO Review*, October, pp. 27–31.

Dismukes, B. and Hayes, J. (1991) US national security in the Mediterranean, *Proceedings of the USNI*, October, pp. 9–14.

Johnsen, W. T. (1993) NATO's Southern Tier: a strategic risk appraisal, *European Security*, 2(1), pp. 44–70.

Kissinger, H. (1982) *American Foreign Policy: A Global View*. Singapore: Institute of Southeast Asia Studies.

Mackinder, H. J. (1904) The geographical pivot of history, *Geographical Journal*, 23, pp. 423–37.

Medvedenko, A. (1999) Mediterranean can participate in NATO operations – SACEUR ITAR–TASS News Agency, 25 February.

Rathbone, T. (1992) TWEU and security of the Mediterranean, *Letter from the Assembly*, July, pp. 17–20.

SIPRI (1996) *SIPRI Yearbook*. Oxford: Oxford University Press.

Van der Wusten, H. (1998) The state political geography is in, *Tijdschrift voor Economische en Sociale Geografie*, 89(1), pp. 82–9.

Vormann, I. (1995) NATO's star rises in the Med, *Proceedings of the USNI*, March, pp. 73–8.

Walters, R. E. (1974) *The Nuclear Trap, An Escape Route*. Harmondsworth: Penguin.

3

Mediterranean economic geography

Michael Dunford and Russell King

The aim of this chapter is to paint a broad-brush picture of the economic geography of the Mediterranean region. The picture has to be drawn in broad outline given the large number of countries which go to make up this region, as well as their internal complexities and different levels and trends of development. We start by re-posing a question that was briefly addressed in Chapter 1: is there such a thing as a specific Mediterranean economy? Once this question has been analysed and answered, we proceed in the main body of the chapter to explore some dimensions of the uneven development of the Mediterranean region. Much of our analysis is built around a series of tables and graphs. The account which we present here moves on from some of our other recent work (Dunford 1997; King 1998; King and Donati 1999) as well as that of others who have described Mediterranean economic patterns (Conti and Segre 1998; Joannon and Tirone 1990; Reiffers 1997; Tovias 1994); it develops these earlier treatments by broadening the framework of analysis and by using the latest available data.

The chapter is organised as follows. First, we attempt to identify the economic specificity of the Mediterranean region. Next, we relate some essential historical background to an understanding of the contemporary economic geography of the Basin. Third, we explore the multifaceted nature of uneven development amongst the countries bordering the Mediterranean Sea (including Portugal), looking at such variables as wealth, human development, employment trends, agricultural change, industry, oil and indebtedness. These variables do not, of course, comprise a complete listing of economic phenomena which have a strong spatial development component – issues relating to trade, migration and tourism are dealt with in other chapters.

Is there a distinct Mediterranean economy?

We saw from Chapter I that many geographers and historians regard the Mediterranean Basin as a meaningful regional unit. What are the economic parameters of this apparently coherent region? According to Tovias (1994, pp. 1–5), the economic commonalities are those linked to geography, climate and factor endowments.

Let us start with geography, and here we need do little more than note the location of the Mediterranean on the world map as a zone of transition between the industrialised (or post-industrial) countries of Europe to the north and the less developed realms of Africa and the Middle East to the south and east. Proximity to north-west Europe, history, climate and demography give rise to northward flows of agricultural produce (especially fresh fruit and vegetables) and labour migrants (especially between 1950 and 1973), and southward flows of European tourists. Another locational feature of the Mediterranean is its strategic position between the energy-rich countries of the Gulf and North Africa on the one hand, and the energy-hungry countries of Europe on the other. Hence the Mediterranean is a sea across which a large quantity of the world's oil is transported.

Second, the Mediterranean climate is a strong conditioning influence on many features of the region's rural economy, and on its attraction for tourists. The principal crops and products of the Mediterranean are found all over the region and have been the same since Roman times: wheat, vines and olives; bread, wine and oil. These are produced in conjunction with the rearing of sheep and goats, and with the production of more locally specialised crops such as tobacco, rice, cotton and citrus fruit. According to Mørch (1999), Mediterranean agriculture is a complex of four components in which the drought of summer is avoided, adapted to or modified:

- rain-fed annual crops such as wheat which are nourished by the winter rain and harvested in spring or early summer;
- permanent tree crops such as olives and figs, which can survive the dry summer;
- irrigated crops such as salads, vegetables and citrus fruit, whereby the summer drought is artificially compensated;
- transhumance, by which migrating livestock avoid the summer drought by ascending to cooler, moister altitudes.

Variations exist around these components, but such variations are often consistently repeated across the region. Almost everywhere there is some agriculture and some pastoralism; however, the proportions vary. The peasant ecotype is predominant in the northern Mediterranean, whilst the pastoral ecotype is more widespread in the southern and eastern Basin countries, where the climate is much more arid. Nomadism survives in regions like Sinai and the Negev which, although quite close to the Mediterranean shore, are in reality true deserts. Another recurrent duality is the polarisation of farm structure between *latifundia* and *minifundia* – very large and very small farm holdings. The former were traditionally geared to the specialised production of cereals or livestock, whereas the latter produced a variety of crops for family subsistence. Despite agrarian structural changes in recent decades – land reform, land consolidation, shifts in inheritance practices and rural depopulation – these polarised landholding structures persist in many countries (Boissevain 1976).

Third, Mediterranean factor endowments are also distinctive: a shortage of fertile, level land, leading to hand tillage based on small plots and terracing; a traditional surplus of labour leading to underemployment in agriculture, labour-intensive and undercapitalised industry, and migration to Europe or overseas; and a favourable climatic endowment and cultural heritage for tourism. However, these production factors, and the types of economic activity they have given rise to, have not always been indigenously controlled, so that externally-driven modernisation rather than locally-embedded development has tended to be the result. Schneider *et al.* (1972) were the first to recognise this syndrome nearly thirty years ago when they analysed the nature of economic and social change in southern Europe. According to these authors, migration, tourism and 'forced industrialisation' – the main agents of change in the region at that time – did not lead to 'autonomous development', in the sense that a backward region could acquire a highly developed and diversified industrial economy by its own efforts alone, but rather they led to 'modernisation' – the process 'by which an underdeveloped region changes in response to inputs, ideologies, behaviour codes, commodities and institutional models emanating from already established industrial centres' (Schneider *et al.* 1972, p. 343).

So, to some extent we can speak of a Mediterranean economy. Drawing on the three parameters above, and including other features which we will develop later in the chapter, its essential elements are as follows:

- an agricultural sector dominated by certain characteristic Mediterranean products, notably oil, wine, fruit and vegetables, grown mainly on small farms and employing a rather large (though falling) share of the total working population;
- an industrial sector which is polarised between a large number of small enterprises, often at the artisan scale, and a small number of large enter-prises, especially in the oil and petrochemical sectors, which are largely controlled by external capital;
- a pattern of employment change away from farming into service activities, with industrial employment not, in general, taking up much of the slack from the massive agricultural exodus;
- rates of development (however measured) which in general have been above the global average, reflecting the Mediterranean region's status as a rapidly-evolving, semi-peripheral region in the world economy;
- a well-developed informal sector which is found in most branches of economic activity – farming, industry, construction, private services, even the professions – and which embraces such practices as the employment of unregistered labour (e.g. casual farm and construction workers), unrecorded family labour (including child labour), and moonlighting

(common amongst white-collar workers such as civil servants and teachers).

This summary qualitative portrait of the Mediterranean economy hides many details and intra-Basin contrasts which will become more evident when we discuss some data-sets. Before we do that, a few historical pointers are necessary.

The historical roots of Mediterranean economic geography

From the outset, the Mediterranean's rich land and sea resources facilitated remarkable early developments in agriculture, trade and urbanisation and allowed it to emerge as the centre of a succession of hegemonic economies. The most complete political and economic unity was accomplished during the Roman Empire when Roman hegemony spread throughout the Mediterranean Basin and beyond. After Roman times, attempts to create a pan-Mediterranean economic empire succeeded only partially, as regionalism reasserted itself. The fall of the Western Empire ushered in an era of Arab domination that spread north from the African shore into Sicily, Malta and much of Spain, while the collapse of Byzantium saw the expansion of Ottoman influence into North Africa on the one hand and north-west to the Balkans on the other. Although the Ottoman Empire maintained its grip on what is now the Islamic Mediterranean until the early part of the twentieth century, it did not extend its hegemony on the northern shore west of the Adriatic. Over the centuries, repeated episodes of Christian-Islamic conflict – the Crusades in the east, the Spanish *reconquista*, Napoleonic expansion in the early nineteenth century – have set the scene for an ongoing cultural confrontation which has powerful echoes in present-day Mediterranean geopolitics, as the previous chapter showed.

In this way, and putting it very simply, two entities emerged: on one side of the Basin the Arab and Muslim world, roughly Turkey to Morocco; and on the other side, from Portugal to Greece, the Western, Christian world. The latter became a core area of the global economy, hungry for raw materials and consumer markets. It has constantly tried to dominate the rest of the Basin by a combination of political, military and economic means, following the Western capitalistic commercial model (Braudel 1979).

In its early forms, this economic imperialism was initiated by the Italian city-republics 400 years ago. Subsequently the Spanish dominated the Western Mediterranean, whilst the Ottomans ruled the Eastern Basin and the Balkans, as noted above. In the late nineteenth century another colonial expansion took place as France, Italy and Spain accomplished their relatively short-lived conquests of North Africa; wealth from agro-pastoral regions was siphoned off to the metropolitan powers and a growing cross-Mediterranean inequality mirrored the hierarchy of political and economic power (Boissevain 1976). European 'outsiders' also gained colonial footholds in the Mediterranean. The British took control of the south-eastern corner in order to construct the Suez Canal to shorten the journey to their East African, Asian and Pacific colonies, and they occupied strategic positions at Gibraltar, Malta and Cyprus. With the end of the Balkan Wars, Turkey was eliminated as a significant actor in the Mediterranean, at least for the time being, while in the Arab realm the actions of rival Western imperialist powers and the creation of the state of Israel in 1948 left a legacy of political fragmentation that remains as a crucial feature of contemporary political economy.

Important events and processes were also unfolding at scales above and below that of the Mediterranean Basin and its constituent countries. Once the Atlantic was opened up, the coasts of the Mediterranean began to lose their primacy as centres of global decision-making. The economic centre of gravity moved north-west to Amsterdam and Great Britain. Later, the nineteenth-century Industrial Revolution largely by-passed the Mediterranean, with the exception of a few restricted areas such as the Basque provinces (not really Mediterranean anyway), Catalonia and Northern Italy. This means that, outside of these few places, there are no heavy industrial agglomerations and hence the social structure does not have a well-developed urban industrial working class. Instead, as Boissevain (1976) relates, social relations in the Mediterranean have always been more personal and subject to moral codes. The use of the term 'moral' implies no value-judgement: indeed, rural regions were inhabited by peasants whose behaviour was often 'amoral', as Banfield (1958) claimed in his controversial book on a village in Southern Italy; whilst cities were inhabited by wealthy landowners, traders, artisans and administrators with patriarchally organised craft and merchant guilds. Thus, on a local level too, the metropolitan centre and the city have generally exploited the countryside: profits from agricultural products have not flowed back to rural areas for investment in roads, irrigation and other farm improvement, but have tended to be spent on the luxurious living of landowners and rentiers resident in the urban centres.

After 1945 several processes have produced a rapidly evolving geography of economic differentiation within the Mediterranean Basin. Industrialisation and modernisation proceeded unevenly and with different trajectories. Manufacturing industry boomed in Northern Italy in the 1950s and 1960s but developed only in a scattered fashion elsewhere. Agrarian reform and emigration affected many rural regions, whilst the arrival of mass tourism in the 1960s started the profound transformation of coastal landscapes that is documented in several chapters of this book (see especially Chapters 9–11). Oil rents have enriched the producer states, most of which lie just beyond the conventional boundary of the Mediterranean region, and have had a differential impact on non-oil states as well as giving the Mediterranean Sea great strategic significance as a corridor for the transport of this vital resource.

Of equal importance has been the establishment of political, economic and military blocs affecting the countries of the region. First, most northern Mediterranean countries, including Turkey, were integrated into the NATO defence alliance created in 1949, although France withdrew in 1966 and Spain did not join until 1986. Second, France and Italy (as founder-members in 1957), and then Greece (1981), Spain and Portugal (1986) were integrated into the European Union, one of the undisputed cores of the global economy. This implied developing intra-EU trade links at the expense of traditional non-Community trade including relations with other Mediterranean countries (Chapter 4). On the other hand, most of the attempts made by the decolonised countries of the southern and eastern Mediterranean to achieve greater political and economic integration, either with each other or with

Europe, have failed. In the Arab countries, military dictatorships and recurrent wars with Israel in 1967, 1973 and 1982 have tended to reorient economic strategy towards arms, large-scale public works and valorisation of reserves of oil and gas. Political and economic rivalry has prevented Arab unity, but it also has to be recognised that fragmentation in this part of the Basin is a legacy of Western imperialism. Meanwhile, development in the Balkans has been sabotaged by the bloody break-up of Yugoslavia and Albania's sudden rupture with its introverted communist past.

Towards a Mediterranean development hierarchy

Previous attempts to place, and divide, Mediterranean countries within a wider European and global hierarchy of development have been made by Lipietz (1993) and Dunford (1997). Lipietz's scale is global and consists of four 'circles of expansion' around the core of the (then) EU 12: first, the EFTA circle; second, the Mediterranean circle (from Turkey to Morocco), linked to the EU by preferential tariff agreements; third, the Lomé countries of Africa, the Caribbean and the Pacific; and fourth, the East European countries recently 'liberated' by the collapse of communism. This framework is already dated, given the reduction of the EFTA circle (Austria, Finland and Sweden joined the EU in 1995), the fast-tracking of EU membership applications for several East European countries, and the redrawing of the map of the Balkans. Dunford's six-fold schema is European and Mediterranean and has a more finely-drawn differentiation: first, Europe's three richest economies (Germany, Switzerland, Luxembourg); second, the 'inner EU/EFTA' countries (France, Belgium, the Netherlands, the United Kingdom, Denmark, Sweden, Norway, Finland, Iceland, Austria and Italy); third, a group of peripheral EU states (Ireland, Spain, Portugal) plus Israel; next another tier, made up of Greece, Cyprus and Malta; fifth, a group consisting of the more developed eastern and southern Mediterranean countries (Turkey, Tunisia etc.); and finally the poorest Mediterranean countries (Morocco, Egypt etc.).

In most of the tables which follow, we will divide the countries of the Mediterranean into four groups. The division has some geographical logic, in that the groups broadly equate the four 'quadrants' of the Mediterranean introduced in Chapter 1, and they also correspond to different levels of development. Of course this begs the question of what we mean by 'development', and we need to recognise at the outset that this is an elusive and multi-dimensional concept. The six tiers of development/underdevelopment referred to above were mapped out by the simple wealth criterion of GNP per capita. For many years now, development specialists have been pointing out that the accumulation of wealth is not a good index of human development: one needs, firstly, to look at how that wealth is distributed and used; and secondly, to acknowledge that other variables such as life expectancy, infant mortality, literacy and food intake are much better indicators of living conditions in the poorer countries of the world. In other words, wealth should be seen as a means to an end, not an end in itself.

Patterns of development and underdevelopment

Population and wealth

Table 3.1 sets out some basic population and wealth data for the Mediterranean countries. Four countries contain around 60 million inhabitants each – France, Italy, Turkey and Egypt. Spain (38m), Algeria (30m) and Morocco (28m) form the next division by population size, followed in turn by a group with 10–15m (Portugal, Greece, Serbia, Tunisia and Syria), and then a group with 4–6m (Croatia, Libya, Lebanon, Israel), and finally several smaller countries with 3m or less. The total population of the countries listed in table 3.1 is 425 million.

The second column of table 3.1 shows the unequal distribution of wealth, concentrated above all in the EU Mediterranean countries. In fact these five countries account for 86 per cent of the Mediterranean's GNP[1] yet only 43 per cent of the region's total population.[2] If the comparison is restricted to the 'big three' (France, Italy and Spain), the percentages are even more contrasting – 80 and 38 per cent.

Table 3.1 Mediterranean countries: population and wealth, 1998

	Population (m.)	GNP ($US bn.)	GNP per capita ($US)
EU Mediterranean			
France	58.8	1,466.0	24,940
Spain	39.3	533.7	14,080
Portugal	10.0	106.4	10,690
Italy	57.6	1,166.2	20,250
Greece	10.5	122.9	11,650
Other Northern Mediterranean			
Malta	0.4	3.6	9,440
Cyprus	0.8	–	–
Slovenia	2.0	19.4	9,760
Croatia	4.6	20.7	4,520
Bosnia-Herzegovina	2.5	–	–
Yugoslavia-Serbia	10.6	–	–
Macedonia-FYROM	2.0	2.6	1,290
Albania	3.4	2.7	810
Turkey	63.5	200.5	3,160
Maghreb			
Morocco	27.8	34.8	1,250
Algeria	30.0	46.5	1,550
Tunisia	9.4	19.2	2,050
Other Southern Mediterranean			
Libya	5.3	–	–
Egypt	61.4	79.2	1,290
Lebanon	4.2	15.0	3,560
Syria	15.3	15.6	1,020
Israel	6.0	95.2	15,940

Source: World Bank (1999).

On the other hand, a large group of southern and eastern Mediterranean countries (Morocco, Algeria, Tunisia, Egypt, Lebanon, Syria and Turkey) contain 52 per cent of the total population but claim only 10 per cent of the Basin's GNP.

Moving to the final column of table 3.1, a series of levels in the ranking of GNP per capita is clearly evident. France and Italy head the ranking, with more than $20,000, then Spain and Israel ($14,000 and $16,000 respectively), followed by a group with around $10,000 (Portugal, Greece, Malta, Slovenia and probably also Cyprus, for which no recent figure is available). Finally, more than half the countries of the region have a per capita GNP of less than $5,000, and most of these have less than $2,000. Some simple cross-Mediterranean comparisons are revealing: Spain's per capita wealth is 11 times that of its close neighbour, Morocco, Italy's is 10 times that of its southern neighbour, Tunisia, and Greece's is 9 times that of Egypt. These comparisons are especially meaningful in the context of the migration flows analysed in Chapters 6–8.

Measuring human development

As the index of GNP per capita does not tell us whether wealth is used to improve human well-being, the United Nations Development Programme (UNDP) has developed a range of alternative measures, the most widely used of which is the Human Development Index, HDI. This is a composite index combining life expectancy at birth, educational attainment (adult literacy and enrolment ratios) and adjusted per capita income measured at 'Purchasing Power Standards/Parities' (PPS/PPP).[3] Various elements of human development are set out in table 3.2.

First it is necessary to make a further comment on the measurement of per capita wealth. In table 3.2 the real GDP per capita figures for the various Mediterranean countries exhibit a closer range than the GNP per capita figures in table 3.1. This is because GDP is measured at Purchasing Power Standards to reflect variations in the cost of living. The low cost of living in poor countries neutralises, to some extent, the low incomes received by these countries' inhabitants. Conversely, in wealthy countries high personal incomes are brought down, in real terms, by the high cost of living. Hence the cross-Mediterranean comparisons we noted above for Spain/Morocco, Italy/Tunisia and Egypt/Greece are reduced from 11, 10 and 9 to their 'real' ratios of 5, 4 and 4 respectively. These are still substantial economic divides, sufficient to encourage cross-Mediterranean economic migration.

Life expectancies at birth usually increase constantly in peace time although the rate of improvement is generally faster in countries where the absolute figure is lower, partly because there is more scope for improvements in diet and health care and partly for purely statistical reasons of measurement against a low base value. Recent improvements in medical facilities in the poorer countries of the Mediterranean have helped to close the gap in life expectancy with the richer countries. As table 3.2 shows, currently that gap stands at 12 years, between France, Spain, Italy and Greece (all with over 78 years) and Morocco and Egypt with 66 years. It is worth noting here that many of the least developed countries of the world have life expectancies of less than 50 years.

Patterns of adult literacy are more wide-ranging, from close to 100 per cent in the European Mediterranean countries to about half that rate in Egypt (53 per cent)

Table 3.2 Mediterranean countries: components of human development, 1997

	Life expectancy at birth (years)	Adult literacy (%)	Real GDP per capita (US$PPP)	Human Development Index	GDP rank minus HDI rank
EU Mediterranean					
France	78.1	99.0	22,030	0.918	4
Spain	78.0	97.2	15,930	0.894	9
Portugal	75.3	90.8	14,270	0.858	3
Italy	78.2	98.3	20,290	0.900	2
Greece	78.1	96.6	12,770	0.867	8
Other Northern Mediterranean					
Malta	77.2	91.1	13,180	0.850	2
Cyprus	77.8	95.9	14,200	0.870	6
Slovenia	74.4	99.0	11,800	0.845	5
Croatia	72.6	97.7	4,895	0.773	18
Macedonia	73.1	94.0	3,210	0.746	28
Albania	72.8	85.0	2,120	0.699	19
Turkey	69.0	83.2	6,350	0.728	−22
Maghreb					
Morocco	66.6	45.9	3,310	0.582	−27
Algeria	68.9	60.3	4,460	0.665	−31
Tunisia	69.5	67.0	5,300	0.695	−34
Other Southern Mediterranean					
Libya	70.0	76.5	6,700	0.756	−6
Egypt	66.3	52.7	3,050	0.616	−14
Lebanon	69.9	84.4	5,940	0.749	−4
Syria	68.9	71.6	3,250	0.663	−11
Israel	77.8	95.4	18,150	0.883	3

Source: UNDP (1999).

and Morocco (46 per cent). These low education rates derive from a structural unevenness in educational provision, with strong rural-urban and gender contrasts. Many Islamic countries suffer from high rates of female adult illiteracy which are commonly twice as high as the figures for males. Furthermore the geographical distribution of the population in some of these countries, with many rural people living in dispersed and remote locations, makes full educational provision very difficult.

The Human Development Index synthesises the first two columns in table 3.2 with a third variable, not recorded in the table – combined participation rates in primary, secondary and third-level education – and a factor based on GDP per head recorded in the third column.[4] The HDI figures are largely self-evident. They range from 0.918 for France to 0.582 for Morocco, and fall into a number of groups which we will comment on presently. Before we do that, attention is drawn to the final column of table 3.2 which expresses individual countries' differences in the world rankings as between their HDI performance and their GDP per capita position. A positive score indicates better performance on the broader HDI (which includes social and health indicators) than for GDP per head; a negative figure indicates a better performance on the narrow wealth-based GDP ranking than for overall human development.

A clear division is evident. European countries (plus Israel) score better on HDI than on real GDP per capita. The southern Mediterranean countries, on the other hand, all score worse on HDI than their wealth would imply, due to their under-performance on indicators of life expectancy and educational attainment. In the case of Algeria, the poor results stem largely from the fact that investment in people has been sacrificed to industrial growth. The low status of women is also important, and this factor is a constant feature, to varying degrees, throughout the Islamic countries listed in the table. It is also interesting to note the countries with the greatest positive difference on the ratings: Croatia, Macedonia and Albania have inherited the Communist system's commitment to universal state education and health care, but their GDP figures have lagged behind in the aftermath of the Balkan crises.

Figure 3.1 provides a framework for an attempt to categorise Mediterranean countries on the dual criteria of wealth and human development: the axes of the graph represent the third and fourth columns of table 3.2. The diagram shows that the countries fall into two main classes, more and less developed, within which there are subclusters. The main cluster of developed nations have HDI values of at least 0.85 and real GDP per capita figures above $12,000; these countries are all located within the top band of high human development at a world level. Slovenia (0.845 and $11,800) is the 'bottom marker' of this group but there is no doubt that it should belong to this category given the large gap on the graph between it and its poorer

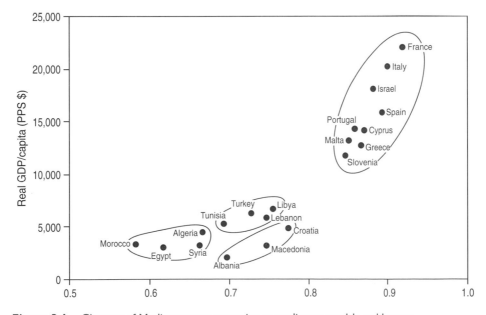

Figure 3.1 Clusters of Mediterranean countries according to wealth and human development, 1997

neighbours. Within the rich and highly developed group of countries France and Italy stand supreme, whilst five other countries – Portugal, Greece, Cyprus, Malta and Slovenia – are tightly bunched.

The large group of poorer Mediterranean countries are defined as those having HDI values of less than 0.8 and per capita GDPs of less than $7,000. Three subclusters can be identified (figure 3.1). Croatia, Macedonia and Albania have HDIs of 0.7–0.8 but rather low GDPs, for reasons noted above. The remaining countries divide into two subgroups, with Libya, Lebanon, Turkey and Tunisia enjoying a more advanced status than Morocco, Egypt, Algeria and Syria. On a global scale, these southern and eastern Mediterranean countries have medium or medium-low levels of human development; they rank decidedly above the Saharan and sub-Saharan African countries, which have a very low level of human development.

Critique and elaboration of the human development approach
The UNDP's decade-long discourse on human development has not gone unchallenged, however, and the UNDP itself draws attention to some of the shortcomings of its indicators, and periodically proposes refinements and new indicators.[5] Amongst the more fundamental criticisms are the accusation that 'human development' is an oxymoron which implies that all previous development has been 'inhuman' (Rist 1996), and the more complex argument that the choice of variables as components of the HDI is subjective and that the indicators themselves are culture-dependent (Benyaklef 1997). Other criticisms have focused on the omission of environmental aspects of quality of life and the scant regard paid to issues of long-term sustainability, whilst important questions remain to be explored about the distribution and quality of the parameters measured within the national territories and populations of the countries concerned (see, for instance, Benyaklef 1997; Doessel and Gounder 1994; Luchters and Menkhoff 1996; Streeten 1995).

The distributional question is fundamental. We have already briefly noted the gender disparity when discussing statistics on literacy. Whilst women generally have greater life expectancy than men, in most underdeveloped and semi-developed countries they receive less education and, often, a poorer diet and less health care than men. Of course gender inequalities are by no means absent in the developed world, and there are some powerful social expressions of patriarchy surviving in southern European countries like Spain and Greece, but generally gender inequality is greatest in the poorer countries (UNDP 1997, pp. 38–40). In recent years the UNDP has introduced a gender-related development index (GDI) which is based on the same variables as the HDI but which takes into account the differences between the sexes in their mean scores on the component variables. The calculations for this are quite complex (see UNDP 1997, pp. 123–4 for the technical details), but Bonavero and Dansero (1998) offer a useful graph (figure 3.2) which summarises the difference between HDI and GDI in percentage terms. The largest percentage – representing the greatest degree of gender inequality – is Libya with over 18 per cent; in this country, for instance, the literacy rate for men is 89 per cent and for women 59 per cent. As figure 3.2 shows, the lowest gender inequality rates, around 2 per cent, are for France, Croatia, Slovenia and, lowest of all, Albania. Also worthy of note is the relatively low gender inequality figure for Turkey: at just over 4 per

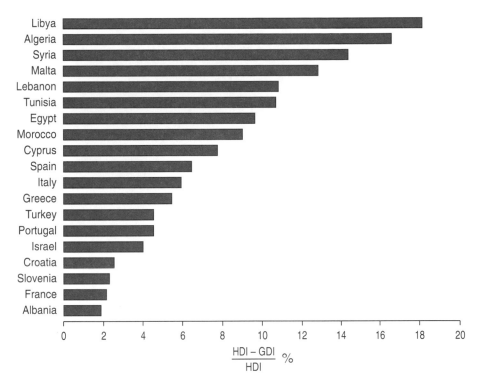

Figure 3.2 Differences between human development (HDI) and gender development (GDI) in Mediterranean countries, 1994
Source: Bonavero and Dansero (1998, p. 306); UNDP (1997).

cent, it is lower than the figures for many richer European countries which, in turn, have only slightly more favourable figures than Morocco, Egypt and Tunisia. These countries, and Turkey, have made considerable efforts to improve gender equality in recent years through education and legislative measures (Bonavero and Dansero 1998, p. 305).

In addition to gender inequality we also have to bear in mind social and spatial variation in development indicators. Regional disparities are of particular importance to economic geographers and here we summarise some key patterns and trends, drawing on the work of Benyaklef (1997), Bonavero and Dansero (1998) and Reiffers (1997).

The strongest regional contrasts are generally expressed between major cities and rural areas, and between the coast and the interior, of Mediterranean countries. The 'littoralisation' of the Mediterranean population, which has been amply documented (see for example Grenon and Batisse 1989; Joannon and Tirone 1990; Reiffers 1997), builds on the concentration of most of the large towns, tourist areas, industrial complexes and transport axes along the coastal regions. Hence it is in these coastal areas that the highest levels of income, education and other forms of social well-being are recorded. Italy is the European Mediterranean country with the greatest regional imbalance in per capita GDP, because of the well-known gap between the less-developed South and the rest of the country. Spain too has a large contrast in per capita wealth: between Barcelona, the Balearic Islands and Madrid on the one hand, and the rest of the country on the other. Marked regional imbalances are also found in many countries of the southern and eastern shores of the Mediterranean, above all in Turkey, Egypt, Tunisia, Algeria and Morocco: in all cases the coastal areas are the richest and fastest-growing, for the reasons stated above.

Regional disparities in the provision of services are also often marked. Figure 3.3 plots the pattern of regional imbalance in access to hospital beds. The graph shows an inverse correlation between the level of health coverage and the regional disparities of provision within each country. European countries have a relatively high number of beds per 1000 inhabitants and this high level of provision is more or less universal throughout the whole country, except for Greece where hospital services are over-concentrated in Athens. Egypt, Turkey and the Maghreb states have much lower ratios of beds to inhabitants but much higher regional contrasts in availability – this variation being mainly articulated between urbanised and rural regions of the countries.

Another example of the danger of reading too much into national averages is provided by a consideration of literacy figures for Tunisia: whilst the overall figure is 67 per cent, this varies from close to 100 per cent for young males living in the city to less than 40 per cent for older females living in rural areas. We also have to be

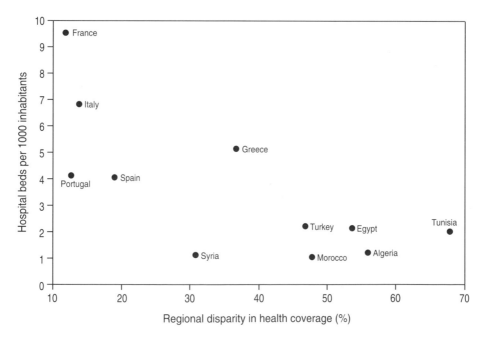

Figure 3.3 Regional disparities in health services (hospital beds) in selected Mediterranean countries, 1992–94
Source: Reiffers (1997, p. 195).

aware of variations in the *quality* of the services recorded in the statistics. Some literate people may be barely able to read and write because of the poor quality of their schooling and lack of practice. Likewise data on access to health services such as doctors or hospitals should ideally take account of the quality of medical services which may be very different in, say, Egypt or Morocco compared to France or Spain.

A different set of issues arises over dimensions of development which are not considered by the conventional indices. Sparrow (1998) presents some interesting data and reflections on the incidence of corruption amongst the Mediterranean countries. Although Sparrow claims that corruption in its various forms is rather prevalent in the southern Mediterranean countries, a graph which plots real GDP per capita against a 'probity index' shows a number of Southern European countries performing poorly. In fact Italy is on a par with Egypt and Turkey, and Spain and Greece are only slightly better.[6]

Yet another issue which should be taken into account when analysing the economic and social characteristics of the Mediterranean is the dynamic role of the informal economy. According to Bonavero and Dansero (1998, p. 305), the informal sector plays an important role in virtually all of the countries of the Basin and can be considered a Mediterranean trait. The activities carried out in the informal sector by definition escape most official statistics and link up with some dimensions of corruption. Typical activities in the diverse world of the informal economy include self-employed individuals and public officials taking second jobs which are not recorded (hence they pay no tax or social security on their earnings), small enterprises which are not registered (e.g. small builders and traders), family firms relying on family and child labour, and individuals engaged in parallel markets dealing in smuggled or stolen goods.

Although some forms of informal economy have probably been endemic to the Mediterranean for centuries, this type of activity appears to have increased since the 1970s due to the contraction of the formal economy in industry, the squeeze on public sector jobs and the implementation of structural adjustment programmes. Estimates which are commonly quoted suggest that 20–30 per cent of total economic activity by value derives from the informal sector in Greece, Italy and Spain (King and Konjhodzic 1996, pp. 43–52), whilst employment estimates for four North African countries indicate a sharp increase of informal employment up to 40–50 per cent of the total by the end of the 1980s (Dunford 1997, p. 150).

Trends through time

So far we have explored various dimensions and indicators of development and underdevelopment for the Mediterranean countries in the 1990s, using a static frame of analysis. Although the developmental status of many countries, especially on the southern and eastern shores, remains weak, the situation is much better than it was 20 or 30 years ago. As tables 3.3 and 3.4 show, scores on HDI and wealth have increased considerably since 1975, although the pattern of improvement varies from country to country (the tables are limited to those countries with complete data sets). The figures indicate (referring also to table 3.2) that by 1997 some southern Mediterranean countries, notably Turkey, Tunisia and Lebanon, were

Table 3.3 Mediterranean countries: trends in human development, 1975–97

	Human Development Index			Change 1975–97	
	1975	1985	1997	Absolute	%
North Mediterranean					
France	0.848	0.875	0.918	0.070	8.3
Spain	0.814	0.851	0.894	0.080	9.8
Portugal	0.735	0.786	0.858	0.123	16.7
Italy	0.824	0.852	0.900	0.076	8.9
Greece	0.792	0.835	0.867	0.075	9.5
South Mediterranean					
Morocco	0.426	0.508	0.582	0.156	36.6
Algeria	0.511	0.605	0.665	0.154	30.1
Tunisia	0.510	0.608	0.695	0.185	36.3
Egypt	0.432	0.531	0.616	0.184	42.6

Source: UNDP (1999).

Table 3.4 Mediterranean countries: trends in income, 1975–97

	GDP per capita (1987 US $)			Change 1975–97	
	1975	1985	1997	Absolute	%
North Mediterranean					
France	12,763	15,324	18,554	5,791	45.4
Spain	6,415	6,992	9,591	3,176	49.5
Portugal	3,117	3,794	5,564	2,447	78.5
Italy	9,629	12,637	15,548	5,919	61.5
Greece	4,552	5,557	6,583	2,031	44.6
South Mediterranean					
Turkey	1,284	1,478	1,940	656	51.1
Morocco	641	822	927	286	44.6
Algeria	2,315	2,966	2,352	37	1.6
Tunisia	980	1,272	1,670	690	70.4
Egypt	467	827	1,015	548	117.3
Syria	998	1,132	1,288	290	29.1
World	2,888	3,174	3,610	722	25.0

Source: UNDP (1999).

getting close to the HDI values recorded by Portugal and Greece in 1975, although the gap remained wider for GDP per capita.

Trends in HDI scores over the period 1975–97 reveal two predominant rates of change: a modest rate of around 9 per cent for four European countries, and a much higher rate of 30–40 per cent for the four North African countries, with Portugal (17 per cent) in between. Certainly, on the criterion of the HDI, the southern Mediterranean countries are catching up with their northern neighbours, although it is also true that, as HDI gets higher and hence closer to its theoretical maximum of 1.0, further increases are likely to be lower both in absolute and percentage terms.

Trends in the evolution of GDP over the period 1975–97 show a rather different set of outcomes. First, there is no ceiling, as with the HDI, and so progress for the richest countries is not 'capped'. Second, whilst the trends for the Southern European countries are rather uniform (though Portugal once again stands out as recording the highest percentage – but not absolute – increase), the pattern of evolution amongst the southern and eastern shore countries is highly variable. Algeria made virtually no progress whilst Egypt, starting from the lowest base level, more than doubled GDP per capita, and Tunisia also performed well. All countries in table 3.4, except Algeria, grew faster than the world average over the period in question, confirming the status of the Mediterranean as an economically dynamic global region.

Two further dimensions of relative economic change are next explored: first the

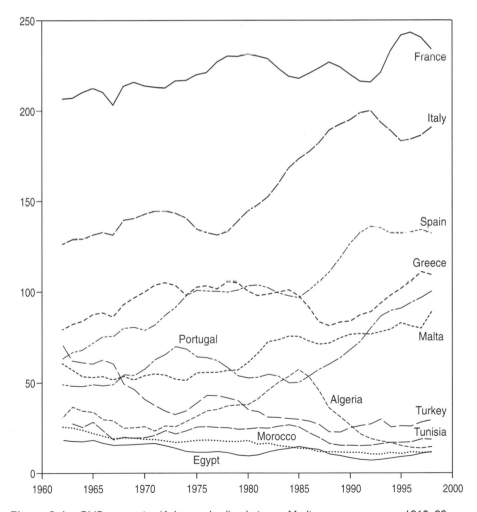

Figure 3.4 GNP per capita (Atlas method) relative to Mediterranean average, 1960–98
Source: Elaborated from World Bank (2000).

changing positions of national economies in relation to the Mediterranean average, and second the trends towards either greater or lesser regional economic disparity within individual countries.

Figure 3.4 plots per capita GNP (Atlas method) for those Mediterranean countries for which a full set of trend data is available for the period 1960–98; the national profiles are measured against the Mediterranean average (100) for each year in the series. A number of different models of change can be identified. First, all EU countries recorded an improvement in their positions. France remained the leading economic power throughout the period, although Italy grew especially rapidly during 1975–90, to almost close the gap, before diverging again in the early 1990s. Spain and Portugal advanced their positions after joining the EU in 1986. Spain's improvement faltered in the 1990s, but Portugal's continued strongly upward to eventually reach the Mediterranean average. Greece developed strongly in the 1960s, then entered a period of fluctuating change before recovering in the late 1980s and early 1990s. Malta is the final European country in this set and has exhibited a general, if unspectacular, upward shift in its relative position *vis-à-vis* the Mediterranean average. A second group of countries, all below the mean line, and therefore poor by Mediterranean standards, saw their position on the GNP criterion generally worsen. For Tunisia, Morocco and Egypt, the curve is relatively flat; the greatest relative decline since 1960 is registered by Turkey. Algeria achieved strong GNP growth during 1973–85, but then changing economic fortunes linked to the price of oil and gas and associated industrialisation led to a sharp downturn during 1985–95. Even more sharply fluctuating trends are recorded by the incomplete data for Libya and Israel but these partial curves are not plotted.

Finally, figure 3.5 sets out the evolution of inter-regional disparities in per capita GDP for several Mediterranean countries over the period 1985–95. Values greater

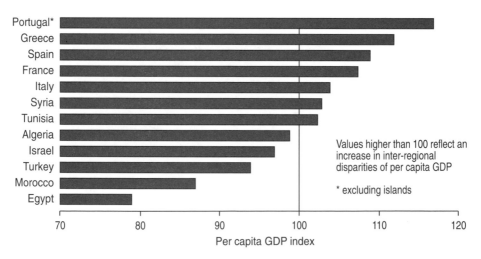

Figure 3.5 Evolution of inter-regional disparities in per capita GDP, 1985–95, selected Mediterranean countries
Source: Reiffers (1997, p. 191).

than 100 indicate an increase in regional imbalance, those less than 100 indicate regional economic convergence within the countries concerned. The five EU countries all experienced increasing regional inequality, whilst several of the Basin's poorer countries – Morocco, Egypt, Turkey (but also Israel) – witnessed a narrowing of the inter-regional income gap. Suggested reasons are as follows (Reiffers 1997, p. 191). In the cases of Portugal and Spain, rapid national development was led by the richest regions which thereby drew further ahead of the less dynamic regions. In Greece, overall national growth was lower, but regional disparities nevertheless increased as the more wealthy regions survived the restructuring process better than the less prosperous parts of the country. Moving to the two cases of marked regional convergence, Egypt found that its better-off governorates reverted towards the norm as the general process of national development was rather evenly spread, whilst in Morocco the dominance of Casablanca and Rabat was diluted by rapid growth in some southern provinces (notably Agadir and Meknes) and others in the northern coastal region (notably Tetouan).

The changing shape of the economy

Patterns of employment
Recent decades have seen profound changes in the structure of Mediterranean economies. A simple way of tracing these changes is through the shifting balance of employment in agriculture, industry and services (table 3.5). In 1965 Albania, Turkey, Morocco, Algeria, Egypt and Syria all had over half their working populations engaged in farming, and another group of countries – Tunisia, Libya, Greece and Cyprus – had over 40 per cent engaged in agriculture. Only in France, Italy, Malta, Lebanon and Israel was agriculture not the dominant employment sector. By 1992 agriculture had become much less important in proportionate terms in all countries.[7] Numbers continue to fall, although in those countries where farming now only employs a residual share of the working population (France, Italy, Malta and Israel) the scope for further agricultural exodus is limited.

In terms of cross-Mediterranean contrasts, what the data in table 3.5 show, in synthesis, is a northern Mediterranean (including Israel) where, by the 1990s, at least 80 per cent of the population were in service and industrial occupations, and increasingly in the former. In these countries, modernisation and urbanisation have reached an advanced stage, and even if large numbers of people continue to live in rural areas, this does not mean that they earn their living from farming. This can be contrasted with the southern and eastern Mediterranean where the employment structure was more evenly distributed amongst primary, secondary and tertiary sectors in 1992, but changing fast. It is also noticeable that whereas in the non-European Mediterranean both industry and services increased their employment shares over the period 1965–92, in the European countries (and Israel), industrial employment declined, leaving a very strong expansion in tertiary sector jobs to compensate for the decline in both the agricultural and industrial sectors. Overall, the Mediterranean economic pattern is for decline of agricultural employment to be linked to a direct switch into tertiary employment, not via the intermediate step of

Table 3.5 Mediterranean countries: percentage distribution of employment, 1965–92

	Agriculture		Industry		Services	
	1965	1992	1965	1992	1965	1992
Mediterranean EU						
France	18	6	39	29	43	65
Spain	34	11	34	33	32	56
Portugal	38	17	30	34	32	49
Italy	24	9	41	32	34	59
Greece	47	23	24	27	29	50
Other North Mediterranean						
Malta	8	3	41	28	51	69
Cyprus	40	15	27	21	33	64
Albania	69	56	19	19	12	25
Turkey	75	47	11	20	14	33
Maghreb						
Morocco	62	46	15	25	24	29
Algeria	57	18	17	33	26	49
Tunisia	49	26	21	34	30	40
Other South Mediterranean						
Libya	41	20	21	30	38	50
Egypt	55	42	15	21	30	37
Lebanon	28	14	24	27	47	59
Syria	52	23	20	29	28	48
Israel	12	4	35	22	53	74

Source: World Bank (1994).

industrial employment as happened in most North European countries (Charmes *et al.* 1993, p. 9).

Three further points can be made about the data in table 3.5 and the above discussion. The first is that the nature of the data overlooks the reality that many people in the region hold multi-sectoral employment profiles. This is especially the case in rural areas where farmers may engage in other activities either on or off the farm – for instance, running an agri-tourism enterprise such as a guest house, or working part-time in the transport sector. But it can also be that an industrial or an office employee works part-time in farming in the evening and at weekends. Such pluri-activity is not recorded in the statistics, which arbitrarily assign individuals to single categories of employment.[8]

The second point is that the distribution of employment does not necessarily match the proportionate shares of agriculture, industry and services in the GDP. A full set of GDP data to compare with table 3.5 does not exist, but partial data shows how the agricultural sector 'underperforms' compared to its weight in employment terms. Compare, for example, the share of agriculture in GDP in the early 1990s for the following countries (employment shares in brackets): France 3(6), Spain 5(11), Portugal 6(17), Italy 3(9), Cyprus 7(15), Turkey 18(47), Morocco 18(46), Egypt 19(42). These figures indicate that farmers are less than half as productive as workers in other economic sectors. However, some of this statistical economic underperformance is explained by an element of self-sufficiency and barter, by which some agricultural produce does not enter the market system and thus is excluded from the GDP.

Third, the nature of work differs considerably from one country to another, again particularly between the northern and southern sides of the Mediterranean. An industrial worker in France or Italy is more likely to work in a modern factory than counterparts in Morocco or Egypt who probably labour in cramped workshops. And whereas in the towns and cities of Mediterranean Europe the expansion of tertiary sector employment is based on the growth of capital-intensive and sophisticated services, in the southern Mediterranean it is more likely to involve petty commercial activities and informal services (Charmes *et al.* 1993, pp. 9–10).

The agricultural sector

Let us look in more detail at the evolution of the agricultural sector in the Mediterranean. As we saw from the beginning of this chapter, agriculture is one of the unifying themes of the Mediterranean economy and landscape. Similar patterns of agricultural production and rural life are observable right across the region. Dry-farming of wheat, permanent tree-crops such as the olive, fig, vine and almond, and the pasturing of sheep are a farming trilogy which is found from the plateaux of Provence to the steppes of Tunisia, and from the rolling hills of Castille to Anatolia (Joannon and Tirone 1990, p. 51). But profound changes have taken place in recent decades. Many of the traditional elements of Mediterranean agriculture are in crisis due to lack of competitiveness on European and other markets, shortage of labour, and the difficulties of rationalising production on land which is often marginal due to its hilly nature, poor soils, unreliable rainfall, and outdated tenures.

Table 3.6 presents five indicators of agricultural change for the period between *circa* 1980 and the mid-late 1990s. The countries selected (on the basis of data availability) are the five Mediterranean EU countries, seven poorer countries of the southern and eastern Mediterranean, and Israel. The data reveal the following main trends and contrasts. First, the proportion of land under permanent crops is static or declining in the European countries, whereas in most of the other countries it increased over the period 1980–96. This contrast is broadly consistent with the food output data on the right-hand side of the table, where static post-1990 indices for the European countries (note how close the 1995–97 values are to the 1989–91 index of 100), are well below the indicators of food output growth for the other Mediterranean countries listed on the table. However, behind these production data lie other statistical trends which, if anything, reverse this interpretation. In the European countries the numbers of people working in agriculture fell considerably over the period 1980–96 (hence productivity increased in terms of output per agricultural worker), whilst the national consumption demand grew very slowly because of falling birth rates and a near-static total population which already enjoyed high standards of nutrition. In the remaining countries, rising food output was achieved against a background of a dense rural population and a rather rapidly rising total population, which had the effect of nullifying the apparent productivity gains. As a result, most southern Mediterranean countries have shown a progressive inability

Table 3.6 Mediterranean countries: agricultural indicators, c. 1980–96

	Land under permanent crops as % of total land		Irrigated as % of total cropland		Tractors per 1000 agricultural workers		Food production index (1989–91 = 100)	
	1980	1996	1980	1995	1980	1995	1980	1996
Mediterranean EU								
France	2.5	2.1	4.6	8.2	737	1,189	93.7	103.6
Spain	9.9	9.8	14.8	17.7	200	513	82.1	99.4
Portugal	7.8	8.2	20.1	21.7	72	203	71.9	99.8
Italy	10.0	9.1	19.3	24.9	370	867	101.5	99.7
Greece	7.9	8.4	24.2	33.8	120	267	91.2	98.4
Southern Mediterranean								
Turkey	4.1	3.2	9.6	15.4	38	57	75.8	106.3
Morocco	1.1	1.9	15.2	13.0	7	10	55.9	94.9
Algeria	0.3	0.2	3.4	6.9	27	43	69.7	118.2
Tunisia	9.7	13.1	4.9	7.5	30	39	67.6	108.3
Egypt	0.2	0.5	100.0	100.0	4	10	68.4	129.8
Lebanon	8.9	12.5	28.3	28.4	28	77	57.8	117.6
Syria	2.5	3.9	9.6	20.4	29	65	94.5	136.7
Israel	4.3	4.2	49.3	45.3	294	336	85.7	114.1

Note: All data in the table refer to three-year averages (hence 1980 = 1979–81, 1996 = 1995–7 etc.).
Source: World Bank (2000).

to feed themselves, and food imports have become a significant part of the trade balance. In Algeria and Egypt food accounts for 30 per cent of the value of all imports. Only in Turkey and Israel do food exports match food imports. Whilst Morocco exports large quantities of citrus fruit and salad products, and Egypt and Syria export cotton, all the Arab countries of the Mediterranean Basin are massive net food importers.

The middle sections of table 3.6 show common trends across all countries, with one or two exceptions. Both mechanisation and irrigation increased strongly, but at very different absolute levels. Irrigation became more widespread in all countries except Morocco, where it registered a small decline, and Egypt where all land is irrigated anyway. Mechanisation of farming has occurred in all countries, although the variation in 'tractorisation' is enormous – there being 100 times more tractors per thousand farmers in France and Italy than in Egypt and Morocco.

Within each country there are marked contrasts between areas of agricultural intensification and areas of abandonment; over time these contrasts have tended to become more marked. Much agricultural dynamism comes from regions of specialised irrigated agriculture, which tend to be based on agricultural regions which have been important for centuries, even millennia, for their fertility and productivity – the Nile Valley, the irrigated coastal plains of the Maghreb, the *huertas* (irrigated gardens) of Valencia and Murcia, the coastal and estuarine plains of Naples and Palermo, the lower slopes of Vesuvius and Etna (Pratt and Funnell 1997). The extension of irrigation has been based both on small-scale individual and cooperative efforts (wells, small lakes and river-water extraction) and on the large-scale engineering of massive dams such as the Aswan on the Nile or the dozen or more barrages on the Tigris and Euphrates in eastern Turkey. As we saw in the previous chapter, conflicts emerge over 'hydraulic politics', as between Turkey, Syria and Iraq, between Israel and Jordan, or between the water-rich and water-poor autonomous regions of Spain.

Despite its overall uniformity, Mediterranean agriculture has always been highly diversified regionally and locally. Sharp contrasts in the physical nature of the terrain give rise to wide variations in the productivity of the land, but also important are the market prices (and price support regimes) for individual products, and the social relations of agricultural production – the distribution of land ownership and the relationships between landlord and tenant, labourer and boss. Most Mediterranean farming is based, still today, on small-scale holdings, *minifundia*. This does not necessarily mean low productivity or poor economic viability, since irrigation and specialisation can yield high levels of profitability on at least some small farms. In fact, it is remarkable that so much commercialisation and modernisation of Mediterranean agriculture has taken place without much change in the agrarian structure which remains dominated by small holdings (table 3.7). On the other hand, large estates or *latifundia* have also existed, notably in southern Spain, southern Italy, parts of Turkey and in Syria, Egypt and the Maghreb. In some of these cases the persistence of large holdings in the face of mounting population pressure and land hunger has been due to the survival and reproduction of indigenous landed elites; in North Africa colonialism played a dominant role. Despite a whole series of land reforms, starting in Italy in 1950 and in Egypt in 1952,

Table 3.7 Size distribution of farm holdings, selected Mediterranean countries, c. 1993

	Size classes (hectares)				
	1–5	5–10	10–20	20–50	50+
France (1993)					
no. of owners (% total)	27	10	13	26	24
area owned (% total)	2	2	5	24	67
Spain (1993)					
no. of owners (% total)	58	16	11	8	7
area owned (% total)	6	9	9	14	65
Italy (1993)					
no. of owners (% total)	77	11	6	4	2
area owned (% total)	14	13	16	20	37
Greece (1993)					
no. of owners (% total)	76	15	7	2	~0
area owned (% total)	32	24	20	16	8
Turkey (1991)					
no. of owners (% total)	68	18	9	4	1
area owned (% total)	24	21	21	20	15
Tunisia (1990)					
no. of owners (% total)	40	25	21	11	3
area owned (% total)	8	13	20	23	36

Source: Médagri (1999).

polarised landholding systems characterise nearly all Mediterranean countries today (table 3.7).

Joannon and Tirone (1990, p. 56) synthesise the current Mediterranean agrarian scene by identifying two broad groups of farmers. First, there are those who are progressive, professionalised and commercially oriented. They tend to live full-time from their farms, which are larger, more mechanised and more likely to be irrigated. Such farms have high yields and are fully integrated into national and international markets. Second, there are the much poorer peasant farmers who tend to occupy the more marginal farming environments (hills, plateaux and mountainsides) and who are unable to specialise in commercially viable products. They live on the edge of the market system and suffer economic insecurity. As well as those who farm their own land we also include in this category farm labourers who receive low wages and who may remain unemployed for parts of the year because of lack of work. The younger members of these marginal rural populations are easily lured away by the prospect of jobs in the industrial and service sectors, or by migration abroad, whilst the older farmers survive, if only just, by having their tiny incomes supported by pensions and welfare payments.

The industrial sector

The Mediterranean does not have a history of large-scale industrialisation, but there are important precursors to postwar industrial expansion to be noted: textiles in

Florence and shipbuilding and glass in Venice in the Middle Ages; eighteenth-century industrial growth in Naples, Palermo and Seville; and nineteenth- and early twentieth-century industrial development both in major inland cities (Milan and Zagreb, for example) and in the larger port cities such as Marseilles, Barcelona and Genoa. Textiles, clothing, food and drink, arms, mechanical industries and shipbuilding were some of the leading sectors during these earlier phases of in-dustrialisation, together with small-scale (in comparison with northern Europe) industrial growth based around deposits of iron and coal in Asturias, the Basque provinces, the western Italian Alps and Slovenia.

Since the early 1960s, the Mediterranean Basin has emerged as an important world industrial region, based on a rather narrow range of mechanical industries (Fiat is the biggest) together with port-based industrial complexes linked for a time to other heavy industries (iron and steel, shipbuilding, processing of imported goods), and then to the strategic location of the region with regard to sources and transport of oil. During the 1960s and 1970s, the oil refinery and the petrochemicals complex became potent symbols of Mediterranean industrialisation, with glinting tank farms and characteristic red-and-white hooped chimneys creating sharp contrasts with the traditional coastal landscapes of fishing ports and terraced olive groves. Iron and steel, cement and aluminium smelting were other important indus-trial developments launched or expanded at this time.

But this phase was short-lived. The economic relapse of the mid-1970s, reinforced by the second oil crisis a few years later, undermined the low-cost platform of the oil-based industrial boom, with the result that, for the past twenty or more years, painful processes of industrial restructuring and closure have hit industrial growth poles which in some cases had only just been launched. The case of Southern Italy, where there was a big push to foster industry-led development in the 1960s and 1970s, is a good example of failed industrialisation, and many European geo-graphers have focused their attention on this regional development experience (for instance Celant 1994; Coppola 1977; Dunford 1988; King 1985 and 1987; Rodgers 1979).

Italian government policy to stimulate the development of the South – *grosso modo* Abruzzo, Molise, Campania, Apulia, Basilicata, Calabria, Sicily and Sardinia – started in 1950 with the land reform and the establishment of the *Cassa per il Mezzogiorno*. Several phases of development policy then ensued:

- 1950–57: emphasis on building up infrastructure (roads, water and elec-tricity) and agriculture (land reform and land reclamation);
- 1957–73: industrial development policies based on tax and grant incentives, on the concept of the growth pole, and on basic industries such as iron and steel, oil and petrochemicals;
- 1973–86: economic crisis and industrial restructuring; establishment of regional governments in all Italian regions; uncertainty over the future of the *Cassa per il Mezzogiorno*;
- 1986 to 2000: abolition of the *Cassa*, concentration on 'ordinary' policies, encouragement of new industrial initiatives (e.g. Fiat at Melfi in Basilicata) as well as small and medium enterprises (SMEs) and vocational training.

Development funds channelled to the South via the *Cassa* and other special agencies were considerable but never rose above 1 per cent of GNP per year. After 40 years of assistance, there was undoubtedly a major improvement in the social conditions of the Mezzogiorno, but the attempt to set in motion an autonomous and self-reproducing pattern of industrialisation failed. Examination of the sequence of ternary diagrams in figure 3.6 shows that, whereas the pattern of employment change 1951–71–91 in the northern Italian regions shifted markedly towards industry before becoming more weighted towards the service sector, in southern regions the evolution was along the horizontal axis leading straight from agriculture to the tertiary sector, with only a slight inflexion towards industrial employment in the 1960s and early 1970s. By 1991 it is true that the distribution of points (one for each region) is much more tightly packed than in 1951, but a north/south divide is still evident, southern regions retaining around 15–20 per cent of their employed populations in agriculture, compared to northern regions with

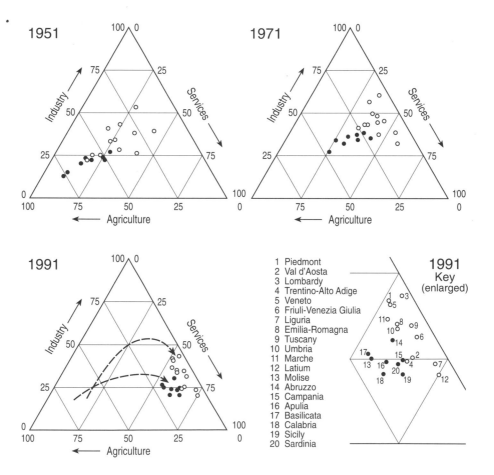

Figure 3.6 Italy: economically active population by sector and region, 1951–91
Source: Celant (1994, pp. 14–15).

less than 10 per cent. However, on closer examination the 1991 graph reveals four clusters (Celant 1994, p. 16):

- at the top of the graph, three regions with the highest levels of industrial employment (Piedmont, Lombardy, Veneto);
- a second cluster just below the first, comprising regions with moderately high levels of industrial employment (30–35 per cent) – Friuli-Venezia Giulia, Emilia-Romagna, Tuscany, Umbria, Marche and Abruzzo (note that the last of these is a 'southern' region);
- a third cluster, located in the bottom-right of the graph, consisting of two regions with very high percentages of tertiary sector employment – Liguria because of its elderly population and tourist vocation, and Latium because of the overwhelming dominance of the service economy of Rome;
- a fourth cluster composed of the remaining southern regions, plus two northern ones (Val d'Aosta, Trentino-Alto Adige) which, because of their mountainous, rural and non-industrial nature, have joined the southern group.

Of course, the employment profile is only one way of measuring structural change in regional economies, and it is necessary to look at other parameters for a more complete evaluation of North/South regional dynamics in Italy. As Celant (1994) shows, the regional income gap has not significantly narrowed: both northern and southern regions have experienced very strong income growth, but along roughly parallel tracks so that the relative divide remains unbridged.[9]

So, what appears to have happened with respect to the industrialisation and development of the South of Italy is this. The South has witnessed a remarkable rise in overall standard of living, in common with all regions of Italy, but in the South this has not been based on solid improvements in economic and industrial potential. Instead, it has been based on some localised developments in agriculture and services and on resource transfers, first through the 'extraordinary' intervention of the *Cassa*, and then on 'ordinary' fiscal transfers equivalent to about 3 per cent of Italian GNP. Those industrial initiatives which have developed in the South since the 1960s have been externally oriented rather than integrated with local regional economic structures. Located near ports, they function on the basis of partial processing of imported raw materials which are then re-exported to northern Italian industrial complexes for final processing and marketing. According to Celant (1994, p. 30) dependence on external industrial systems is great: the Mezzogiorno has many industrial plants which use the 'just-in-time' system, with their respective decision-making centres located in the centre or north of Italy or even abroad. The result is that the industrialisation of the South of Italy has been, and still is today, 'insufficient, fragmentary, episodic, poorly coordinated, not very widespread and with poor staying power'. This description also sums up the experience of industrialisation of many other countries of the Mediterranean as well, including those privileged by possessing their own oil reserves.

Oil, development and indebtedness in the non-European Mediterranean
In 1998 the Middle East and North Africa accounted for two-thirds of the world's proven oil reserves and one fifth of gas reserves. Table 3.8 spells out some of these data, placing the figures on production and reserves for Mediterranean countries within the global context. Because of the dependence of the major world economies (North America, Western Europe, Japan and East Asia) on imports of Arab oil, the production and transport of this key resource have tremendous geopolitical significance. However within the Arab world, the distribution of oil production is extremely uneven. Generally oil is found in sparsely populated countries, whereas the densely populated states lack large reserves of oil and gas (Algeria is an exception to this rule). Amongst the countries of the Mediterranean Basin, the key producers are Libya, Algeria, Egypt, Syria and less so Tunisia. Algeria is also a major producer of gas – 56 billion cubic metres per year in the mid-1990s. But the really big oil producers lie just to the east of the Mediterranean region: Saudi Arabia, the Gulf States, Iraq and Iran. Middle East oil exports depend on land and sea routes that include strategic passages through the Straits of Hormuz, the Suez Canal and the Mediterranean Sea. North African oil is less affected by these strategic bottlenecks, whilst Algerian gas is conveyed north to Europe via the Transmed pipeline to Sicily and a new pipeline under construction via Morocco and Spain.

Oil from Libya and the Middle East is abundant, easy to extract and of high quality. Its costs of production are lower than anywhere else in the world – North Sea oil, for example, costs fifteen times as much to produce. However, there are

Table 3.8 Oil reserves and production: Mediterranean countries and the world context

	Proven reserves (thousand billion barrels)		% world total	Production (thousand barrels daily)		% world total
	1978	1998	1998	1978	1998	1998
North America	57.1	85.1	8.1	14,640	14,165	19.4
South/Central America	25.3	89.5	8.5	4,110	6,730	9.2
Europe	27.4	20.7	2.0	4,520	6,885	9.4
Former Soviet Union	71.0	65.4	6.3	12,595	7,360	10.1
Middle East	369.6	673.7	64.0	15,420	22,795	31.2
Syria	2.1	2.5	0.2	270	565	0.8
Africa	57.9	75.4	7.0	5,715	7,525	10.3
Algeria	6.3	9.2	0.9	1,250	1,385	1.9
Egypt	3.2	3.5	0.3	875	860	1.2
Libya	24.3	29.5	2.8	1,060	1,445	2.0
Tunisia	2.3	0.3	<0.1	105	60	0.1
Asia-Pacific	40.0	43.1	4.1	6,275	7,645	10.5
World	648.3	1,052.9	100.0	63,275	73,105	100.0

Source: BP Amoco (1999).

other costs as well as dangers. Supplies have been threatened by a series of Middle Eastern conflicts – the Six Day War in 1967, the Iran–Iraq conflict of the 1980s and the Gulf War (1990–91) – which have altered prices and upset supply patterns and transport routes. Other price increases have been produced by the actions of the OPEC countries in October 1973, after the Yom Kippur War, and in 1979 following the revolution in Iran: the first of these 'oil crises' resulted in a quadrupling of oil prices almost overnight, whilst the second led to a further tripling of prices over the years 1979–81. After these oil price increases, GDP per capita in the oil-producing states rose to levels which were on a par with, or even exceeded, the advanced industrialised countries of the world.[10]

The growth of oil revenues generated large trade surpluses that were available for

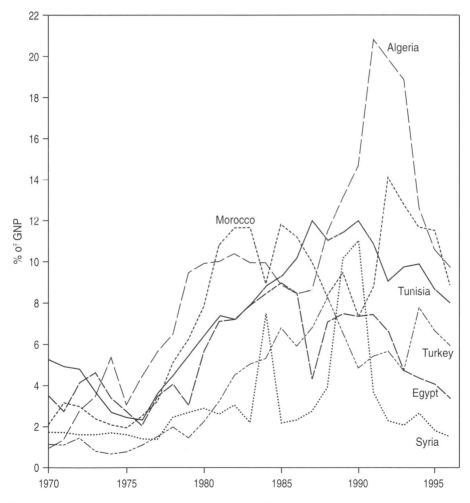

Figure 3.7 Total debt service as a percentage of GNP in some southern Mediterranean countries, 1970–97
Source: Elaborated from World Bank (2000).

investment in other sectors, as well as for spending on consumer goods and welfare services. A large share of new national wealth was diverted into arms expenditure, partly to satisfy national aggrandisement and partly as a reflection of the tense political situation. Most petrodollars, however, were invested in financial assets managed by international financial institutions, creating immense flows of interest payments to the asset owners and their countries. The international financial institutions holding the petrodollars recycled them as loans to less developed countries, including some of the poorer countries around the Mediterranean which used the loans to finance industrialisation and other projects of economic modernisation, including the development of export industries which, it was hoped, would generate earnings to repay the loans. However, the countries concerned did not reckon with the impact of monetarism which had devastating effects for them: first, the profound international recession meant that exports grew more slowly than expected; and second, higher interest rates dramatically increased the cost of debt servicing. Figure 3.7 shows how this cost escalated during the late 1970s and 1980s for a large number of countries which consequently saw their reserves depleted relative to their outstanding debts. In 1986, for example, 59 per cent of Algeria's export earnings were required simply to service its debt; the figures for Morocco (51 per cent) and Egypt (43 per cent) were almost as high.

Faced by mounting economic crisis, many countries had little alternative but to comply with the programmes of structural adjustment imposed by the International Monetary Fund. The consequences were numerous, especially for the North African countries and others in the same boat such as Jordan (Gizard 1993; Valmont 1993). First, reductions in expenditure on health and education threatened the population's quality of life, especially long term. Second, reduced recruitment into public administration added to the burden of unemployment. Reduced subsidies on food and other essential goods were a third element of policy with strong hardship implications. Fourth, there were reductions in the public funds used to finance development and public enterprises. It is not hard to see in these effects of structural adjustment and in the perceived 'failures' of Western-style international economic management the root causes of dissatisfaction and escalating unemployment that often lead to alternative aspirations, either ideological, as with radical Islamist 'solutions', or pragmatic, as with migration to richer countries.

Conclusion

This has been a long chapter, so the conclusion will be short. In the face of powerful processes which enhance the sharpness of the geographical patterning of uneven development within and across the Mediterranean, other processes and institutional initiatives attempt to create a more coherent Mediterranean economic system. The role of trade as an integrating force for the region is considered in the next chapter, where it will be seen that trade links across the Basin have strengthened and expanded, but in an asymmetrical fashion.

The main institutional initiative to 'bridge the development gap' across the Mediterranean has been the Euro-Mediterranean Partnership, launched at the

Barcelona meeting of 1995 which brought together the 15 EU member states and 12 non-member states.[11] The proclamation of the Partnership arose out of a European concern over the political instability of North Africa and a southern Mediterranean concern at the EU's growing support for East European countries, which were perceived as being favoured over relations to the south. The Partnership had three arms: economic aid and trade; political and security issues; and social and cultural development. The economic arm of the Partnership was its most clearly defined feature and involved the establishment of a Euro-Mediterranean free trade area by 2010, supported by nearly Ecu 5 billion of financial aid to the southern countries. The ambition was that these economic measures would stimulate development, create employment, reduce the pressures for migration, and foster social stability.

This is not the place for a full appraisal of the Euro-Med Partnership, not least because, despite its bold rhetoric and path-breaking agenda (or perhaps precisely because of them), relatively little has happened and the initiative has stalled.[12] Indeed, in the view of Joffé (1997), the Partnership has done little more than reflect the ongoing hegemony that Europe has established over the Mediterranean region. Although there is a clear rhetoric of providing 'trade and aid' to the southern Mediterranean countries, there are hidden dimensions to this policy. The first is that trade and aid will buy security from Islamic terrorism and from an 'invasion' of political refugees and economic migrants. The second is that trade liberalisation – designed to facilitate access of southern Mediterranean products to European markets and hence promote employment and development in the southern countries – is very partial and is not extended to those agricultural products which would most benefit North African farmers and challenge the livelihood of European farmers. Meanwhile, according to Joffé, the people of the southern Mediterranean feel that the EU had misunderstood the real significance of Islamism, whilst the neoliberal economic agenda imposed by the West has unwittingly helped radical political forces by diminishing the quality of life for many poor people. Joffé concludes (1997, p. 12) that 'the Partnership is paternalistic, full of contradictions, and offers little hope of resolving the social, economic and political problems of the region'.

Notes

1   GNP is a measure of the value of goods and services a country's residents produce in any one year plus net receipts of primary income (employee compensation and property income) from non-resident sources. To enable comparison across economies GNP, calculated in national currency, is usually converted to US dollars at official exchange rates. To smooth fluctuations in prices and exchange rates, the World Bank Atlas method applies a conversion factor that averages the exchange rate – adjusted if the official rate is considered to diverge by a large margin from the rate applied to international transactions, and for differences in rates of inflation between the country and the G–5 countries (France, Germany, Japan, the United Kingdom, and the United States) – for a given year and the two preceding years.

2   For this comparison Bosnia, Serbia, Libya and Cyprus are omitted from the totals because of missing data on GNP.

3   'Purchasing power parity' refers to the number of units of a country's currency that is required to purchase a standard representative basket of goods and services that a US dollar would buy in the United States. See UNDP (1990) and subsequent UNDP

Human Development Reports for detailed discussions on this and the Human Development Index. Note also that the HDI is based on GDP not GNP per capita; the difference is that GDP refers to the value of the goods and services produced within the territorial confines of the country in question, excluding financial flows such as emigrant remittances.

4   In calculating the index, income above a cut-off point of world average per capita income was traditionally discounted. In 1997 a new method of discounting, involving taking the logarithm of income throughout, was introduced to discount incomes in a more gradual way. For this reason and also because of the use of improved life expectancy data, revised data on adult literacy and combined gross primary, secondary and tertiary enrollment rates and updated data on purchasing power parities – following the more comprehensive 1997–8 surveys of the International Comparison Programme – the figures reported in this chapter differ from those published in the past, including our own analyses (Dunford 1997; King 1998).

5   Amongst these are modifications in the measurement of the basic HDI as well as attention paid to gender (Gender Development Index), empowerment (Gender Empowerment Measure) and the condition of the world's poorest people (Human Poverty Index).

6   This graph, on which not all Mediterranean countries are marked, is in Sparrow (1998, p. 231). Top of the probity chart, with values close to 10, are New Zealand and Denmark; at the bottom, with values of less than 1, are Pakistan and Nigeria. Italy, Spain and Greece are the main cluster of countries which exhibit lower levels of probity (or higher corruption) than their 'development' (real GDP per capita) would suggest.

7   Despite the marked proportionate decline, high rates of population increase in the southern and eastern Mediterranean countries mean that absolute numbers of people working in agriculture remain very high, in some cases even increasing since 1965.

8   According to Papayannaky (1991, p. 192) this explains why the percentage of economically active people working in agriculture in Greece is so high in comparison to other Southern European countries. The figure should be reduced by about 8 percentage points according to this author.

9   This oversimplifies the complexity of economic change at the scale of Italy's 20 regions; in particular it overlooks the strong performance of several north-central regions such as Tuscany and Emilia-Romagna which have tended to displace the main regions of earlier industrial development (Lombardy and Piedmont) in the hierarchy of Italian regional prosperity, due above all to the dynamism of SMEs in the so-called Third Italy.

10 In 1980, when the French per capita GDP figure was $12,680, the figures for Libya and Saudi Arabia were $10,640 and $14,250 respectively. Highest of all were figures for Qatar and the United Arab Emirates – $33,420 and $32,210. Egypt's was just $500.

11 The non-member countries of the Partnership are Turkey, Israel, Cyprus, Malta, Syria, Lebanon, Palestine, Jordan, Egypt, Tunisia, Algeria and Morocco.

12 For literature on this see Pierros *et al.* (1999) and the special issues of *Mediterranean Politics*, 2(1), 1997 and the *Journal of North African Studies*, 3(2), 1998.

References

Banfield, E. C. (1958) *The Moral Basis of a Backward Society.* Glencoe, Ill.: The Free Press.

Benyaklef, M. (1997) Socio-economic disparities in the Mediterranean, *Mediterranean Politics*, 2(1), pp. 93–112.

Boissevain, J. (1976) Uniformity and diversity in the Mediterranean: an essay in interpretation, in Peristiany, J. G. (ed.) *Kinship and Modernization in Mediterranean Society.* Rome: Center for Mediterranean Studies, American Universities Field Staff, pp. 1–11.

Bonavero, P. and Dansero, E. (1998) Human development in the Mediterranean, in Conti, S. and Segre, A. (eds) *Mediterranean Geographies.* Rome: Società Geografica Italiana (Geo-Italy vol. 3), pp. 297–316.

BP Amoco (1999) *Statistical Review of World Energy 1999.*
   http://www.bpamoco.com/worldenergy/oil
Braudel, F. (1979) *Civilisation Matérielle et Capitalisme, XV–XVII Siècles. Les Jeux d'Echange.* Paris: Armand Colin.
Celant, A. (1994) Eliminating the gap: public policies and the development of the Italian South, in *Restructuring Processes in Italy.* Rome: Società Geografica Italiana, pp. 11–38.
Charmes, J., Daboussi, R. and Lebon, A. (1993) *Population, Employment and Migration in the Countries of the Mediterranean Basin.* Geneva: International Labour Office, Mediterranean Information Exchange System on International Migration and Employment, Working Paper 93/1.
Conti, S. and Segre, A., eds (1998) *Mediterranean Geographies.* Rome: Società Geografica Italiana (Geo-Italy vol. 3).
Coppola, P. (1977) *Geografia e Mezzogiorno.* Florence: La Nuova Italia.
Doessel, D. P. and Gounder, R. (1994) Theory and measurement of living levels: some empirical results of the human development index, *Journal of International Development,* 6(4), pp. 415–35.
Dunford, M. (1988) *Capital, the State and Regional Development.* London: Pion.
Dunford, M. (1997) Mediterranean economies: the dynamics of uneven development, in King, R., Proudfoot, L. and Smith, B. (eds) *The Mediterranean: Environment and Society.* London: Arnold, pp. 126–154.
Gizard, X. (1993) *La Méditerranée Inquiète.* Paris: DATAR.
Grenon, M. and Batisse, M. (1989) *Futures for the Mediterranean Basin: the Blue Plan.* Oxford: Oxford University Press.
Joannon, M. and Tirone, L. (1990) La Méditerranée dans ses états, *Méditerranée,* 70(1–2), pp. 1–70.
Joffé, G. (1997) Southern attitudes towards an integrated Mediterranean region, *Mediterranean Politics,* 2(1), pp. 12–29.
King, R. (1985) *The Industrial Geography of Italy.* London: Croom Helm.
King, R. (1987) Italy, in Clout, H. D. (ed.) *Regional Development in Western Europe.* London: David Fulton Publishers, pp. 129–63.
King, R. (1998) The Mediterranean. Europe's Rio Grande, in Anderson, M. and Bort, E. (eds) *The Frontiers of Europe.* London: Pinter, pp. 109–34.
King, R. and Donati, M. (1999) The 'divided' Mediterranean: re-defining European relationships, in Hudson, R. and Williams, A. M. (eds) *Divided Europe: Society and Territory.* London: Sage, pp. 132–62.
King, R. and Konjhodzic, I. (1996) Labour, employment and migration in Southern Europe, in Van Oudenaren, J. (ed.) *Employment, Economic Development and Migration in Southern Europe and the Maghreb.* Santa Monica: RAND, pp. 7–106.
Lipietz, A. (1993) Social Europe, legitimate Europe: the inner and outer boundaries of Europe, *Society and Space,* 11(5), pp. 501–12.
Luchters, G. and Menkhoff, L. (1996) Human development as a statistical artefact, *World Development,* 24(8), pp. 1385–92.
Médagri (1999) *Annuaire des Economies Agricoles et Alimentaires des Pays Méditerranéens et Arabes.* Montpellier: Institut Agronomique Méditerranéen.
Mørch, H. F. C. (1999) Mediterranean agriculture – an agro-ecological strategy, *Geografisk Tidsskrift,* Special Issue 1, pp. 143–56.
Papayannaky, M. (1991) Greek agriculture: current problems and exports, in Montanari, A. (ed.) *Growth and Prospects of the Agrarian Sector in Portugal, Italy, Greece and Turkey.* Naples: Edizioni Scientifiche Italiane (Collana IREM, vol. 1), pp. 189–205.
Pierros, F., Meunier, J. and Abrams, S. (1999) *Bridges and Barriers: The European Union's Mediterranean Policy, 1961–1998.* Aldershot: Ashgate.
Pratt, J. and Funnell, D. (1997) The modernisation of Mediterranean agriculture, in King, R., Proudfoot, L. and Smith, B. (eds) *The Mediterranean: Environment and Society.* London: Arnold, pp. 194–207.
Reiffers, J.-L. (1997) *La Méditerranée aux Portes de l'an 2000.* Paris: Economica.

Rodgers, A. (1979) *Economic Development in Retrospect: The Italian Model and its Significance for Regional Planning in Market-Oriented Economies.* New York: Winston Wiley.

Schneider, P., Schneider, J. and Hansen, E. (1972) Modernisation and development: the role of regional elites and non-corporate groups in the European Mediterranean, *Comparative Studies in Society and History*, 14(3), pp. 328–50.

Sparrow, O. (1998) Framework scenarios for the Mediterranean region, *Journal of North African Studies*, 3(2), pp. 229–46.

Streeten, P. (1995) Human development: the debate about the index, *International Social Science Journal*, 143, pp. 24–37.

Tovias, A. (1994) The Mediterranean economy, in Ludlow, P. (ed.) *Europe and the Mediterranean*. London: Brassey's, pp. 1–46.

UNDP (1990) *Human Development Report 1990.* New York: Oxford University Press.

UNDP (1997) *Human Development Report 1997.* New York: Oxford University Press.

UNDP (1999) *Human Development Report 1999.* New York: Oxford University Press.

Valmont, A., ed. (1993) *Economie et Stratégie dans le Monde Arabe et Musulman.* Paris: EMAM.

World Bank (1994) *World Tables.* Washington: International Bank for Reconstruction and Development.

World Bank (1999) *World Development Indicators.* Washington: International Bank for Reconstruction and Development.

World Bank (2000) *World Tables.* Washington: International Bank for Reconstruction and Development.

4

Spatial patterns of trade as indicators of regional unity in the Mediterranean Basin

W. Jan van den Bremen

The Mediterranean and its rimlands as a regional entity

According to Braudel (1979), the Mediterranean Sea is not one but many seas. Nevertheless, this 'sea between the lands' has been more enabling than discouraging with respect to communication between the peoples of its rimlands and hinterlands: '*mare nostrum et mater nostra*' is a very old and well chosen aphorism which encapsulates this situation. The relative ease of maritime communication within the Mediterranean Basin has contributed to high levels of migration and mixing of human beings, to the diffusion and assimilation of cultural characteristics, especially religion and related ideas about societal structure, and finally to the spatial extension of political power and of exchange and trade. These processes have usually acted to strengthen the regional coherence of the Basin. A high level of unity is perceived by many writers, most notably by Braudel (1966) but also more recently by other authors (Grenon and Batisse 1989; Regelsberger and Wessels 1984).

The central research question addressed in this chapter is simply this: how real is the perceived regional unity of the Mediterranean Basin? This leads to further, related questions. How wide is the gap between 'perceived reality' and the 'factual reality' of regional unity today? In which direction are the various dimensions of regional integration and unity developing? This leads in turn to a discussion about selection of indicators. According to Shmueli (1981) such indicators should be of two kinds:

- homogeneity in characteristics like landscape, culture, economic development;
- spatial interaction and coherence in politics, economics etc.

Chapter I made a general survey of some of these indicators pertaining to the contested interpretations of unity versus diversity in the Mediterranean Basin. The present chapter makes a more detailed analysis of one indicator, commodity trade. At first glance, this may seem a narrow indicator of spatial interaction and regional coherence. On the other hand trade both reflects, and is dependent on, national policies, levels of economic development, socio-cultural affinities, and trade agreements or conflicts between states. The choice of commodity trade for analysis is also prompted by the availability and comparability of appropriate data sources, usually standardised in US dollars.[1]

Background contexts to Mediterranean trade

Against the background of the globalisation and regionalisation of the world economy, and within the context of the strengthening regional dominance of the EU Mediterranean economies during the 1980s and 1990s, the developing countries of the southern and eastern shores of the Mediterranean Basin try in their own way, as independent states, to participate in trade-based political and economic development. Unfortunately, most of these initiatives have been short-lived; more success is being hoped for from the Euro-Mediterranean partnership, referred to at the end of the previous chapter and elsewhere in this book. Although trade is theoretically one of the best ways to profit from the localised comparative advantage that derives from regional differentiation, in practice it seems that, amongst the non-EU countries of the Mediterranean Basin, there are too few economies (with the exception of Israel) with strong comparative advantages and with sufficiently high levels of development and organisation to cooperate effectively in trade stimulation.[2] However, there are sufficient differences in natural factor endowments to stimulate trade from Algeria, Libya, Morocco and Turkey in primary commodities such as crude oil, natural gas and minerals. Furthermore, there is always, according to Krugman (1991, p. 13), 'a strong arbitrary, accidental component to international specialisation', and therefore to trade.

In the Mediterranean region, as in some other parts of the world, geographical distance can hamper trade, but 'transaction problems' seem to be more important as barriers to trade. These problems originate from bad communication and time costs caused by differences in language, customs, regulations and procedures, which are often underlain by deeper and diverging trends in political outlook and socio-cultural endowment. Such divergences exist particularly between the European, historically Christian countries of the north-western quadrant of the Basin and the mainly Arab-Muslim countries of the other three quadrants. But these transaction problems also occur between countries within these groupings. Summing up, although there is some 'nearest neighbour' effect, proximity costs are no longer a crucial control on trade and regional integration; the main constraints are related to transaction problems.

Before concentrating on an analysis of the spatial pattern of trading relations, a short description will be given of the regional differentiation within the Mediterranean Basin, based on individual countries. This will provide some necessary background to understanding patterns of economic and trade integration.

Only a few of the circum-Mediterranean states and their populations are entirely Mediterranean in character: Cyprus, Malta, Albania, Israel and Lebanon fit into this category. The islands of *Malta* and *Cyprus* are left out in the analysis which follows because of their small populations and economies, data problems and (in the case of Cyprus) the political situation. *Albania* also has a modest economy, but more critically has behaved as a political and economic 'outsider' until the early 1990s. Even during the 1990s its economy and civil society were in a state of repeated shock and chaos. Here, too, there are very limited data available. This country will therefore be disregarded in the analysis below.

The once-rich and well-developed *Lebanon* has gone through many years of turmoil within its borders as well as in relation to its nearest neighbours Syria and Israel. It was and still is a real 'trader' and has an 'openness to trade' rather like the island economy of Hong Kong. It maintains strong trade relations outside the Mediterranean Basin, notably with the oil-producing states of the Gulf.

Fifty years as an independent nation-state in the eastern Mediterranean have brought *Israel* few political and economic partners within the Basin, a situation which is the result of serious transfer problems, as defined earlier. However, as was amply demonstrated in the previous chapter, Israel's levels of social and economic development are those of a modern Western economy, and relatively rare to find in this part of the Basin. It maintains strong political and trade relations with countries outside the Basin, especially with the United States. Only these two countries – Lebanon and Israel – sustain export relations with the EU which account for less than half their total export value.

Next, there is a group of countries which have at least half of their population living within the Mediterranean region: Greece (90 per cent), Italy (73 per cent), Tunisia (70 per cent), Libya (63 per cent) and Algeria (53 per cent). *Italy* is a politically young country, but with a long and important history in the Mediterranean Basin. It is one of the four largest circum-Mediterranean countries in terms of population – nearly 60 million – together with France, Turkey and Egypt. In the postwar period it rapidly built a modern economy which is now the fifth largest in the world and second to France among the Mediterranean states. It occupies a central geographical position in the Basin, connecting its eastern and western parts and bridging Europe to North Africa. As a result it enjoys the best locational proximity to all countries of the Basin. As a founding member of the Common Market/EU, it maintains a key position and role in the Mediterranean policies of the Union.

Greece is one of the smaller countries of the Basin. Its economic position is roughly at the median of the 14 countries considered in the analysis. Its location is somewhat comparable to Italy's regarding the possibility of connecting up the eastern and western Mediterranean and it has further links with North Africa (especially Egypt) and the newly-democratised states of Central and Eastern Europe, many of which are candidates for future membership of the EU. However, the recent turbulence in the Balkans has prevented the realisation of this northward-linking role for Greece, while its historical conflictual relationship with its 'large' neighbour Turkey has hampered potential eastward initiatives.

Algeria and *Libya* have very arid hinterlands; population is concentrated in Mediterranean coastal regions rather than the deserts. However, they are rich in oil and natural gas. These assets make them rather one-sided economies, but ready partners in trade with their rich neighbours across the Mediterranean Sea. Their political structures and behaviours tend to frustrate sound international economic relationships; ideologically they align with other Arab states, although some of these alliances are volatile and intermittent. Their export trade with the EU takes a large share of their total exports – 60 per cent for Algeria, 80 per cent for Libya – very high compared to other Basin countries.

The final group of countries are those with a low percentage of their populations living in the Mediterranean sphere, yet which have special interests there. This category contains Spain (36 per cent Mediterranean), Egypt (35 per cent), Turkey (20 per cent), Morocco (15 per cent), the former Yugoslavia (13 per cent), Syria (11 per cent) and France (10 per cent). It is understandable that this group of countries has a character and an orientation which are as much connected to the 'exterior worlds' of the Mediterranean region as to the Basin itself. Nevertheless, their easy access to the Mediterranean means that they have ample opportunity to take part in the economy of the Basin.

France has a dominant role within Western Europe, especially with its Common Market co-founders, Germany and the Benelux countries. Its former colonial ties to the Maghreb and its early political and cultural influence in the Near East have given France a major place in the wider economic and political life of the Mediterranean Basin, and this position is continuing within the frame of recent EU Mediterranean policy. For over 25 years France has accounted for a stable one-third of all intra-Mediterranean exports by value.

Syria is highly involved politically in the eastern Mediterranean realm, especially in Lebanon and eastwards towards the Tigris-Euphrates region. It is rich in raw materials. For more than 10 years the EU has accounted for more than 50 per cent of Syrian export trade.

Morocco is for the main part a country open towards the Atlantic, but it keeps long-established and strong relations with its Maghreb neighbours and with France and Spain, former colonial powers in the country. About two-thirds of its export trade is directed towards the EU, especially France.

In addition to its position at the north-eastern quadrant of the Mediterranean, *Turkey* has a special relation with regard to regions outside the Basin – to the continental Middle East and to the Tigris-Euphrates river valleys and, more recently, to the southern parts of the former Soviet Union near the Black and Caspian Seas. Turkey's industrial output is only just behind that of Spain. However, compared with some other Mediterranean countries, trade relations with the EU are at a moderate to low level, accounting for around half of total export value.

Egypt is historically and culturally highly interconnected with many Mediterranean Basin countries. It holds a strong political position within the Arab-Muslim world, and has a special relationship with both Israel and the Palestinians. A large and fast-growing population is combined with a large but still poor economy; it is one of the heaviest receivers of bilateral aid in the world. The oil-rich Middle East is its nearest and most influential neighbouring bloc.

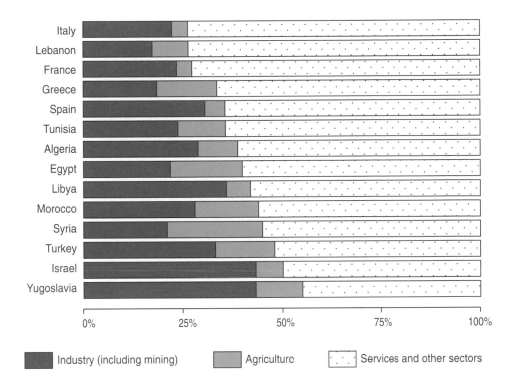

Figure 4.1   Mediterranean countries: division of GDP, 1989

Nevertheless the EU is an important trade partner, receiving more than half of Egypt's exports.

Yugoslavia and its successor states are a region still in turmoil. Whilst first Slovenia and then Croatia attained a measure of economic independence and prosperity during the 1990s, other parts were prevented (or prevented themselves) from partaking in the normal ways of economic life for parts of the decade. Before the early 1990s the country as a whole had an important role in the Mediterranean economy, in manufacturing as well as in trade. Its total GDP equalled the sum of Greece's and Israel's, whilst its export trade virtually matched the combined total of Israel and Turkey, although much of Yugoslavia's exports were directed outside the Mediterranean. Because of data problems, Yugoslavia is included in the following analysis only up to 1990.

Figure 4.1 provides a final piece of background for the trade analysis. It shows the 14 countries' economic structure in terms of percentage shares of GDP contributed by agriculture, industry (manufacturing, mining and quarrying) and services. The year chosen – 1989 – is the last for which Yugoslav data are available.

Globalisation and regionalisation: the EU and the Mediterranean Basin

Nowadays, two directions in the economic relations between countries prevail: globalisation and regionalisation. For trade, globalisation implies the establishment of mutual economic relations between any two countries of the world with regard to the location of business and the exchange of goods (Anderson and Norheim 1994). This orientation of economic life should make it feasible to use the comparative advantages throughout the world in the most efficient way, and by doing this attain the highest prosperity for all countries and the world population in the future. This is the goal of the World Trade Organisation (WTO 1995a, 1995b). The other side of the coin is regionalisation, which implies that countries are more inclined to do business with their near neighbours and with states possessing a similar, and therefore familiar, societal and cultural environment.

For most countries in the world, the Dutch expression 'the skin is nearer than the skirt' describes what seems to happen in reality. Bilateral and specific multilateral trading relations develop with countries which are nearby, or at least not too far away, in geographical and in political and cultural terms. Lack of knowledge of more distant countries combines with travelling/transport costs to boost transaction costs to a level where profitable economic ties are not, for the time being, developed.

Many countries seem to choose a strategy for economic development which lies somewhere between self-sufficiency and globalisation: which means some kind of regionalisation. This situation can be seen to apply to the EU and the Mediterranean countries: starting from a regional trading bloc and moving to an appropriate extension in number and quality of national economies to build up trading links over time. In this way the EU functions as a starting-point for the further Mediterranean neighbours. The target of globalisation can be reached in the future from this more or less 'natural' situation and expansion. But it is a long process during which many tariff and non-tariff barriers to trade have to be dismantled, and transaction thresholds cleared.

Since the end of the Second World War three regional blocs have shaped the global economy: Western Europe, North America and Pacific Asia, dominated in turn by Germany, the United States and Japan. The most emblematic of these regional poles developed in Western Europe during the 1950s. The Common Market included two countries with Mediterranean characteristics as founding members – Italy and France. As the European Economic Community developed, purely economic considerations concerned with production and trade combined with other issues to create a political union regulating all kinds of social, economic and cultural aspects of a growing number of countries – first six, then nine, then twelve and now fifteen. The political boundaries have vanished to promote an open market and a free exchange of people, money and goods. A key overarching objective is to develop a 'territorial unity' with minimal differences in prosperity, a democratic way of government, and stable security for its inhabitants against inside and outside disturbances.

Since 1957 many Mediterranean countries have tried to become full members of the European Community or to obtain a special associate status which would secure trading access for their goods to a large, prosperous and diverse market. However, in 1973 only three 'northern' countries joined as full members; the social, cultural and economic differences between the founding members and the United Kingdom, Ireland and Denmark were only slight. Meantime the political climate was changing in Southern Europe with the transformation to democratic government in Greece, Spain and Portugal, and full membership was accorded to these countries in the early 1980s. But this meant the temporary end to the process of extension of membership in the Mediterranean Basin: Turkey stayed out. With a further 'northward' extension of the EU in the 1990s (Sweden, Finland, Austria), the main thrust for new membership now is towards the more developed economies of Central and Eastern Europe, not towards the Mediterranean.[3]

However, the non-member Mediterranean countries were not left aside. From the early 1970s on, most countries got the possibility, in a framework of bilateral cooperation and association agreements with individual or groups of EC/EU countries, to export their manufactured products to the larger market. Of course, manufactured goods were not their main commodities available for export; and even within the manufactures sector wage-intensive products such as clothing and textiles did not benefit from full access to the European market. Barriers were maintained against Mediterranean agricultural and horticultural products because of competition with similar products from Mediterranean member states of the EU (Featherstone 1993; Pomfret 1986). On the other hand there have never been admission problems with regard to vital raw materials like oil and oil products, even from non-associated countries like Libya.

A 'renewed policy' toward the Mediterranean Basin was introduced in 1989 (Jones 1997). One of its objectives was a better economic integration of the Mediterranean countries, both amongst themselves and with the EC/EU as a whole. Another was maintaining (and improving where possible) access for Mediterranean products to the Community market. At the historic meeting in Barcelona in 1995, the decision was taken to strengthen the trading partnership for the period 1995–99, with the ambitious target of the establishment of a Euro-Mediterranean free-trade area by 2010. This new policy is in harmony with the principle of non-discrimination in the field of trade promoted by the General Agreement on Trade and Tariffs (GATT) and the World Trade Organisation.

In sum, the period from the late 1960s to the mid-late 1990s is an excellent timeframe within which to examine the economic and especially the trade relations within the Mediterranean Basin and between Basin countries and outside trading partners, notably the economic poles of Germany, the United States and Japan.

Trends in export trade

For most Mediterranean countries the value of export trade[4] increased steadily over the thirty years 1966–96, although there were two slow-downs in the early 1980s and early 1990s. Figure 4.2 demonstrates these trends, and is split into two for ease of display of the data. Figure 4.2a shows the European Mediterranean countries,

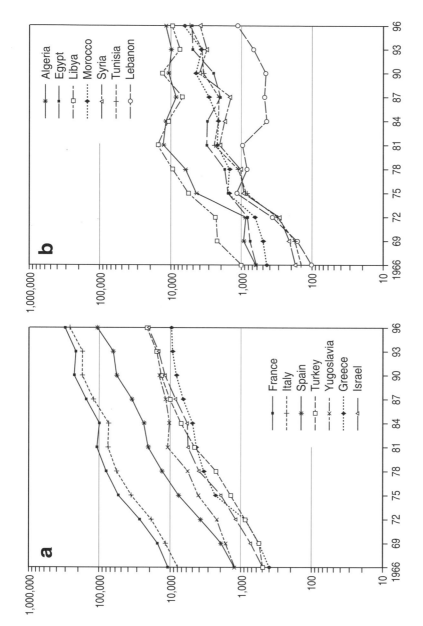

Figure 4.2 Commodity exports, 1966–96 ($ million): (a) European Mediterranean countries plus Turkey and Israel; (b) non-European Mediterranean countries

including Turkey and Israel; figure 4.2b shows the mainly weaker and smaller economies of the southern and eastern Basin countries. World development trends were followed the most closely by the larger and stronger economies of the Mediterranean Basin such as France, Italy, Spain and Israel. Export trade from Greece and Yugoslavia lagged behind after the early 1980s (figure 4.2a). The oil exporters, Algeria and Libya, showed problems with export value growth after the onset of the second oil crisis in 1979. Similar profiles are displayed by other countries in figure 4.2b (Egypt, Morocco, Syria and Tunisia) with the exception of Lebanon, whose diminishing export trade over the period 1975–90 is a result of the country's turbulent political situation.

At the beginning of the officially-stated Mediterranean Policy of the EC in the early 1970s (when only France and Italy were EC members), intra-Mediterranean exports accounted for one fifth of all exports from the Basin countries. By the mid-1990s this proportion had increased to one quarter. During the same period Germany's share of the export value of the Basin countries decreased from 18 to 15 per cent, the USA's from 11 to 7 per cent, and Japan's remained stable at only 2 per cent. Here, then, is some initial evidence for the functional integration of the Mediterranean as a regional economic unit.

Moving to figure 4.3, which presents the proportional shares of all Mediterranean export trade for four sample years over the period 1969–96, it can be seen that France retains a dominant, if slightly diminishing, two-fifths share over the entire period. Over the same period France accounted for one third of all intra-Mediterranean trade. Figure 4.4 shows that the share of France's total export value destined for other Mediterranean countries exceeded 20 per cent in 1969–70 and 1994–5, and over 25 per cent during 1979–80. The picture for Italy is rather similar: a stable, and slightly increasing, one-third share of total Mediterranean Basin country exports (figure 4.3); a 30 per cent share of total intra-Mediterranean trade; and roughly one-quarter of Italian national trade oriented to the Mediterranean (figure 4.4). The third EU Mediterranean country, Spain, saw major developments in its pattern of trade over the period in question. Its share of total Mediterranean commodity exports almost tripled from 5 per cent in 1969 to 14 per cent in 1996; it accounted for an even larger proportionate increase in the share of total intra-Mediterranean trade, 5 per cent to 17 per cent; and figure 4.4 shows that the Mediterranean has accounted for a progressively larger share of Spanish trade over time, rising to one third by 1994–5. However, much of this 'Mediterranean' growth of Spanish trade is with its EU Mediterranean partners, France and Italy. This pattern is reciprocal. Half of France's intra-Mediterranean exports are to its Mediterranean EU partners. For Italy, half of intra-Mediterranean exports by value are directed to France alone, and nearly a quarter to Spain. Figure 4.3 shows that these three economically strong countries are responsible for about 80 per cent of total Mediterranean exports (to all destinations) over the period 1969–96, the percentage increasing from 78 in 1969 to 86 in 1996, largely because of the impact of Spain's export growth. In terms of intra-Mediterranean trade this triumvirate increased their share from 70 to 80 per cent between 1969 and 1996.

The impact of the other 11 Mediterranean countries, with lower levels of development and smaller, less industrialised economies, is much more modest. Just a few

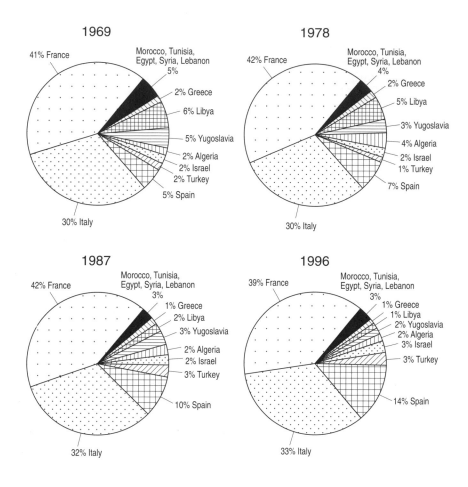

Figure 4.3 Percentage shares of total commodity exports amongst Mediterranean countries, 1969–96 (in 1996 the 'Yugoslavian' share is the aggregated shares of Slovenia, Croatia, Serbia, Bosnia-Herzegovina and Macedonia-FYROM)

noteworthy cases stand out from figure 4.3: Turkey and Israel have increased their shares of total export value, whilst Greece, Libya and (ex-)Yugoslavia have lost ground.[5]

Figure 4.4 shows that, for all Mediterranean countries taken together, the intra-Mediterranean share of the total export trade has increased only slightly to 26 per cent in 1994–5. The highest shares – above 50 per cent in 1994–5 – are held by the Maghreb countries, Libya, Egypt and Syria. Several countries are very unstable as regards the Mediterranean orientation of their trade – for instance Morocco, Syria, Egypt and Lebanon – whilst others (Tunisia and Greece) have a much more regular pattern. Israel is the only country with a very low intra-Mediterranean trade relationship for all three sample years.[6]

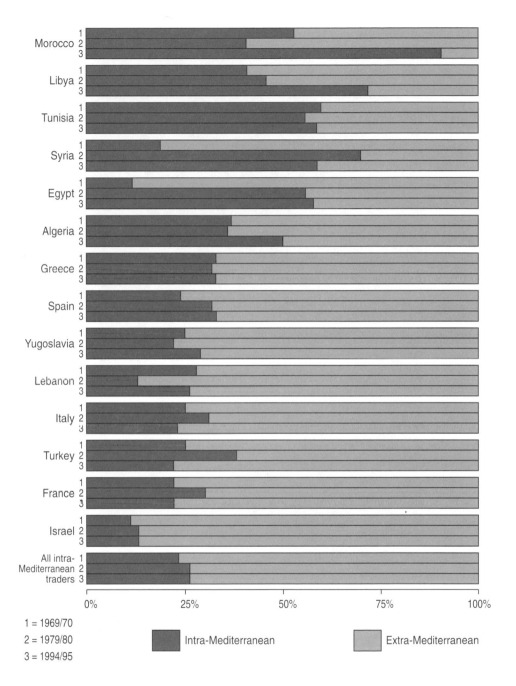

Morocco 1 2 3
Libya 1 2 3
Tunisia 1 2 3
Syria 1 2 3
Egypt 1 2 3
Algeria 1 2 3
Greece 1 2 3
Spain 1 2 3
Yugoslavia 1 2 3
Lebanon 1 2 3
Italy 1 2 3
Turkey 1 2 3
France 1 2 3
Israel 1 2 3
All intra-Mediterranean traders 1 2 3

0%   25%   50%   75%   100%

1 = 1969/70
2 = 1979/80
3 = 1994/95

Intra-Mediterranean   Extra-Mediterranean

Figure 4.4 Mediterranean countries: division of their trade by intra- and extra-Mediterranean origin and destination (by value), 1969–70, 1979–80 and 1994–95

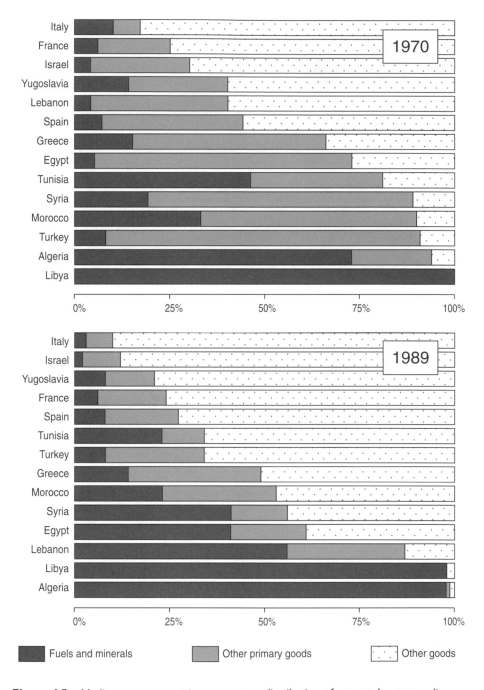

Figure 4.5 Mediterranean countries: percentage distribution of exports by commodity group, 1970 and 1989

The development of export trade can to a considerable extent be explained by changes in the types of commodity traded, which in turn reflect the changing production structures of the exporting countries. Figure 4.5 shows the percentage distribution of exports by commodity group for the 14 countries in 1970 and 1989. Exported goods are collapsed into three simple groups: fuels and minerals, other primary goods (chiefly deriving from agriculture), and other goods (mainly manufactures). In 1970 many Mediterranean exporters had rather one-dimensional export profiles: Algeria and Libya relied mainly on fuel and mineral exports, whilst Turkey, Syria, Egypt and Greece relied heavily on other primary goods. Only Italy, France, Israel, Yugoslavia, Lebanon and Spain derived more than half their export revenue from non-primary exports in 1970. By contrast Turkey, Tunisia and Morocco diversified their exports away from primary products to a significant degree, as a comparison between 1989 and 1970 shows.

This kind of change over time in the composition and 'quality' of commodity trade is an indication of the transformed nature of trade in general and within the Mediterranean Basin in particular. In the next section of the chapter a model will be applied to analyse the spatial pattern of trade over a period of 25 years, with special reference to trade structure and volume of the Mediterranean Basin countries.

Assessment of export trade relations by country: the intensity index

The indicators of world trade discussed above are only fully valid in their description of the levels of interaction and dependence of countries in cases where the economics involved have similar status and shares in the world trading economy. The countries of the Mediterranean Basin are very different in these respects, so there is a problem of appraisal and interpretation, for instance with respect to figures 4.2 and 4.3. To overcome this difficulty a practical but rather complex index is available: the so-called intensity index, a kind of double quotient. This index uses an actual and an expected or hypothetical value for dyads in import/export trade. The hypothetical value is the status or position of a country in the world trading system. Briefly, the index expresses a standardised comparison of the percentage share in country j of the exports of country i to the share of the total exports of country i in world trade. If the value of the quotient is higher than 1 it means that there is an intensive relationship: country i has a more than average dependency on country j for its export trade. If the relation in the reverse direction – exports of country j to country i – is calculated and this also produces a value above 1, this mutual relationship demonstrates bilateral dependence.[7] This can be interpreted as an indication of regional integration or unity, the more so if there is also spatial proximity, as there is amongst Mediterranean Basin countries. It is self-evident that such a regionalisation only has significance if there are several countries involved, if the relations have some stability over time and if these connections have higher values for contiguous than for non-contiguous trade relations. Very high levels of intensity are attained by countries with a mono-structural trading pattern, for instance oil exporters: their dependence on the economy of the importing country is very high. Above-unity trade relationships are sometimes highly asymmetrical, the index in one direction being more than double that in the other trading direction. In such cases there is a certain dependency in the relationship between the two countries. The mere values of the index do not of themselves constitute an explanation for trading relationships and mutual dependency; they are only a descriptive measure. As Anderson and Norheim (1994, p. 21) point out, 'both history and geography, in addition to government policies, play fundamental roles in shaping the pattern of world trade'. Hence we need to bear in mind some of the contextual discussion presented earlier in the chapter about the size and history of the Mediterranean countries, their factor endowments, transaction costs and so on.

The intensity indices were calculated for the 14 countries considered in this analysis, and for four two-year sample averages:

- 1969–70, to coincide with the end of postwar economic growth, the closure of the Suez Canal (1967–75), but before the first oil crisis (1973), and before the start of the European Community's Mediterranean economic policy;
- 1979–80, just after the second oil crisis and a peace treaty between Israel and Egypt, and just before the Mediterranean enlargement of the EU to include Greece (1981) and Spain (1986);
- 1989–90, after another economic recession and before the beginning of the economic growth of the 1990s, and coinciding with the Union of the Arab Maghreb;
- 1994–95, reflecting the growth of the mid-1990s and recent available data.

Yugoslavia could not be integrated into all calculations for 1994–5 because of its break-up into five independent states in 1993.

Internal and external intensity indexes for the Mediterranean Basin

First we present the intra-Mediterranean trade intensities for the Basin as a whole. Over the sample years the overall intensity index has remained below 2 but has increased progressively from 1.37 in 1969–70 to 1.46 in 1979–80, 1.70 in 1989–90 and finally 1.77 in 1994–95. These figures exclude Yugoslavia. If Yugoslavia is included (possible only for the first three sample two-year periods), the figures are 1.30, 1.39 and 1.62. This evidence clearly shows a steady increase in overall trade integration of the Mediterranean countries over the 25-year period examined.

Extra-Mediterranean trade relations will be illustrated by looking at the indices with Germany (representing the European economic core), the USA (representing the North American core) and Japan (the East Asian core). Japan is the least connected of these three to the Mediterranean Basin; only sporadically does the intensity index attain a value of 1.0 or more, but there is no consistent trend or pattern. The USA maintains more systematic trade relations with Mediterranean countries. Israel is the main partner involved, with an intensity index for imports and exports rising from 1.1 in 1969–70 to 2.0 in 1994–95. The same trend holds for trade from Egypt, but at a lower level. Oil-rich Algeria and Libya record indices of above 1.0 respectively for the last three and first two sample years of the series.

Spain, Turkey and Tunisia also sustain indices slightly above 1.0. Overall, however, there is no evidence over the 25 years of analysis that the trade linkages with the North American core have changed relatively in value or intensity.

Trade relations with Germany are more important and systematic in character. Over all four sample years the index of intensity was above 1.0 for imports and exports for each of the four EU partner countries, France, Italy, Spain and Greece. These dual-intensive relationships with Germany increased for the other Mediterranean countries from one – Turkey – in 1969–70 to three (Turkey, Tunisia, Libya) in 1994–95. For all countries the number of one-sided relations with a value exceeding 1.0 increased from five to nine. So, by 1994–95 all Mediterranean countries were intensively connected with Germany, at least in a one-sided relationship.

But this growth in number of countries linked to Germany does not hold when the intensity index is calculated for the relationship between Germany and all the Mediterranean countries counted together. This index hardly changes over the period examined. Aggregate Mediterranean exports to Germany record indices of 1.96 in 1969–70, 1.72 in 1979–80 and 1.84 in 1994–95, while the import relationship expresses figures of 1.52, 1.35 and 1.84 respectively. The increase in imports from Germany reflects the strengthening of trading ties with France, Italy and Spain. If the Mediterranean EU countries are excluded from the calculation, the export relationship to Germany also remains rather stable over the three two-year periods quoted above: 1.74, 1.60 and 1.74. Imports from Germany to non-EU Mediterranean countries are much lower, respectively 0.83, 0.97 and 1.1. This discrepancy between intensity of exports and of imports can be explained by the low purchasing power of these non-EU Mediterranean states *vis-à-vis* high-value exports from Germany.

The conclusion from this part of the analysis must be that extra-Mediterranean trade relations have not changed very much, with the exception of links to nearby European countries like Germany. In this case, the number of intensive (>1.0) trade relations has increased considerably. There has been a steady increase in high-value exports from Germany to the Mediterranean Basin countries, whilst Mediterranean exports to Germany remain at a higher, but stable, level compared to goods moving the other way.

The changing intensity of the spatial pattern of trade within the Mediterranean Basin

With 14 countries in the Basin, each country has the possibility of 13 two-way trading relations. If these trade linkages have an intensity value of 1.0, the existing trade exchanges equate to the expected level if all relationships exactly match the proportions in world exports and imports. Inevitably, dyad relationships will deviate from this hypothetical norm. Higher indices indicate a higher trade and economic reliance on the partner country; and as we have seen, this can be either a mutual or a one-sided relationship. The existence of mutuality means a higher level of economic integration between these countries than between them and their external environment. For countries at the same level of economic and techno-logical development, some symmetry in trade exchanges and intensity indices is to be expected – as is the case, for instance, between France and Italy. Between countries with differing levels of economic development, the mutuality may be very unbalanced, as between Italy and Libya, the latter being a mono-structural oil exporter.

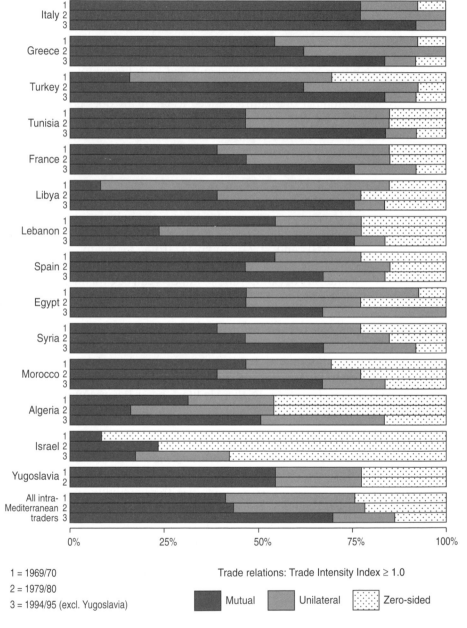

1 = 1969/70

2 = 1979/80

3 = 1994/95 (excl. Yugoslavia)

Trade relations: Trade Intensity Index ≥ 1.0

Mutual   Unilateral   Zero-sided

Figure 4.6 Intensity levels of intra-Mediterranean trade: 1969–70, 1979–80 and 1994–95

Figure 4.6 plots the distribution of the intensity of trade dyads for each country and its potential 13 trading partners for three pairs of years: 1969–70, 1979–80 and 1994–95 (for the last of these dates the trade dyads are reduced to 12 because of the non-availability of data for ex-Yugoslavia). For all Mediterranean countries taken together, the percentage of dyads with one-sided or mutual intensity indices of more than 1.0 increased from 75 per cent in 1969–70 to 85 per cent in 1994–95. The proportion of high-intensity (>1.0) mutual dyads increased more sharply, from 40 to 70 per cent.

Three countries with central locations within the Mediterranean Basin – Italy, Greece and Tunisia – recorded consistently high intensity trade linkages throughout the three sample periods. Italy, the most highly developed and centrally located of these three, displayed the highest level of mutual relationships of at least 1.0 – 10 out of 13 in 1969–70 and 1979–80, and 11 out of 12 in 1994–95. Most other countries intensified their trade relationships with other Mediterranean countries, as figure 4.6 shows. Turkey and Libya made spectacular progress in this respect, starting from a very low level in 1969–70. Lebanon is an atypical, fluctuating case, due to the political troubles mentioned earlier. Israel is a special case too, because of its low trade integration with other Mediterranean countries (only Egypt, Greece and Turkey were linked at above-average intensity). For several countries (e.g. Lebanon, Spain, Morocco, Algeria), the tendency for the middle period to record lower intensity than the first or third sample periods is probably related to the effects of the second oil-price recession on trade.

From the above, it can be concluded that the network of high-intensity trade dyads has increased substantially over the 25 years analysed, and especially over the 15 years from 1979–80 to 1994–95.

The re-orientation of the spatial pattern of trade within the Mediterranean Basin
In order to answer the research question about the changing direction of export trade relations within the Mediterranean Basin, an analysis will be made of the changes in the orientation of the spatial distribution of high-density dyads. This means the selection of those mutual dyads which exceed the average value of the intensity index for the Basin as a whole for each given two-year period. The overall Basin intensity index increased progressively from 1.4 (1969–70) through 1.5 (1979–80) to 1.7 (1989–90) and 1.8 (1994–95), representing an increase of 25 per cent over 20 years and 29 per cent over 25 years, excluding the former Yugoslavia from the calculation to maintain the validity of comparison across all four two-year periods. For the sake of the record, the index is only slightly higher if Yugoslavia is included where possible.

From these data the six polygons of figure 4.7 were constructed to illustrate and visualise the analysis of the orientation of the spatial pattern of high-intensity trade relations. In these polygons the solid black line portrays trade relations amongst EU countries (and Yugoslavia) and between EU states and the other Basin countries. The dotted lines represent trade relations between the other Mediterranean countries, which means, with the exception of Israel, the so-called Arab-Muslim countries.

The first series of polygons in figure 4.7 presents the analysis for the first three

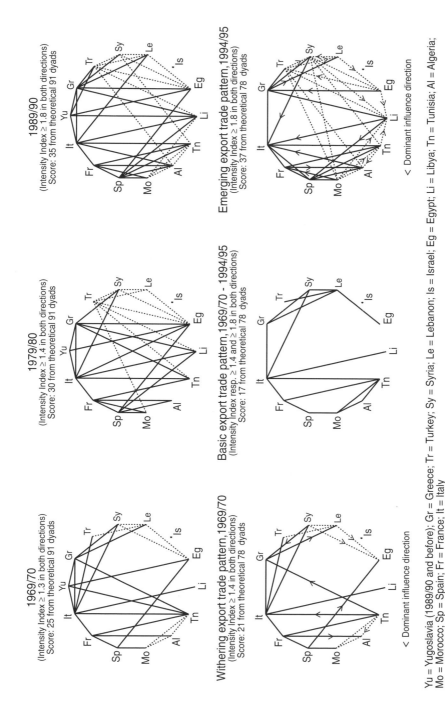

Figure 4.7 Diagrammatic representation of reciprocal and dominant trade relations amongst Mediterranean countries, various dates

Yu = Yugoslavia (1989/90 and before); Gr = Greece; Tr = Turkey; Sy = Syria; Le = Lebanon; Is = Israel; Eg = Egypt; Li = Libya; Tn = Tunisia; Al = Algeria; Mo = Morocco; Sp = Spain; Fr = France; It = Italy two-year periods with a ten-year interval for all Basin countries, including former-Yugoslavia. Several characteristics emerge from a scrutiny of these networks:

- an increase in the number of north-south reciprocal relations represented by solid black lines and thus of relations between Southern Europe and North Africa;
- a substantial increase in the number of dotted lines, indicating intensifying trade relations between the countries of the north-eastern, south-eastern and south-western quadrants of the Basin, or between the so-called Arab-Muslim countries;
- an increase in the diversity of the trade relations of France and Spain with the Maghreb countries;
- an increase in trade relations of the Maghreb countries themselves and with Libya.

The second series of polygons in figure 4.7 pictures the trade relations of the Basin countries excluding the former Yugoslavia, but for a period of 25 years. The central polygon in this series is the representation of those trade relations satisfying the condition that they were present in all four two-year periods with an above-Basin average for the trade intensity in the respective two-year period. This polygon represents the so-called Basic Export Trade Pattern. North–south directions in trade relations and economic interdependence dominate in this diagram. On one side the Western Mediterranean – the EU countries (Italy, France and Spain) are intensively related to the Maghreb countries, and on the other side – the Eastern Mediterranean – Greece is the main trader with or together with Turkey, Syria, Lebanon and Egypt.

The left polygon of the second series of diagrams of figure 4.7 has a so-called 'withering' structure, which means a partly-disappearing pattern of trade relations and directions. However, the earlier-mentioned 'basic structure' dominates in this polygon. The right polygon portrays the most recent pattern of trade relations and directions. Compared to the 'withering' pattern, the following trends may be observed:

- an increasing density and diversification of the overall pattern of trade relations;
- a further strengthening of the 'traditional' north–south pattern in the Western Mediterranean;
- the same strengthening in the north–south pattern in the Eastern Mediterranean;
- the development of a north-east to south-west pattern between the Arab-Muslim countries.

So a basic and continuing north–south structure of the trade pattern with an eastern and a western cluster seems to have developed, together with a rather new and emerging network of relations between the eastern Mediterranean and the Maghreb.

Concluding remarks

This chapter has tackled the problem of measuring the nature and scale of regional unity within the Mediterranean Basin. Trade intensity and interdependence have been chosen as plausible indicators of this unity. By calculating a trade intensity index for dyads of Mediterranean countries and all countries together for four two-year periods from 1970 to 1995, it has been possible to visualise and analyse the spatial patterns and changes in the trade interdependence within the region. The conclusion is that since 1970 the level of intensity and the number of trading dyads with higher levels of intensity have increased, and so therefore has the general level of regional unity within the Mediterranean Basin. However, this overall unity over-lays a still-diversified regional patterning, which means more or less separate clusters with their own regional unity. But over the years the pattern of these clusters has changed and there are now four noteworthy ones:

- the trade relations between the Mediterranean EU countries (especially Italy, France and Spain) and the Maghreb countries (Morocco, Algeria and Tunisia) have intensified and diversified, forming one of the main clusters of the Basin;
- the trade relations between the Maghreb countries themselves and with Libya have increased, contributing to the formation of a kind of secondary cluster;
- the trade relations of the Arab-Muslim countries of the Middle East (Turkey, Syria, Lebanon and Egypt) with the Maghreb countries and Libya have developed to such an extent that a new regional unity or cluster can be discerned;
- the trade relations of Greece with the coastal countries from Turkey to Libya have increased in such a way that an eastern cluster of trade inter-dependence and regional unity has developed.

It is very difficult to explain the existence of these clusters as regional entities; this is an avenue for further research. For the time being, two plausible hypotheses can be advanced:

- the western and eastern clusters, both strongly connected with EU countries, have developed and are developing further as a result of the economic and regional policies of the EU, especially the Euro-Mediterranean policy directed at the establishment of a free-trade area by 2010;
- the cross-Mediterranean east–west cluster is more likely to have originated from the influence of two factors: a more general one, the strong development of world trade and the greater openness of countries to trade as such; and a special one, the diminishing influence of transaction thresholds between countries with a more or less comparable cultural/religious and political background, as is the case with the Arab-Muslim countries of the Mediterranean.

Hence, the original physical division between the eastern and western halves of the Basin still persists; there is a tendency for more unity, but the ambitions of the Euro-Mediterranean partnership are still not fulfilled.

Notes

1   See Eurostat (1988), UNCTAD (1992), WTO (1995a, 1995b) as well as the *Direction of Trade Statistics Yearbooks* issued by the IMF (1956–97).
2   Cyprus might also be considered an exception, although its spectacular growth rates in the fields of tourism and financial services are based on trade in 'invisibles'.
3   Except, of course, that Slovenia can be considered a part-Mediterranean country.
4   Volume (weight) is a relevant measure of trade if the analysis is based on transport means and transport nodes like seaports. However, when the spatial integration of national economies is the focus of analysis, as in this chapter, monetary value is a much more appropriate indicator.
5   The 1996 figure for Yugoslavia is the aggregate export values of Slovenia, Croatia, Bosnia-Herzegovina, Serbia-Montenegro and Macedonia.
6   Of course, the choice of just three (or four) sample years to cover a 25-year period leaves open the possibility of wide fluctuations outside the chosen years.
7   The double quotient has been taken from Savage and Deutsch (1960) who developed a so-called 'null model' for politics that was later successfully applied to the analysis of trade and transport by economists and economic geographers (Sautter 1974). See these sources for technical details and mathematical renderings of the model.

References

Anderson, K. and Norheim, H. (1994) History, geography and regional economic integration, in Anderson, K. and Blackhurst, R. (eds) *Regional Integration and the Global Trading System*. New York: Harvester/Wheatsheaf, pp. 19–51.

Braudel, F. (1966) *La Méditerranée: La Part du Milieu*. Paris: Armand Colin.

Braudel, F. (1979) *Civilisation Matérielle et Capitalisme XV–XVII Siècles: Les Jeux d'Echange*. Paris: Armand Colin.

Eurostat (1988) *Trade: EC–Mediterranean Countries, Theme 6 Foreign Trade*. Luxembourg: Statistical Office.

Featherstone, K. (1993) The Mediterranean Challenge: cohesion and external preferences, in Lodge, J. (ed.) *The European Community and the Challenge of the Future*. London: Pinter, pp. 186–201.

Grenon, M. and Batisse, M. (1989) *Futures for the Mediterranean Basin: The Blue Plan*. Oxford: Oxford University Press.

IMF (1956–1997) *Direction of Trade Statistics Yearbooks*. Washington DC: International Monetary Fund.

Jones, A. R. (1997) The European Union's Mediterranean policy: from pragmatism to partnership, in King, R., Proudfoot, L. and Smith, B. (eds) *The Mediterranean: Environment and Society*. London: Arnold, pp. 155–63.

Krugman, P. (1991) *Geography and Trade*. Leuven: Leuven University Press.

Pomfret, R. (1986) *Mediterranean Policy of the European Community: A Study of Discrimination in Trade*. London: Macmillan.

Regelsberger, E. and Wessels, W. (1984) European concepts for the Mediterranean region, in Luciani, G. (ed.) *The Mediterranean Region: Economic Interdependence and the Future of Society*. London: Croom Helm, pp. 239–66.

Sautter, H. (1974) Tendencies of regionalisation in world trade between 1938 and 1970, in Giersch, H. (ed.) *The International Division of Labour: Problems and Perspectives* Tübingen: Mohr, pp. 483–540.

Savage, R. and Deutsch, K. W. (1960) A statistical model of the gross analysis of transaction flows, *Econometrica*, 28(3), pp. 551–72.

Shmueli, A. (1981) Countries of the Mediterranean Basin as a geographic region, *Ekistics*, 290, pp. 359–69.

UNCTAD (1992) *Handbook of International Trade and Development Statistics 1991*. New York: United Nations Conference on Trade and Development.

WTO (1995a) *Trading into the Future*. Geneva: World Trade Organisation.

WTO (1995b) *Regionalism and the World Trading System*. Geneva: World Trade Organisation.

------------ *5* ------------

Cultural representations of urbanism and experiences of urbanisation in Mediterranean Europe

Lila Leontidou

▬▬▬▬▬▬▬

The proletarian dictatorship will preserve this magnificent apparatus of industrial and intellectual production, this driving force of civil life, from the ruin which is looming so threateningly over it. Bourgeois power . . . is now revealing the progress of its decay in the cities, which are steadily declining in comparison with the countryside . . . The proletarian dictatorship will save the cities from ruin. . . In this way, it will prevent those miraculous engines of life and civil progress which are the cities of today from being destroyed piecemeal by the landholders and usurers of the countryside who, in their uncouth way, hate and despise modern industrial civilisation.

Gramsci (1994 p. 136)

It was January 1920 when Gramsci wrote this polemic text and, at that time, on the other side of the Atlantic, the Chicago human ecologists Park, Burgess and McKenzie were busy constructing a model for their own cities which would change the direction of urban geography. Both sides considered the innovative potential generated in the city, but were antithetically coloured by the attitudes of their own societies: Gramsci saw the city as a workshop for culture, progress and innovation; the human ecologists introduced the distinction between Nature and Culture, and saw the city as an ecosystem balancing out the disruptive tendencies for crime, marginality and erosion of the moral order, embedded in early industrial settlements (Wirth 1938). For them, urbanisation was linked with, even 'determined by', industrialisation. For Gramsci, as well as Max Weber (1966) before him, there were other forces creating urbanisation besides industrial development.

This chapter explores such alternative forces of urbanisation in one of the world regions where urban restructuring has not been centred on the industrial revolution and economically-motivated urban growth: Southern Europe. It will be argued that, in this case, urbanisation, urban landscapes and life patterns are culturally rather than economically conditioned and are related with representations of urbanism. *Urbanisation* is used here in the demographic sense: as the process of population concentration into cities. The term is also employed in the structural sense of social change related with the growth of cities. Note that the usual definitional linkage between cities and industrial growth is lacking here, and is considered Eurocentric. Current urban geography keeps taking it for granted that industrialisation must lead to urban growth and extends this by linking counter-urbanisation with deindustrialisation. In this framework, South European urban development is explained away as 'retarded' in Anglo-centred convergence theories. As Mediterranean cities do not grow because of industry, they are considered to be at a stage prior to 'proper' development, and have often been labelled 'precapitalist' cities (White 1984). According to the analysis presented in this chapter, this renders urban theory normative and a-historical, even where it discusses change, as in 'urban life-cycles' theory (van den Berg *et al.* 1982). The definitional linkage between urbanisation and industrialisation is especially inadequate for under-standing Mediterranean development dynamics which have been based on cities since antiquity, not on industrial capitalism.

As for the definition of *urbanism* employed in this chapter, it is not just a refer-ence to a way of life associated with residence in an urban area, where relationships forged by size, density and heterogeneity replace community ties. My definition of urbanism refers to an intersubjective construction and representation of the city-idea: a *cultural attitude* towards the city and urban life. Urbanism refers to the positive light shed on the city-idea, the 'city as virtue'; it is opposed to the 'city as vice' (Schorske 1998, pp. 37–55), which is referred to as *anti-urbanism*. This chapter explores the linkages between urbanism (as culture and way of life) and urbanisation (as a form of demographic and economic restructuring), in order to investigate an event which is known, and goes beyond the city as virtue: the city as a *magnet* in Mediterranean Europe. This has often been contrasted with the type of urban-isation led by economic forces, and industrial capitalism more particularly; but it has not been adequately contrasted with representations of Anglo-American anti-urbanism as in the present chapter.

It is important, at this preliminary juncture, to stress in-between cases and the diversity of the 'urban': the poles presented here are intercepted by thousands of formations around the world. Mainland European cultures are hardly homogeneous. Paris and Vienna contrast with Berlin and Brussels, the com-plexity of Celtic cities such as Dublin, Edinburgh or Cardiff dilutes the picture, and Afro-American urban cultures in the USA present a contrast with 'white' representations of the city. This chapter basically focuses on Southern Europe, but here again the construction of urbanism presents a multitude of articulations. These will be touched upon here, in an attempt at a (re)interpretation of different facets of Mediterranean urban-oriented identities and representations which relate to a mosaic of urbanisation trajectories reflected in the popularity of the inner city, the inverse-Burgess model, spontaneous urbanisation, spaces of immigrants, and urban competition.

Representing the city as vice: industrial anti-urbanism and rural nostalgia

Schorske (1998) has depicted the 'city as vice' on the basis of literary sources, which attribute anti-urbanism to urban squalor during the industrial revolution and its aftermath:

> The eighteenth century developed out of its philosophy of Enlightenment the view of the city as virtue. Industrialism in the early nineteenth century brought to ascendancy an antithetical conception: the city as vice. Finally there emerged, in the context of a new subjectivist culture born in the mid-nineteenth century, an intellectual attitude which placed the city beyond good and evil. No new phase destroyed its predecessor. Each lived on into the phases that succeeded it, but with its vitality sapped, its glitter tarnished. Differences in national development, both social and intellectual, blur the clarity of the themes. (Schorske 1998, p. 37)

And yet the Enlightenment intellectuals, Voltaire, Adam Smith and J. G. Fichte, seem less than influential in the subsequent Northern European and American city-ideas analysed by Briggs (1968), Williams (1973), White and White (1977) and Wiener (1985). Even in the eighteenth century, secular intellectuals, the French Physiocrats, the familiar romantic poets, and then a large current of nineteenth-century artists, writers and reformers, archaistic and futuristic, pushed the view of the city as vice: squalor, crime, and disappointment of hopes raised by the Enlightenment project (Schorske 1998, pp. 43–9).

Northern intellectuals would concede that 'city air makes man free', but, simultaneously, the city was painted in black. Saunders' (1986) summary of their work and taken-for-granted premises led him to the most extreme anti-urbanism of all: there is no 'urban' sociology, there is no 'urban' realm, he concludes. However, there are variations in the theme of anti-urbanism dramatised by industrial squalor. Marx and Engels passed from ethical rejection of the modern city as the scene of labour's exploitation and alienation, to an historical affirmation of its liberating function; Durkheim's urban anomie can raise fewer doubts about his idea concerning the city; but Oswald Spengler's decline of the West is considered by Schorske (1998, p. 53) as a view of the city as a fatality beyond good and evil.

Pessimism, criticism and disappointment are not peculiar, given the squalor associated with urban life during the industrial revolution and early industrialism in Northern Europe and America. However, anti-urbanism in much of Northern Europe predates the industrial revolution. It has been discovered back in Roman times. While the Romans, like the Greeks before them, believed that participation in a real, natural city constituted the precondition of civilisation, the English public life did not need the city: the elements of civilisation already existed in everyone's home (Briggs 1968, p. 72).

According to Williams (1973), the English gentry has tended to idealise nature and live near the countryside, visiting their urban properties only for a season. The Victorian industrial elite adopted a rural orientation particularly reflected in literature, with Thomas Hardy's 'country' books as leading examples (Williams 1973). 'Our England is a garden', observes Wiener (1985, p. 46), and the 'pretense that the

Englishman is a thatched-cottager or country squire at heart' is noted by Hobsbawm (1968, p. 142). Explicit anti-urbanism was not just a reaction to rapid urbanisation. Wiener (1985, pp. 47–8) distinguishes two types of Victorian anti-urbanism. The distaste for city growth and the protection of landed interests, which in any case lost their appeal as farming lost its importance in the national economy, form only part of the story. There was in parallel 'the criticism of city life by those whose way of life was essentially urban' (Wiener 1985, p. 48). This was a cultural phenomenon and did not fade away with economic restructuring; on the contrary, according to Raymond Williams (1973, p. 248), there was almost an inverse relationship in the twentieth century between the importance of the working rural economy and the cultural importance of rural ideas.

We cannot possibly go into the particular relationships among peasants, rural landed elites and urban groups, which led to differences among Britain, France and the USA (see Weber 1976; Wiener 1985, pp. 47–59). However it should be pointed out that, in contrast with continental Europe where Marx's rural 'barbarism' and 'idiocy' was a widespread attitude, the English cultural tradition exalted the countryside. Anti-urbanism was also evident in the USA. The flight from the city, the role of the countryside and the highway as a refuge, have constituted a constant theme from Thomas Jefferson to Frank Lloyd Wright (White and White 1977). We encounter this in novels, film and music, from Ernest Hemingway to Bruce Springsteen (Leontidou 1990, pp. 257–8). Those stressing the liberating potential of the city, from Marx to Wirth, did so hesitantly and with ambivalent notes about the ills of urban living. Rural nostalgia was usually played in imagination. The city in these cultures became a scapegoat for the ills of modern life, while at the same time the countryside was 'remade' as a rural idyll (Williams 1973).

Experiencing the city as virtue: Mediterranean urbanism

From antiquity and the Enlightenment until today, the strongest antithesis to the above cultural orientation is to be found in Mediterranean urbanism. The city-state was a very important focus of life around the Mediterranean Sea, and the collective memories of those glory days have never faded (Leontidou 1997). Since a very early period, the city represented progress and civilised cultural life, while the country-side was considered the domain of ignorant peasants. In Turkish tradition 'urban life, because of cultural values rooted in history, is considered a higher form of organization' (Karpat 1976, p. 232). In Greece, 'one can not conceive of civilisation at all outside urbanity' (Kayser et al. 1971, p. 195), and the experience of urbanisation has been named with a word which means 'friendliness to the city', astyfilia. In fact, the postwar process of migration has always been considered as a common fact of life, and there was hardly any rural-urban conflict in popular culture since the nineteenth century (Mouzelis 1978). Social scientists have agreed that urbanisation disentangled villagers from traditional authority. Many kept their bonds with their place of birth by keeping their property and attracting chain migrants (Leontidou 1990, pp. 258–61). This, however, along with the strong regional consciousness found in Greek urban migrant settlements, did not imply any hostility for the city. It meant a permanent affiliation with family and place of birth, which permeated many spheres of life, including urban spatial patterns.

On the western edge of Mediterranean Europe, Iberian countries also developed urban-oriented cultures with a global impact, since these were diffused not only at home but also in colonies. Urban orientation was not a novelty in those countries which had experienced the remarkable urban civilisations of pre-Columbian cultures (Hardoy 1973). In any case, already in the sixteenth century, the urbanism of the Spanish and Portuguese *conquistadores* is reflected in colonial customs in urban planning in Latin America (Morse 1976). The *conquistadores* established dual land rights in cities and the rural periphery, which led to certain similarities between Third World and Mediterranean urbanisation, such as the inverse-Burgess model and the colonisation of the suburbs by the poor (Leontidou 1990, pp. 250–1).

The 'aversion to nature in traditional Italian upper class culture, in contrast to the culture of Northern Europe' (Fried 1973, p. 106), is echoed in writings reflecting the concerns of broader social strata. Gramsci eulogises Italian urbanism in the quotation at the beginning of this chapter and in many others, among which is the following humorous remark referring to the Italian *Risorgimento*.

How would Italy of today, the Italian nation, have come into existence without the formation and development of cities and without the unifying influence of cities? 'Supercountrymanism' in the past would have meant municipalism, as it meant popular disarray and foreign rule. And would Catholicism itself have developed if the Pope, instead of residing in Rome, had taken up residence in Scaricalasino? (Gramsci 1971, p. 288)

Many years after Gramsci, the attraction to the city has spurred the massive migrations, internal and international, for which the Mediterranean is well known. People have been 'voting with their feet' and expressing their viewpoint on urbanism by migrating and struggling for a stake in the city through illegal building and informal work practices. They have built dense and compact communities, so that in Rome, for instance, 'the lack of parks and recreation facilities in the city is attributable . . . to the recent urbanization of many inhabitants and their continued association of green with rural misery and of cement with civilization' (Fried 1973, p. 106).

Southern urbanism had an important discontinuity during Fascism, as experienced in other European countries: from the proto-Nazi *litterateurs* in Germany assaulting urban people as vicious, to the neo-rightist French by the end of the nineteenth century calling for a return to the provinces. The racist Germans called 'Back to the soil where blood runs clear', and the Nazis, while excoriating the 'pavement literature' of the 1920s, and branding urban art as decadent, 'brought out in their city-building all the elements which the urban critics had most strongly condemned' (Schorske 1998, p. 53).

In Southern Europe, anti-urbanism similarly peaked during the Italian Fascist regime, combined with impressive city-building operations in Rome and spectacles and parades to go with them. Mussolini idealised 'peasant Italy' and relocated urban residents from the inner city in order to resurrect the glory of ancient Rome. Planning, *urbanistica,* received official recognition, and grand access routes through the centre of Rome indicated the concern of planners with monumentality rather than amenity (Fried 1973, p. 166). Historical buildings were rehabilitated, medieval institutions were resurrected and several new towns were built (Mariani 1976). Rome is a typical case of 'Haussmanisation', resulting in massive relocation of the poor as well as the artisan and small merchant class from the centre. During the monumentalisation of the city centre, antiquity was loudly recuperated by Mussolini and reconstructed via public works as massive as the flattening of the city's hills. Large-scale planning coexisted with speculation and the exclusion of the poorer inhabitants from cities. In 1924–40, 12 *borgate* (peripheral villages) were built in a punitive spirit to rehouse those evicted. Then the Roman periphery filled with spontaneous and jerry-built *baracche*, shanty settlements (Fried 1973, pp. 30–8).

Mussolini also attempted to check migration through anti-urbanisation laws and legislation on forced domicile (Allum 1973, p. 27; Fried 1973, p. 80; Gabert 1958). The Fascist laws of 1931 and 1939 aimed at controlling rural–urban migration with the introduction of work permits in cities. However, in order to obtain a work permit, it was necessary to hold a residence permit! These laws were circumvented, of course, as always in Southern Europe, through an illegal system of recruitment of low-wage illegal migrants, especially in Northern Italian cities (Gabert 1958). Later, during the 1950s and 1960s, urban social movements in Rome, Milan and Naples struggled for the abrogation of these laws and for 'freedom of residence' (della Seta 1978, p. 308).

In Spain after the civil war, the same contradictory combination of anti-urbanism with urban planning was dominant. Franco strengthened Madrid (Salcedo 1977), while the city was portrayed as the centre of vice and evil (communism, crime, divorce) by state-controlled institutions. Occasional threats to control domestic migration to Athens were also voiced by the Greek military junta, in the context of a rhetoric about 'parasitic' Athens, but were never realised (Carter 1968). In any case, the set of policies against popular house-building and the facilitation of de-industrialisation underlines the dictators' apprehension about the social threat presented by urban concentrations. However, these policies were combined, in the most contradictory manner, with an active boost to inner-city multi-storey construction by loosening relevant legislation, thereby creating high-rise develop-ment and the current urban congestion (Leontidou 1990, pp. 212–15).

In Greece this rhetoric against urbanisation has lasted longer than the dictator-ship. It has constituted the undercurrent of middle-class suburbanisation waves, and is currently intensified by urban congestion and environmental pollution. But this is scarcely anti-urbanism: it is no more than a rhetoric against the 'problems' which are invading middle-class space, familiar from other examples, such as Istanbul, where environmental deterioration is blamed on the migrants. An aristo-cratic family descendant remarked indignantly about *geçecondu* (shanty-town) residents: 'it is a disgrace that learned men devote their time to studying these wretches, instead of fighting to preserve what is left of the civilization of our great peers' (quoted by Karpat 1976, p. 63).

Greek planners and policy makers have also tended to stigmatise migrants as 'immoral' and blame congestion on them. However, overbuilding was not practised exclusively by migrants. Most of Athens' affluent residents have been busily erecting multi-storey apartment buildings, *polykatoikies,* clinging stubbornly to the inner city, even gentrifying some enclaves such as Plaka. They have been moving to 'suburbs' which are so close to the centre and so dense, that they do not deserve the name in its Anglo-American sense (Leontidou 1990, p. 244). The popularity of the inner city cannot possibly fade in the Mediterranean. Despite current stabilisation of urban population and trickles of inner-city decline in Madrid or Rome, or more strongly in the Italian Northern 'Triangle', it is possible that Southern Europe will not see a counterurbanisation trend to the extent that this was experienced in the North.

The inner city as a magnet: landscapes of street life and unhurried leisure

Urban life revolves around the city and the home, and there is another North/South divide here since Roman times (Briggs 1968, p. 72). The Mediterranean tradition sees the whole city as the context for civil life and the housing units as small private enclaves. Anglo-American cultural stereotypes, by contrast, see houses as the essential setting for everyday life, and the city as a context for movement toward houses and leisure spots. This is extended to a contrast between street life in the South (which spreads to continental cities such as Paris), and 'hurried leisure' in America (Karapostolis 1983, p. 262), where at least in the period of modernism the private car has superseded the pedestrian (Jacobs 1961). The antithesis extends to leisure patterns such as eating out, and to many cities in Northern Europe. 'Hurried leisure' is evident in restaurant queues, in the practice of waiters of keeping glasses full, and in their pressure to clear the table for the next round of customers as soon as a meal appears to be finished. This is not experienced in Mediterranean cities, nor is anybody expected to leave the table or the cinema as soon as the meal or the film ends. People are allowed to enjoy their meals and performances in unhurried leisure.

Leisurely enjoyment of public spaces and the tendency toward street life cannot be attributed to the mild Southern climate alone. They reflect urban-oriented cultures in compact and enclosed cities, where home entertainment is difficult in restricted housing spaces. Certain aspects can also be found in French urban 'café societies', in Vienna, and in tourist squares elsewhere. The popularity of the city centre and the inner city more generally is also not exclusive to the Mediterranean. However the tendency of the more affluent social groups to live in the centre contrasts with inner-city poverty in Anglo-American and several Northern European cities, where the more affluent classes have suburbanised from a very early period. It is not spacious living (as in the Burgess and Alonso models), but accessibility to the city centre which is highly valued and preferred in Mediterranean cities. Theirs is an inverse-Burgess model.

Values for central living are combined with a rural second home for days of leisure, usually located at the place of origin of migrants to cities, which a lot of the Mediterranean urban populations are. Second-home access throughout continental

Europe has been important in the creation of compact cities: between 8 and 20 per cent of urban residents had access to a rural second home in the early 1980s (White 1984, pp. 163–4), and the rate has gone up considerably since then, with the spread of residential North–South tourism lining the Mediterranean shores. The widespread practice of renting in middle-class areas of many mainland European cities contrasts with the high rates of owner-occupation in Britain and the USA. Wealthy Europeans may rent a central apartment but will often own a rural second home. In the Mediterranean, owner-occupation does not correlate with social status. It is location which correlates, and the affluent homes in the centre make nonsense of the term 'gentrification' in Mediterranean cities: there can be no such 'urban life-cycle' as re-urbanisation in cities without any experience of counterurbanisation to begin with.

The popularity of the city centre combined with street life is evident in the urban landscape and land use. Inner cities undergo *continuous gentrification* in Southern Europe. The Central Business District is not dominated solely by economic activity, but also by residence and cultural and leisure spaces. Zoning is not strict: mixed land use throughout the urban fabric creates a combination of residence with economic and leisure activities at walking distance – a postmodern collage (Leontidou 1993). Zoning is actively undermined by spontaneous urban development. Even the social space is mixed, and segregation is milder than in the Anglo-American city. In the densely-built Southern urban neighbourhoods, vertical differentiation in multi-storey apartment buildings overlaps with mild neighbourhood segregation (Leontidou 1997). The urban landscape is interspersed with small squares in both central and peripheral areas rather than large parks. It is difficult to imagine a Hyde Park or a Central Park in any city of the South, or a Green Belt for that matter. Piazzas are for seating and eating out, leisure spots are dispersed in neighbourhoods within walking distance, and lined with the outdoor tables of cafés and taverns. Harbours and inner-city archaeological parks are enjoyed for strolling. This creates a striking contrast with the London Thames waterfront, which is rather unfriendly for strolling, though recently outdoor city life has been adopted in the North too, as detailed in the last part of this chapter.

The metropolis as a magnet: spontaneous urbanisation and the informal city

Urbanisation in the Mediterranean has been generally rapid and spontaneous, combined with the experience of squatting (or, rather, semi-squatting). These have been informal cities, not industrial cities. With exceptions such as Turin, Milan and Bilbao, South European cities did not attract migrants because of their industrial development. Analysis until the early 1980s stressed that people have been repelled by rural poverty and attracted by opportunities of informal work in the large agglomerations (Williams 1984). However, the forces were not just economic. The culture of urbanism was strong in the massive rural exodus, enhanced by the availability of cheap peripheral land for settlement. In this, linkages between urbanism and urbanisation are now explored.

Most Southern urban networks 'distort' Anglo-American settlement hierarchy models, defying the proverbial neat rank-size rule, sometimes with primate distributions, sometimes with urban bipolarity (Leontidou 1990, p. 259). Most of the large metropolitan growth centres are on the Mediterranean coast: Athens, Istanbul, Beirut, Tel Aviv, Alexandria, Tripoli, Tunis, Algiers, Barcelona, Naples. The tendency for primacy is accentuated when capital cities are also ports, while in other cases bipolar urban networks evolve (Madrid and Barcelona, Rome and Milan, Ankara and Istanbul). The South is still the most dynamic crescent of the European urban network.

Unauthorised, spontaneous popular settlements, springing up around all Mediterranean cities until at least the mid-1970s, have been a stark contrast with controlled suburbanisation in Anglo-American cities. In Britain and the USA, peripheral urban land was always strictly safeguarded against popular invasions and control of illegal building was effective. The dominant classes kept central urban land for economic purposes, and protected peripheral land for residential purposes, including upper-class living and the controlled development of working-class housing estates (Leontidou 1990, pp. 247 8).

Evidence of this abounds. As urbanisation rates mounted during the transition to capitalism, the 'crowd' of London did try to squat; but the movement was curbed from the outset. Every kind of illegal building was suppressed at birth (George 1966, p. 79). The few self-help associations starting in Birmingham in the mid-nineteenth century (Briggs 1968, p. 19) were soon to be integrated into the market. Shanty towns also appeared in the periphery of German and Swedish towns and in Eastern Europe during the nineteenth century, but they were controlled at once (Niethammer and Bruggenmeier 1978, pp. 125–7). The same process was observed in other European cities (Braudel 1973, p. 205). As for the USA, many cities were encircled by shanty towns during the industrialisation phase. There were 40–50 acres of shanties on the west side of Chicago in the 1850s, where the population rose from 57,000 to 214,000 during 1862–72 (Karpat 1976, p. 11). However, these were displaced 'after the civil war when streetcar systems opened new areas to middle-income development' (Ward 1973, p. 297).

This contrasts with loose control of peripheral urban land in the Mediterranean. Even today some Southern countries have no land registration systems. It might be that Mediterranean dominant classes were apprehensive of dangerous masses in the city core, after the negative experiences of Northern industrial conurbations. At the same time, however, they wanted to preserve the style of central living for themselves. There are references to the discouragement of industrialisation in papal Rome, to avoid the 'restless proletariat' (Fried 1973, p. 21). In fact, Italian dominant classes consistently relocated the poor from the inner city or, occasionally, from 'miserable boats near the quays . . . or under the bridges on the canals' in Genoa and Venice (Braudel 1973, p. 205). The experience of Rome during the Fascist period (noted earlier) was repeated at a smaller scale in other Italian cities.

The predominance of cultural over economic forces in spontaneous urbanisation can be effectively demonstrated by the negligible impact of property structures on urban expansion. The uniformity of popular land colonisation in

South European cities until the 1970s contrasts with the diversity of land-ownership patterns in them. There are several types of landlordism in Southern Europe (private, large and small, state, church), but basically two models can be discerned between Greece and Portugal on the one hand, and Spain and Italy on the other (Leontidou 1990, p. 248). In the former, properties are generally small, urban landownership is fragmented, and urban redevelopment is piecemeal. In the latter two cases, large real-estate entrepreneurs have emerged in the context of a generally fragmented urban space. In Rome, the extent of concentration of landownership in the Agro Romano already in 1913 was such that 11 families owned 40 per cent of the land and in 1954 six families together owned more land than the city government (Fried 1973, p. 115). This, however, did not impede spontaneous popular settlement. *Borgate* were mushrooming in this urban fringe already in the early 1900s. Apparently, some of this land was subdivided and sold in anticipation of popular pressure for infrastructure expansion. There are even references to illegal dwellings built with subsidies from the state (Fried 1973, pp. 29, 120). The contrast with British and North European Green Belt policy can hardly be over-emphasised.

The remarkable homogeneity in peripheral land colonisation, despite diversity of landownership patterns, can be generalised more widely by a comparison of Greece with Latin America. Land fragmentation in the former, and in the Balkans more generally, prevailed in the nineteenth century, whereas Latin America experienced trends towards concentration of landownership (Mouzelis 1978, p. 49). In both cases, however, there was always overwhelming popular land colonisation at the urban fringe.

Illegality has been an important aspect of spontaneous urbanisation, not restricted to peripheral settlements. Within approved city plans, illegal additions are a standard speculative venture, giving the middle classes extra space. Rooftop additions create a couple of supplementary dwellings for exploitation, as in Roman apartment buildings (Fried 1973, p. 59); the adding of *áticos* and *sobre-áticos* retreating from the facade gains Barcelona residents some height; and balconies become rooms and glass houses in Athens apartment buildings. Illegality within the city plan is most obvious in the case of Naples, where the building boom during the 1950s was invaded by people related with the *Camorra* and was partly controlled by city bosses, the virtual economic empires of Lauro and Gava (Allum 1973, pp. 36–9, 296–7). Neapolitan speculators were immortalised by Rossi in his film *I mani sulla città*, reminiscent of American rather than Mediterranean corruption. The role of the mafia in the city-building process of Palermo is less known. Elsewhere in the South illegality is much more 'democratically' spread, and a whole population is well trained in contravening regulations and devising informal strategies of living and working.

Beyond the border: Southerners in Northern cities

Mediterranean urban-oriented cultures have tended to overflow towards global space through international migration. The urban ideal was reaffirmed beyond the boundaries of Southern Europe by emigration, which for the best part of the postwar period was economically motivated. Bilateral agreements brought *Gastarbeiter* ('guest-workers') from Southern to Northern Europe especially during the 1960s (King 1994). Their communities, as well as those of the diaspora more generally, have often been as large as Southern towns and as relevant, linked with them by chain migration. Greek, Italian and Jewish diasporas have been forming lobbies in cities of the USA, Canada and Australia. Their positive construction of identity and difference transforms host cities, while their 'topophilia' and 'sense of place' have a lasting impact on home towns and villages. This ranges from the financing of monumentalisation and development by the Greek 'Great National Donors', to the modest remittances to rural families sent by *Gastarbeiter*.

Southern Europeans are dispersed globally but, by clustering together in large cities abroad, render certain places their own. This cannot be dismissed due to its generality: Chinatowns, Muslim communities, 'little Italies' and ghettos indicate that migrants universally cling together. Their concentrations are solidly urban: large agglomerations and global cities are renowned for their large migrant communities, which have a profound impact on them. Third World migrants, Greeks, Turks, Yugoslavs and Iberians are mentioned in studies as populations which 'interfere' in Northern 'urban life-cycles'. They counteract inner-city decline and counterurbanisation. High immigrant fertility has had the same effect, though declining in the new setting. Examples abound of European urban cores kept from decaying by migrants who sustain schools and services there, while native populations move out to the suburbs and villages. This is especially felt in Brussels and in German cities such as Stuttgart and Munich (King 1994, p. 230).

There are important variations in these patterns. North Africa seems to send its migrants to the peripheries rather than the inner cities of France. Already in 1560, Balkan migrants were 'living a shanty-town existence in holes in the earth' and huts appeared at the edges of forests in the suburbs of Paris. These were demolished after a French ordinance of 1669 (Braudel 1973, p. 205). However, the building of peripheral shanty towns continued, by the rag-pickers in the nineteenth century, by the residents of the Parisian fortification belt in the early twentieth century, and by the interwar *pavillons* (White 1984, p. 45). But it was the massive influx of North African labour in the 1960s which transformed the Parisian periphery, crystallising the inverse-Burgess urban model in the French capital. It is interesting here that it was Mediterranean populations, mainly Algerians but also Portuguese, who demanded their rights in the city by building peripheral *bidonvilles* (Leontidou and Afouxenides 1999). By the late 1970s these were gradually reduced, reabsorbed into 'normal' housing after a 1964 law and the erection of *grands ensembles* (Borde and Barrere 1978). They never disappeared, however, and can still be seen today on a smaller scale, notably around Marseilles.

Recent migration waves have become very complex, composed of several different streams. After the mid-1970s, repatriation of the *Gastarbeiter* was sought by governments in host countries and also chosen by many emigrants. These people have sometimes returned to their rural areas of origin, where they have engaged in the tertiary sector, for instance as small entrepreneurs in shops, taxis, cafés and tourist enterprises. Usually, however, they settled in small towns and larger cities. Other streams originate in Africa and Eastern Europe. Now Southern European cities have joined those of Northern Europe as destinations, passing from emigration to immigration (King 1994). The North–South divide persists, but its axis has shifted to a new parallel: now 'Southerners' are Africans, and, increasingly, Eastern Europe is characterised as the 'new South' of the EU.

An emergent migration stream which highlights our argument about Northern representations of the city and the country is international residential tourism. If Third World migrants and Southerners invaded Northern cities in the past, since the 1980s but especially the 1990s, with the removal of European internal boundaries, Northerners have started to invade the Southern coasts and countryside. These contrasting migration destinations, urban versus rural/coastal, bring to the fore the spatial dimensions of the contrast between Southern urbanism and Northern anti-urbanism. International residential tourism and retirement migration are stepping up and creating migration waves opposite to those of the 1960s (Williams *et al.* 1997). Population movements in the past were overwhelmingly urban-oriented. Now Northern Europeans invade the rural South.

The 'Mediterraneanisation' of European urban landscapes

Population movements nowadays indicate a revival of urban living. Gentrification swings Northern European and American 'urban life-cycles' towards a re-urbanisation trend (van den Berg *et al.* 1982), currently enhanced by urban competition. Mediterranean cities have been constantly gentrifying and are therefore not facing any major restructuring in this area. They are, however, affected fundamentally by urban competition: production is rapidly outstaged by consumption in strategies to attract entrepreneurial interest in postmodern Europe. A strong component of place marketing is developing, most usually located in cities. Cities compete as city states once did, though with new methods and peaceful means. In this race the Mediterranean has a peculiar ambivalent advantage (Leontidou 1997).

Urban rivalries have been more pronounced in the South than in other European regions. The bipolar urban networks, but also urban primacy, have long created a tension between the two largest cities: Milan and Rome, Barcelona and Madrid, Thessaloniki and Athens represent the antithesis between industry and administrative power. The fierceness of such urban rivalries before and during political consolidation in Iberia and Greece, and during and after the Fascist regime in Italy, suggests a relationship with government propaganda. Mussolini's rhetoric on behalf of Rome reinforced anti-Roman sentiments among other cities, and created the Northern slogan 'Turin produces, Milan sells, and Rome consumes' (Fried 1973, p. 71). Similarly, during the 1970s, other Mediterranean capital cities were fiercely stigmatised as 'parasitic' and moralised against, especially Madrid and Athens (Leontidou 1990; Salcedo 1977). The latter has experienced a rhetoric which facilitated policies of the dictatorship for the control of spontaneous urban growth and industrial decentralisation.

All capital cities were antagonised by the second largest cities, but also by smaller towns with a strong sense of place, originating in local histories over the *longue durée*. In a sense, many Mediterranean towns and cities never lost city state or capital city status in the popular imagination. As in ancient times, urban identity is now reaffirmed, often aggressively, in all corners of Southern Europe.

This can be turned into an advantage in postmodern culture, where the re-invention of tradition is sought for the sake of place marketing. The past is valorised in urban design and authenticity sought in the urban landscape. The local development of strategies for visibility is developing in place of the welfare state, reform and redistributive planning (Jensen-Butler *et al.* 1996; Kearns and Philo 1993; Leontidou 1995). These neoliberal strategies are based on strong cultural identities and the capitalisation of heritage. There is plenty of this potential in the pride of Mediterranean cities for their ancient and medieval heritage. There is also the resilience due to the experience of constant urban competition within the respective nation states. Corporate mobilisation and the culture of urbanism led Spain to provide three star examples in urban competition in 1992, hosting international events: Barcelona for the Olympics, Madrid as the European city of culture, and Seville for EXPO 92. Since then, the international spotlight has often turned on the three cities throughout the 1990s. More generally, after the important administrative reform of 1978, cities in the various regions of Spain (Catalan, Castillian, Basque, Andalusian) autonomously reaffirmed their identities in Europe and claimed a section of EU funding, centrality and entrepreneurial interest. Support was plentiful and is easing out past rivalries (Garcia 1993; Leontidou 1995).

The city state may never be resurrected again in Europe, but the nation state is now questioned. Moving towards a 'Europe of the regions' means increasing local autonomy, but also escalating urban competition, to the point of the actual personification of towns and cities. Networking leads to the reconciliation of co-operation and competition, as in the economic sphere. Hundreds of different realities emerge and speeds of urban and regional development are multiplied. Repolarisation and spatial divisions in some corners, but also socio-spatial homogenisation in others, tear the European urban network apart and at the same time lead to its depolarisation: Northern cities have to follow the culture of urbanism. Southern cities are *not* developing towards Northern models. By contrast, I venture to say, anti-urbanism is moderated by urban competition and by 'selling places', mostly cities, and diluted by gentrification (Jensen-Butler *et al.* 1996; Kearns and Philo 1993). The 'Mediterraneanisation' of inner-city landscapes such as that of central Liverpool exemplifies ways in which, in postmodern times, Southern urban-oriented cultures have explicitly influenced and literally penetrated the North.

Conclusion

Neoliberal developments are swinging Mediterranean cities from pre- to post-modernism, without full-scale modernisation in between (Leontidou 1993).

Convergence theories, 'urban life-cycle' models and evolutionist perspectives of the past are thus discredited in postmodern Europe. Mediterranean urban trajectories are not 'prior' to those of Anglo-American cities, nor do they constitute their exact reversal, for that matter. The building up of positive Mediterranean urban identities *preceded the industrial revolution* and the latter is not a relevant event in their own particular cultural development or urbanisation trajectories or urban landscapes, except as a global constraint. If the city has been cherished in Mediterranean Europe, it may be because the ills of industrial squalor and modern life never had to be attributed to any scapegoat, as in Anglo-American cultures.

In this chapter, a *spatial* dimension has been introduced in the long-standing debate on the city-idea, by contrasting North/South representations of urbanism and highlighting some in-between spaces. Schorske (1998) has elaborated the *temporal* dimension, leading from the Enlightenment's 'city as virtue' to industrial capitalism, which portrayed the 'city as vice'; Schorske also tackled the in-between spaces of the 'city beyond good and evil'. In this chapter, this temporal focus is embedded into European *space*, combining Braudel's (1975) '*longue durée*' with the theory for the Mediterranean city, where the past (antiquity to Enlightenment) is constantly revived and incorporated into the present. The fact that modernism, industrialism and Fordism did not take root in Southern Europe does not mean that these cities are pre-industrial, pre-capitalist, pre-modern, as often portrayed in Northern urban theory. This grand narrative is tragically wrong and unconvincing. It may be that these cities were rather long ago 'post'-modern (Leontidou 1993). My analysis here has aimed at rejecting evolutionist and convergence models, the grand narratives which have kept Mediterranean cities captive of anti-urbanist perspectives and locked in unsuitable theory for so long.

References

Allum, P. (1973) *Politics and Society in Post-War Naples.* Cambridge: Cambridge University Press.

Borde, J. and Barrere, P. (1978) Les travailleurs migrants dans la communauté urbaine de Bordeaux, *Revue Géographique des Pyrénées et du Sud-Ouest*, 49(1), pp. 29–50.

Braudel, F. (1973) *Capitalism and Material Life 1400–1800.* London: Fontana/Collins.

Braudel, F. (1975) *The Mediterranean and the Mediterranean World in the Age of Philip II.* London: Fontana/Collins.

Briggs, A. (1968) *Victorian Cities.* Harmondsworth: Pelican.

Carter, F. W. (1968) Population migration to Greater Athens, *Tijdschrift voor Economische en Sociale Geografie*, 59(2), pp. 100–5.

della Seta, P. (1978) Notes on urban struggles in Italy, *International Journal of Urban and Regional Research*, 2(3), pp. 303–29.

Fried, R. C. (1973) *Planning the Eternal City: Roman Politics and Planning since World War II.* London: Yale University Press.

Gabert, P. (1958) L'Immigration italienne à Turin, *Bulletin de l'Association de Géographes Français*, 276–277, pp. 30–45.

Garcia, S. (1993) Local economic policies and social citizenship in Spanish cities, *Antipode*, 25(2), pp. 191–205.

George, M. D. (1966) *London Life in the Eighteenth Century.* Harmondsworth: Penguin.

Gramsci, A. (1971) *Selections from the Prison Notebooks.* New York: International Publishers.

Gramsci, A. (1994) *Pre-Prison Writings.* Cambridge: Cambridge University Press.

Hardoy, J. E. (1973) *Pre-Columbian Cities.* New York: Walker.

Hobsbawm, E. (1968) *Industry and Empire: An Economic History of Britain since 1750.* London: Weidenfeld and Nicolson.

Jacobs, J. (1961) *The Death and Life of Great American Cities.* New York: Random House.

Jensen-Butler, C., Shachar, A. and van den Weesep, J., eds (1996) *European Cities in Competition.* Aldershot: Avebury.

Karapostolis, V. (1983) *Consumption Behaviour in Greek Society 1960–1975.* Athens: EKKE (in Greek).

Karpat, K. H. (1976) *The Geçecondu: Rural Migration and Urbanisation.* Cambridge: Cambridge University Press.

Kayser, B., Pechoux, P. Y. and Sivignon, M. (1971) *Exode Rural et Attraction Urbaine en Grèce.* Athens: EKKE.

Kearns, G. and Philo, C., eds (1993) *Selling Places: The City as Cultural Capital, Past and Present.* Oxford: Pergamon Press.

King, R. (1994) Migration and the single market for labour: an issue in regional development, in Blacksell, M. and Williams, A. (eds) *The European Challenge: Geography and Development in the European Community.* Oxford: Oxford University Press, pp. 218–54.

Leontidou, L. (1990) *The Mediterranean City in Transition: Social Change and Urban Development.* Cambridge: Cambridge University Press.

Leontidou, L. (1993) Postmodernism and the city: Mediterranean versions, *Urban Studies*, 30(6), pp. 949–65.

Leontidou, L. (1995) Repolarization in the Mediterranean: Spanish and Greek cities in neoliberal Europe, *European Planning Studies*, 3(2), pp. 155–72.

Leontidou, L. (1997) Five narratives for the Mediterranean city, in King, R., Proudfoot, L. and Smith, B. (eds) *The Mediterranean: Environment and Society.* London: Arnold, pp. 181–93.

Leontidou, L. and Afouxenides, A. (1999) Boundaries of social exclusion in Europe in the 1990s, in Hudson, R. and Williams, A. M. (eds) *Divided Europe: Society and Territory.* London: Sage, pp. 255–68.

Mariani, R. (1976) *Fascismo e 'Città Nuove'.* Milan: Feltrinelli.

Morse, R. M. (1976) The city-idea in Argentina: a study in evanescence, *Journal of Urban History*, 2(3), pp. 307–30.

Mouzelis, N. (1978) *Modern Greece: Facets of Underdevelopment.* London: Macmillan.

Niethammer, L. and Bruggenmeier, F. (1978) Urbanisation et expérience ouvrière de l'habitat dans l'Allemagne impériale, in Murard, L. and Zylberman, P. (eds) *L'Haleine des Faubourgs.* Fontenay-sous-Bois: Recherches, pp. 103–54.

Salcedo, J. (1977) *Madrid Culpable: Sobre el Espacio y la Población en las Sciencias Sociales.* Madrid: Tecnos.

Saunders, P. (1986) *Social Theory and the Urban Question.* London: Hutchinson.

Schorske, C. E. (1998) *Thinking with History: Explorations in the Passage to Modernism.* Princeton University Press: New Jersey.

van den Berg, L., Drewett, R., Klaasen, L. H., Rossi, A. and Vijverberg, C. H. T. (1982) *Urban Europe: A Study of Growth and Decline.* Oxford: Pergamon Press.

Ward, D. (1973) The making of immigrant ghettoes 1840–1920, in Callow, A. N. Jr (ed.) *American Urban History: an Interpretative Reader with Commentaries.* New York: Oxford University Press, pp. 296–307.

Weber, E. (1976) *Peasants into Frenchmen: The Modernization of Rural France, 1870–1914.* California: Stanford.

Weber, M. (1966) *The City.* New York: The Free Press.

White, M. and White, L. (1977) *The Intellectual Versus the City: From Thomas Jefferson to Frank Lloyd Wright.* Oxford: Oxford University Press.

White, P. (1984) *The West European City: A Social Geography.* London: Longman.

Wiener, M. J. (1985) *English Culture and the Decline of the Industrial Spirit 1850–1980.* Harmondsworth: Pelican.

Williams, A. M., ed. (1984) *Southern Europe Transformed.* London: Harper and Row.

Williams, A. M., King, R. and Warnes, A. M. (1997) A place in the sun: international retirement migration from Northern to Southern Europe, *European Urban and Regional Studies*, 4(2), pp. 115–34.

Williams, R. (1973) *The Country and the City.* London: Chatto and Windus.

Wirth, L. (1938) Urbanism as a way of life, *American Journal of Sociology*, 44(1), pp. 1–24.

6

Demography, international migration and sustainable development in the Euro-Mediterranean region

Armando Montanari

This chapter offers an overview of three phenomena which are of crucial relevance for the future development of Europe and the Mediterranean Basin countries: demographic change, international migration across the Mediterranean Sea, and sustainable development. The extent to which these three processes are intimately interlinked will also be explored. In providing this overview, the chapter also sets the scene for several of the more specific analyses which follow in other chapters, especially the regionally-specific case studies of migration across the Straits of Gibraltar and into Portugal which are the topics of the next two chapters, the nature of coastal landscape change analysed by Elio Manzi in Chapter 11, and the important question of Mediterranean environmental crisis documented and evaluated by John Thornes in Chapter 14.

The central question which forms the fulcrum of the present chapter is trans-Mediterranean migration, seen as a product of cross-Mediterranean contrasts in both demographic and economic structure, and as a stimulus for geopolitical initiatives most recently expressed in the Barcelona Agreement of November 1995. However, there are also wider-scale linkages to models of development and resource use; these connections are explored in the final part of the chapter. Table 6.1 sets out a range of data for the various groups of countries which will be referred to during the remainder of the account. Reference can also be made to several of the tables in Chapter 3, especially tables 3.1–3.4.

For some years now, the Mediterranean region has been the centre of immigration flows towards the rest of Europe. Since the mid-1980s, the migration balance for the 15 countries of the European Union (EU) has become the most

Translated from Italian into English by David Katan, University of Trieste; subsequent editing by Russell King.

Table 6.1 Mediterrranean inequalities: population and GDP change, 1985–2010

Group of countries	Population (m.)		Annual Population change (%)		GDP (billion $)	GDP change (%)
	1995	2010	1985–95	1995–2010	1995	1985–95
Non-Mediterranean EU	197	188	0.1	0.0	7002	3.4
Mediterranean EU	175	180	0.0	0.0	3375	2.1
Likely future EU Mediterranean member countries	64	80	0.2	0.3	–	–
Other MEDA countries	155	210	0.3	0.4	264	0.7
Other Mediterranean Balkan countries	25	28	0.1	0.1	–	–
Libya	5	9	0.4	0.6	–	–
Euro-Mediterranean region	621	695	0.1	0.1	–	–

Sources: UNDP (1995); World Bank (1997).
Note: Likely future EU member countries from the Mediterranean are considered to be Cyprus, Malta, Slovenia and Turkey.
Other MEDA countries are Algeria, Egypt, Israel, Jordan, Lebanon, Morocco, Palestinian Authority, Syria, Tunisia.
Balkan countries are Albania, Bosnia-Herzegovina, Croatia, Federal Republic of Yugoslavia (Serbia-Montenegro), and FYROM (Former Yugoslav Republic of Macedonia).

important component in the increasingly sluggish population increase figures. Leaving aside any discussion of specific social, economic or foreign policies, the EU member countries have both individually and collectively attempted to implement all possible immigration controls to reduce the number of immigrants. This situation has now left the way open for an increase in irregular or clandestine immigration.

Since 1995, the EU has introduced a common peace and stability agreement with the Mediterranean countries; this has been seen as the starting point for the economic and social development of the whole Mediterranean region. Immigration control is a central plank within this agreement, based on the thesis that existing social and demographic differences will continue to encourage immigration, whatever future form this human mobility might take. The principal geopolitical problems to be tackled can be sketched out through an analysis of the recent changes in migration patterns and of the existing Mediterranean immigrant communities in the EU. The figures used here refer to those residing legally in EU countries. It is widely known that there are also numerous irregular or clandestine immigrant communities not included in these figures. Indeed, the complexity of the problems arising from the various types of immigrant, coupled with the difficulties this complexity adds to the public authorities' ability to manage the situation, seriously affect the host countries' capacities to accommodate these migrant influxes. Sometimes the host countries treat the phenomenon with sympathy; on other occasions with misinformation, disinterest and with a degree of opposition which can lead to xenophobia. The problem of integration is clearly a key area. The cultural and religious differences are such that assimilation can be considered anything but an easy process. Equally, the state cannot be held uniquely responsible for solving these complex issues. What is needed is collaboration from both the host and the emigrant states, and hence the involvement of the local administrations which have much more of a grassroots contact with the problems at a human level. In the Euro-Mediterranean area all three principal monotheistic religions can be found, which in itself could provide a useful framework for discussing the problem.

The EU already has policies in place to improve the management of migration and to reduce present imbalances. These are based on economic collaboration and the earmarking of structural funds. These policies must take account of the fragile natural and cultural environment which is possibly the most important resource both for the countries of the North as well as for those of the South of the Mediterranean Basin.

The European Union and Mediterranean migration: the demographic background

At the end of 1995 the EU (with its three new members) counted 375.6 million inhabitants. In 1986 the population of these 15 states had been 360 million. In 1986, the population of the five EU Mediterranean countries (France, Greece, Italy, Portugal and Spain) was 171 million, slightly less than that of the rest of the EU: 189 million. During the decade 1986–95 the population of the EU Mediterranean countries increased by 2.6 per cent while the remaining 10 countries increased by 4.3 per cent. Hence, contrary to earlier demographic history, the European countries with the slowest population growth are now the Mediterranean ones. The countries with well below average growth rates were Portugal, Italy, and Ireland, followed by Denmark and Spain.[1] However, since 1989 the migrant balance for the 15 countries has been above the natural balance and is now the most important population growth factor (figure 6.1). Germany (including ex-East Germany) has had a constant negative natural population balance over the period in question, as has Italy since 1993. In both cases this has been compensated through a positive migration balance. Meanwhile, the EU population continues to age while there is a net excess of young workers in the non-EU Mediterranean countries (Montanari and Cortese 1993).

Just as there are national-level contrasts in demographic trends amongst the various EU countries, as noted in the previous paragraph, so there are – in some countries at least – strong regional differences in demographic dynamics, as illustrated in the following *intermezzo* on the Italian case.

The population of Italy was about 57.3 million at the end of 1995, around 55,000 above the previous year. However the number of births was 29,000 less than the

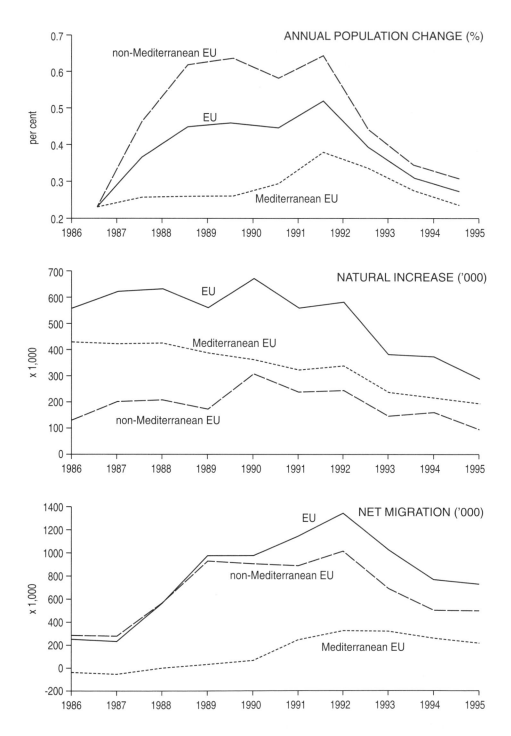

Figure 6.1 Elements of population change in EU, Mediterranean EU, and non-Mediterranean EU countries, 1986–95

number of deaths. The population increase, then, was due entirely to new immigration. At a regional level, there was a natural increase in the population in Campania (4.5 per thousand), Apulia (3.4), Calabria (2.5), Sicily (2.4), Trentino-Alto Adige (1.9), Basilicata (1.2) and in Sardinia (0.7). All these regions, with the exception of Trentino-Alto Adige, are part of the Mezzogiorno, Italy's poor South. These regions form part of the EU's Structural Funds for 'late developers' (Objective 1 regions) as they all had a per capita GDP of less than 75 per cent of the EU average for 1988–90. Traditionally two other regions are associated with the Mezzogiorno: the Abruzzi and Molise. The Abruzzi was excluded by the Commission as its GDP was 89 per cent, while Molise was actually included as an Objective 1 region even though its GDP was 79 per cent. The most significant population decreases were to be found in Liguria (–7.2 per thousand), Friuli-Venezia Giulia (–5.2), and Tuscany, Emilia-Romagna and Piedmont (each around –4). The 1991 Census revealed a Northern Italian population of about 25 million, which could decline to around 20 million in 30 years if the current very low birth rate is maintained. At that point, the number of old people will be equal to the number of those of working age, and this will risk bringing with it a number of predictable economic and social problems.

Returning to the European scale, migration towards the EU countries began to stabilise after 1992. This was due to a number of factors dating back to 1988–91, when the political situation engendered a boost in the flow of immigrants. After this period there was the inevitable decline. In Germany, for example, the net migration balance dropped from 782,000 in 1992 to 330,000 in 1995. In the same period France dropped from 90,000 to 45,000, and in Spain the numbers dropped from 32,000 to 27,000. It should also be noted that over the last 10 or so years a number of EU countries still have a net emigration. Ireland is a case in point, except in very recent years. Other Mediterranean EU countries have moved from net exporters to importers of migrants: Italy (since 1987), Spain (since 1990) and Portugal (since 1992).[2]

The major reason for this stabilisation of immigration has been restrictive EU legislation. Between 1994 and 1995, the Europe of Schengen, which was to have encouraged the free flow of people after that of goods, has in reality resulted in further restriction: the policy considered to be the only antidote to the waves of clandestine immigration. In fact, and in particular during 1995, security checks and controls were quickly reintroduced. This was the result of a series of terrorist attacks and general unrest in France coupled with the pressure of clandestine immigration in Britain, Italy, Spain and Greece. Some have defined these measures as more worthy of the medieval period than the period under the largesse of the free market, internationalism and globalisation. Thus, migration is currently being considered under the same light as terrorism and international crime, against which countries have to defend themselves – using every means at their disposal, including walls and fences. As a result, according to Pugliese (1995) among others, the Imperial Roman idea of the '*limes*' has been introduced. Originally intended as a means to defend the Empire against the barbarians, these fortress-like barriers are now being used to defend the EU countries from new potential invaders. In this scenario of total clampdown on immigration the only option left is probably the illegal route, and this is inexorably under the control of international crime. The director of the Catholic charity 'Caritas Diocesana di Roma' also voiced a number of concerns in his presentation of the 1995 *Dossier Statistico*.[3] Regarding how immigration had been viewed by individual European states and the European Commission, he said:

> After the Treaty of Europe there has been a tendency to frame the phenomenon rigidly as a problem of public order and not to give due weight to the importance of integration. We might expect a little more from an enlightened Europe – not just a simple shutting of the door on work and cultures ... where the increase in xenophobia is often fought with words alone. (Di Liegro 1995)

In the midst of Europe's crises today – crises of ideas, ideals and political leadership – we should add the need to take a lead in the process of globalisation and to compete effectively with the other emerging world macro-regions. Europe has been able to solve its Eastern border problems unexpectedly quickly, but now finds itself having to face problems on its Southern flanks without the necessary tools or strategies. Without the rigid division of the world into its previous ideological blocs, this region is fast becoming increasingly unstable and uncontrollable. The Mediterranean, even those parts visited by around 200 million European tourists who arrive every year, has become a battle-ground like never before. This civil and urban guerrilla war is being played out not only in far-flung corners of the Mediterranean but also in the slums of European towns, where the clandestine, and mainly Mediterranean, immigrants have gathered. As Liauzu (1996) points out, Mediterranean migration is a good illustration of the paradox in all human relocation whether it be economic or touristic. In both cases the movements of people constitute factors of cultural unification while, at the same time, creating elements of alienation in the societies involved. Europe has entered in a new 'cycle of difference', a cycle, without doubt, no less easy to tackle than those of previous centuries. But to fully understand the migratory flows towards Europe we need to piece together the important changes that have occurred in the Mediterranean area within their cultural, economic, religious and social contexts.

Mass migration in the Euro-Mediterranean region

The demographic imbalance between North and South within the Mediterranean Basin is bound to increase over the coming decades (table 6.1). In the North, low fertility rates are contributing to a reduction in birth rate and an increase in the number of elderly. Some regions are already beginning to feel the reduction in the labour force. In the South, on the other hand, the population might even double within 30 years. This demographic problem adds to, and clearly worsens, the present economic problem in these southern Mediterranean countries. With an increase in unemployment the social situation deteriorates, thus creating the conditions for civil tension and conflict. Demographic, economic and social imbalance are primary conditions for a South–North migration flow. On one hand, the European countries are having difficulty in managing and controlling this migration while, on the other, the countries of origin lack an adequate development policy, thus preventing any effective or long-term initiatives (Montanari 1998, 1999).

The migratory flows from the Mediterranean, though important both in themselves and for their knock-on effects, have been considered secondary problems for the EU governments in the wake of the fall of the Berlin Wall and the breaking up of the Soviet regime. Priority seemed to be given, above all by the key player Germany, to financial aid to Eastern European countries with the Phare and Tacis programmes to support their economic reconstruction and to prevent mass migration – which had already reached worrying levels towards the end of the 1980s (Blotevogel *et al.* 1993; Kemper 1993). Yet in reality, notwithstanding the Yugoslavian war, migration from Central European countries has always been considerably less than that coming from the Mediterranean Basin. The management of these migration flows, and of the associated international relations, has revealed itself to be far more complex in the Mediterranean countries, with the result that even French and German cities have directly felt the effects. In fact, at the end of the French presidency of the EU (January–June 1995), the Council of Europe at Cannes passed an aid package of 4.7 billion Ecu for the period 1995–99. This sum was only slightly less than the 6.7 billion Ecu proposed by the Commission for Central and Eastern Europe during the same period (Commission of the European Communities 1995).

After Cannes, the Barcelona Conference (27–28 November 1995) took another step forward, as we saw at the end of Chapter 3. The Barcelona Declaration clearly showed the desire of the 27 governments of the Euro-Mediterranean region to 'give their future relations a new dimension, based on comprehensive cooperation and solidarity, in keeping with the privileged nature of the links forged by neighbourhood and history'. It is the political dialogue which today seems to be the most difficult area to develop, and the most difficult to make into concrete reality. Yet, in the immediate future, the EU needs political harmonisation on its Southern borders as it has achieved to the East. This is essential to ensure a common area of peace and stability in exchange for an offer of financial aid through the MEDA (Mediterranean Development) Programme. However, on approval of the Regulation the package was reduced to just over 3.4 billion Ecu.

Management of migration will have a central role in any Euro-Mediterranean political partnership. The Ministers who signed the Barcelona Declaration agreed on the need to:

- strengthen cooperation to reduce migratory pressure, which should also include the creation of new jobs in the countries of origin;
- guarantee legal rights to those who immigrate legally;
- combat illegal immigration;
- prepare an executive programme to put the partnership principles into practice.

The Work Programme attached to the Barcelona Declaration states the following:

> Given the importance of the issue of migration for Euro-Mediterranean relations, meetings will be encouraged in order to make proposals concerning migration flows and pressures. These meetings will take account of experience acquired, inter alia, under the Med-Migration programme particularly as regards improving the living conditions of migrants legally established in the Union.

The Euro-Mediterranean region will have to deal with the problems of international migration for a long time yet. The best estimates suggest that, even with the injection of substantial direct private investment, the creation of an area of free economic exchange will not reach sufficient levels of regional economic integration over the short term. It has already been suggested for some time now (Montanari and Cortese 1993) that, though a general improvement in the economic situation could contribute to an increase in job opportunities, quality and economic conditions, these will certainly not be enough to resolve the general situation of unemployment and poverty. In fact, any improvements in income, in production or changes in cultural habits may well provoke increasing motives for migration. A further motive for migration, acting as a pull factor, will be the changes in the demographic characteristics of the EU countries (Khader 1997).

The communities of migrants from the Mediterranean countries are not evenly distributed throughout the EU. In order to understand the importance and the development of migration within the context of the Euro-Mediterranean partnership we will consider flows between the 15 countries of the EU, the 12 non-member countries which have countersigned the Barcelona Declaration, and the six Balkan states (Albania, Bosnia-Herzegovina, Croatia, the Former Yugoslav Republic of Macedonia, Slovenia and the Federal Republic of Yugoslavia). The Balkan states did not attend the conference as they were involved, and benefiting from, other EU initiatives due to the civil war and their phase of political and economic transition. We will consider the last 10 years so as to include and check for any effects due to East–West political relations and to encompass the close relationships the EU has made with the countries of Central and Eastern Europe. The use of the Eurostat data bank, SOPEMI publications (various years) and the Council of Europe bulletins (various years) has allowed us to identify the community sizes of the citizens from these 18 Mediterranean countries resident in the EU for the years 1986, 1991 and 1995.[4] These statistics are not available for France, so we have relied exclusively on national census results.

In 1986 there were 13.6 million foreigners resident in the present 15 countries of the EU. Only 5.4 million of these foreigners came from another EU member country, and these were mainly resident in France, Germany and Belgium. The remaining, non-EU, migrants amounted to just over 8.1 million, of whom 60 per cent came from Mediterranean countries. In Austria, those coming from the Mediterranean, Yugoslavia and Turkey represented almost 90 per cent of the non-EU population. In France, Holland, Germany and Belgium the percentage of Mediterranean migrants oscillated between 70 and 80 per cent. In the Southern EU countries, the percentage of Mediterranean migrants accounted for 38 per cent of the total in Greece, around 10 per cent in Italy and Spain, and was almost non-existent in Portugal.

By 1991, the number of foreigners present in the EU had risen to 15.8 million, amongst whom 9.7 million were non-EU citizens, 61 per cent of whom came from Mediterranean counties. The host country proportions remained more or less constant with the exception of Spain and Italy, which saw their Mediterranean quota rise to 26 and 40 per cent respectively, largely as a result of migration from Morocco.

By 1995, there were 18.1 million foreigners present, of whom 12.3 million or 68 per cent were non-EU migrants. This latter figure was now so large that it represented the seventh EU country in terms of population size. Turkish citizens made up the largest group with 2.6 million people, followed by Moroccans (1.1 million), Yugoslavians (800,000), Algerians (700,000), Bosnia-Herzegovinians (400,000), Tunisians and Croatians (300,000 each), to name just those nationalities with more than 100,000 residents.

Between 1986 and 1991 Mediterranean nationals (non-EU) living in the EU increased by 26 per cent, while during the period 1991–95 the increase (6 per cent) was much more limited. During the first period the most significant increase (by 43 per cent) came from Moroccan nationals; also important were the increases in Yugoslavs (by 30 per cent) and Turks (28 per cent).

The non-EU Mediterranean countries have been grouped into three categories to analyse any possible relationship between migration and EU foreign policy. A first category includes those countries close to EU membership, such as Cyprus, for whom there was already a positive vote for entry in 1993, and Slovenia. Also included in this group are countries which already have had some form of membership in the past, and are likely to become members in the future, such as Malta and Turkey. Turkey has had an association agreement since 1964 and requested entry in 1987 (although this was refused as recently as 1997). A second category (the MEDA group) is made up of the remaining nine countries which signed the Barcelona Declaration, while the third group contains the remaining five Balkan countries.

Table 6.2 clearly shows the strong overall growth in the number of immigrants from the Mediterranean during 1986–91, about four times that recorded for the following period. In the first period, the growth rate is rather evenly spread among the three categories. Between 1991 and 1995 the MEDA immigrant number remained constant, the 'next-round' EU countries increased a few percentage points, while the Balkan group increased by more than 25 per cent. Within the

Table 6.2   'Foreign' population in EU countries by citizenship, 1986–95

Citizens of countries belonging to the following groups	1986	1991	1995
Candidates for EU membership	2,004,918	2,536,893	2,641,240
MEDA	1,876,857	2,295,905	2,261,516
Balkan	937,993	1,239,263	1,552,937

Sources: Council of Europe, *Demographic Reports*, various years; SOPEMI, *Annual Reports*, various years.

framework of a general reduction of immigration we can see how the EU has 'given in' to those who were fleeing the civil war. This confirms the thesis that the EU lacks an immigration strategy and that reaction to migration pressure is the result of an inadequate foreign policy. Not without reason, a number of scholars have pronounced that Europe risks drowning in its own sea (Garcin 1994).

Mass migration: an economic, geopolitical and cultural problem

Mass migration presents a number of specific characteristics. Firstly, it is not easy to measure economic migration due to the difficulty in defining 'migrant'. An immigrant may have a variety of statuses: legal, illegal, or clandestine. In the first case the statistics may be reliable, though this will vary considerably from country to country, both in terms of definition and in survey method. Hence cross-national comparisons are hard to make. In the second case, we have immigrants who have entered the country legally, but who have continued their sojourn without the necessary documents. However, in this case, an estimate is still possible. In the third case, though, we have migrants who have entered illegally and remain in that country or move from country to country without any possibility of statistical control. For this third category there are several ways to estimate the number but none of them can give any certainty of accuracy.

The migratory phenomenon is often considered negatively in Europe. With the active support of extreme right-wing parties this provides a fertile environment for xenophobic behaviour (Vandermotten and Vanlaer 1993). Moreover the success of the far-right in mobilising electoral support through the immigration issue makes centre-right, centrist and even left-of-centre parties also harden their policies and their rhetoric *vis-à-vis* the immigration 'problem'. In Spain, public opinion surveys have revealed that at least 40 per cent of the population manifests at least partial xenophobic behaviour, though only 7 per cent is explicitly so. Those with xenophobic tendencies are mainly people who are over 65 years old, conservative and of a low socio-economic status (Ministerio de Trabajo y Asuntos Sociales 1996). Clandestine immigration nurtures this negative perception. It creates insecurity and alarm in the host countries due to the assumption of involvement with international criminality, and the fact that the public perception is that the administration is clearly unable to manage or control its own borders. The problem is not, though, only that of policing and public order, but rather more one of international politics. In a world dominated by a culture of globalisation this situation seems, in fact, to be paradoxical with a return to the culture of borders and the *limes*.

Italy also has its fair share of negative perceptions of non-EU immigrants. There (as in Spain), the immigration is relatively recent and on a smaller scale compared to the rest of Europe, though it is illegal immigration which is seen by many as the real problem. Rampini (1994) suggests that the communities of immigrants present in Italy, who are generally poor and with few civil rights, have created such public alarm precisely because Italian citizens' rights are themselves rather modest in reality. This attitude is not, on the whole, found in the United States, where immigration is still today considered by the dominant culture to be a fundamental value of the society (Collinson 1993). We should, however, make a distinction here, as before, between immigrants from different origins: European, African, Asian and those from other parts of the Americas. Nevertheless, Millman (1997) has collected a large number of immigrant success stories from Indians, Mexicans, Chinese and Caribbeans. As small entrepreneurs they have contributed greatly to the economic and cultural renewal of the USA. It is certainly not easy to quantify the precise role or importance of the immigrants in the process of the development of the American economy and society, but there are clear indications that the immigrants bring with them the seeds of innovation. More than the native-born Americans, the immigrants tend to be autonomous workers and small entrepreneurs, to have large families and to practise intense solidarity towards their own community.

According to economic models, migrations are determined by 'push-pull' factors, though not always does an expulsion factor in one place coincide mirror-like with an attraction factor in another. Collinson (1993) suggests that the expulsion–attraction equilibrium theory operates in practice with illegal immigration rather than with migration deriving from State agreements, which was the main type of migration in Europe up to the mid-1970s. Clearly, in the change from the Fordist to the post-Fordist model of production, so too have the migration flow attraction models changed. The attraction is now temporary work, without protection and unstable, regardless of local levels of unemployment (Blotevogel and King 1996). This altered relationship can be seen in the distribution of the non-EU immigrants in Italian municipalities (Montanari 1993a), as well as illustrated by detailed case studies based on field work. For instance, the results of an in-depth study of the Bangladeshi community in Rome by Knights (1996) showed that the Bangladeshis contribute to the Italian economy in three ways: as employees, as street-hawkers and as entrepreneurs. The second and third groups are particularly sensitive to any social or economic changes, and are susceptible to continuous migration. This precariousness in itself creates general unease and social prejudice amongst the native Roman population, who are inclined to consider the immigrants as 'an unemployed mass of delinquents'. According to Di Liegro (1995), Italians are all too ready to forget the fact that some jobs are only done by immigrants, and often without a contract; and that Italian employers are only too happy to save on social security payments and union minimum wages. 'Even if there are delinquents among them, it is out of place to allow generalisations that have no scientific basis' (Di Liegro 1995).

Migration flows force the EU countries to redefine their own national and political identity (Camilleri 1997). For example, in 1995, more than 33 per cent of the Luxembourg population was foreign, while in Belgium and Germany the figure was around 9 per cent.[5] In the Southern countries of the EU – Greece, Italy, Portugal and Spain – this figure falls to between 1 and 2 per cent.[6] The EU, on the whole, is fast becoming a multi-ethnic society. In all probability, the phenomenon of a pluralistic integration (of some kind – various models are extant) will take the place of the typically North American history of assimilation. The Mediterranean migration flows have a mainly Islamic cultural and religious identity, which means a particularly strong and deeply-rooted culture, and also one which is very far from the host countries' mentality (Khader 1995).

One sign of the lack of integration can also be found in the percentage of foreigners in a country's prison population. In Belgium, France and the Netherlands roughly one-third of the prisoners is foreign. In Italy, where the proportion of foreigners is much lower, the numbers held in detention in 1994 nevertheless accounted for over a quarter of the whole prison population. It is not difficult to infer that international crime operates in the shadow of immigration regardless of the size or type; and that international crime is ready to exploit the weaker elements, and hence the most vulnerable. The percentage of foreigners in Italian prisons has increased as follows: 1970: 3.5 per cent, 1971: 10 per cent, 1972: 16.2 per cent, and 1994: 26 per cent (Caputo and Putignano, 1995). Todisco (1995) has constructed an index which is obtained by summing the number of people reported, arrested, detained, expelled or to be expelled in relation to the number of foreigners legally staying in the country. The highest rate goes to Algerian immigrants (145 per cent). Other Mediterranean countries also have high rates, such as Yugoslavia (36 per cent), Morocco (23 per cent), Tunisia (22 per cent), Albania (12 per cent) and Croatia (11 per cent). By comparison, for the United Kingdom and Germany, the equivalent rates are 2.6 per cent and 0.9 per cent respectively. Though these data are only very approximate, they are a valuable indicator of the extent to which foreign communities have (not) integrated. Even more important is the fact that the data also correspond to how much the various ethnic groups appear in the crime pages or on radio and TV, and consequently they reveal the degree of negative impact that this creates in the public opinion.

Building the Euro-Med region: the role of local organisations

The beginning of the 1990s saw an acceleration in the globalisation of markets and economic interdependence. This was characterised by increased competition and the primacy of financial strategies over the more traditional productive ones. When world customs barriers are lowered, as with the creation of the World Trade Organisation (WTO), transport costs are also reduced, while communication capacities increase enormously. At the same time, we witness an effort to create common markets, customs unions, and free-exchange areas based on geographical proximity. Hence new regional entities are emerging on the world stage, as we saw in Chapter 4. The Euro-Mediterranean Conference held in Barcelona, in fact, planned for the institution of a free-market area by the year 2010 where all the Mediterranean and EU countries could participate. The Euro-Mediterranean area will consequently become another global player to be added to NAFTA (North American Free Trade Agreement), MERCOSUR (Southern Cone Common Market Treaty), CIS (Community of Independent States), and APEC (Asia-Pacific Economic Co-operation).[7]

National economies today are now dominated by the world scale. They are based on multinational capital and industrial production, cross-border information exchange and international consumption patterns and ways of life. Within this context, it seems evident that we will also need to 'think globally' about demographic change and its associated international migration. Nevertheless, the problem remains as to what level of government is the most useful to facilitate access to the benefits of the world economy. In the last two centuries, the traditional reference has been that of the nation-state and its varyingly effective and charismatic representatives. These individuals often identified societal ideals with their own interests. Today, many nation-states in the Euro-Mediterranean are still operating independently and autonomously. As they find themselves incapable of freeing themselves from the myths and the ideologies of the nineteenth century, so they create more of an obstacle to growth and integration. The contemporary national economy has neither the scale nor the power nor the local-level links of the old national administrations. Yet, within the changing dynamics of the world economy, local administrations have been acquiring much more importance both at regional and at metropolitan level. These are perhaps the only administrative levels capable of ensuring the necessary impulse for development and welfare in the decades and centuries to come.

For those Mediterranean countries which are not members of the European Union, there is the justified worry that an area of free trade will expose their economics to the risk of unfair competition from the more advanced sales and distribution methods to be found in the Northern countries. There is also a concern that decisions regarding the Mediterranean will be taken outside the region without the participation of Mediterranean peoples themselves (Philip Morris Institute 1995). As a result it would seem necessary that such a free-trade area be built around popular consensus. Plans and projects should mobilise public opinion through the exchange of culture, science and technology to solve concrete problems such as the safeguarding of natural resources, the restoration of rational planning, the creation of infrastructures, employment, demographic balance and sustainable growth. The Mediterranean is an urban world, as was demonstrated in the previous chapter, and its cities are the ideal administrative level to link the necessity to 'think locally' with the tendency to globalise and internationalise. During the 'Mediterranean Local Agenda 21' Conference, held in Rome during 22–24 November 1995, only a week before the Barcelona Euro-Mediterranean Conference, there was already a clear general awareness of the role and responsibilities of local administrations in the management of local sustainable development in a globally dominated economy. The preparatory documents for the Rome conference show the need to discuss how the local administration could contribute to the objective of sustainable development at national government levels. However, the positions taken by the many city mayors present underlined the fact that reference to national interests was an obvious, though not the most important, issue in relation to the collaboration that the mayors themselves felt was needed to reach their local sustainability objectives. The speeches given by the Mayor of Rome (host) and by the Mayor of Barcelona (host to the Euro-Mediterranean conference), in fact, marked a significant political change and the beginnings of a new cooperation between regions which had been too firmly tied to past stereotypes.

The role of local authorities has been a great innovation within the European Union too. In fact, while national governments were putting the brakes on applying the Maastricht Treaty, a more positive reaction from the local authorities resulted in the relative success which has been obtained in policy areas such as the urban environment. An early reference to the role of local authorities came in 1993 from Jacques Delors, President of the European Commission at the time, during his speech at the seminar on 'Environment and Development: Toward a European Model of Sustainable Development' (Brussels, 9–10 November, 1993), the fore-runner to the White Book on 'Growth, Competitiveness and Employment'. This was the first time that it was officially noted at an authoritative level that two-thirds of the population of Europe and 70 per cent of the employed workforce live and work in urban conglomerations with over 100,000 people. Logically, this is where most of the major EU economic, social and environmental problems are to be found, and this is the setting in which they have to be solved. Amongst these issues there is, of course, immigration and its associated cultural ramifications. Moreover, the new line of thinking initiated by Delors was a recognition of the right for local authorities to have direct relations with the European Commission. These author-ities had long been hostage to the member states as a result of an erroneous interpretation of the concept of subsidiarity.

A further new element was the role of the great world religions. The three monotheistic world religions in the Mediterranean have shaped the process of globalisation in a number of ways. Firstly, there has been difficulty in obtaining political agreement over birth control, as noted during the Cairo Conference (1994). Secondly, concerns have been expressed over the management of environmental resources. Pope John Paul II, in a speech in June 1996,[8] mentioned the problems concerning sustainable development and inter-religious dialogue in the Mediterranean area. He pointed out how contemporary man is led to ask some fundamental questions: 'How can we avoid this rapid development turning against man?… How can we prevent the catastrophes which destroy the environment, threatening all forms of life? And how can we remedy the negative consequences that have already occurred?' He concluded, 'no person can be in a position to deter-mine matters of the environment according to his or her own needs and ignore the rest of humanity'.

Abdallah El Maaroufi (1996), director of the European Office of the World Bank, recognises the fact that, although developing countries on the edge of the Mediterranean have substantial reserves of petroleum and natural gas, their exploitation is compromised due to the lack of, or simply bad management of, two resources: drinking water and productive land. The lack of a clear policy to defend natural resources is creating a risky situation. Many countries consume more water than they actually have. Demographic growth has encouraged migration towards the metropolitan areas, which in a few decades has resulted in a massive increase in urban population. Yet these areas have been unable to provide adequate services or infrastructure (Montanari 1993b). The urban populations are consequently exposed to atmospheric pollution and water shortages, while rural areas are hit by ever-worsening soil erosion, deforestation and biodiversity loss. The artistic heritage is also exposed to pollution, neglect and rapid degradation.

The Fifth Programme, a political, environmental action and sustainable growth programme published by the European Commission in 1993, introduced the concept of responsibility sharing. This concept presupposes the participation of every element of society, which means the public administration, public and private companies, and individual citizens. The sharing of responsibility is essential if we are to reach a happy equilibrium between short-term advantages for individuals, companies and administrations on the one hand, and the long-term benefits for society as a whole. Local communities cannot just be left as mere executioners of what has been decided at national or international level, for the following reasons:

- local community collaboration is essential for the effectiveness of planning;
- it is advantageous if local authorities can find their own role; in this way they can increase the effectiveness of what has been decided at a more general level;
- at a local level it is easier to overcome any particular obstacles and to identify where economic sectors and social interest may be integrated;
- the concept of partnership is being gradually substituted by that of governance, where the citizen may verify, on the ground, the effectiveness of political decisions;
- information and training programmes are more effective if delivered in concrete terms, in a way that the participants can feel involved, and with a clear, verifiable idea of the advantages to be found at local level.

In many EU countries the foreign communities contribute to the maintenance of the demographic levels and to the economic development of those areas where they are concentrated.

Conclusions

In conclusion I wish to sum up and tie together the major themes of this chapter: migration, the nature of Mediterranean development, and the need for partnership amongst the Mediterranean Basin countries. First, regarding migration, the three possible tracks for the EU policy-makers to follow over the next few decades are as follows. Firstly, to continue the closed-door policy, intensifying border controls where necessary. This policy has not been effective over the last few years and has left a space for clandestine immigration and its parasitic associate, crime. Secondly, to allow in limited and selected migrant groups in relation to job needs. This could develop through temporary immigration policies which would exclude the arrival of other family members and any social or cultural integration with the host country. The third track is a more active migration policy combining selection and integration of the immigrants in the host country. These three possibilities, though theoretically logical choices, are unlikely to work practically in a region where the interests of the European countries dominate. Would it actually be possible to draw a form of 'Maginot Line' from east to west through the Mediterranean that would block migration from south to north when the idea is to intensify the movement of people and goods from north to south within the context of a free Euro-Mediterranean exchange? This obvious contradiction between the ideology of an open market on the one hand and the migration policy measures of closed borders on the other is also discussed with reference to Spain and Morocco in the next chapter.

My second conclusion is that the development of a Euro-Mediterranean space should only take place on the basis of economic, social and, above all, environmental sustainability. We have seen – and we will see further in Chapter 14 – that the natural environment of the Mediterranean is extremely fragile and particularly sensitive to any changes carried out by man, both in quantitative and in qualitative terms. The rapid population increase, the change in life-styles and the substantial growth in the movement of both goods and people cannot but have serious consequences for the environment. Reference to the availability and use of drinking water would seem to be especially significant for the Mediterranean. This resource more than any other can identify areas of pressure and shortage, and indicate national strategies of buying up and exploitation. At times these strategies will also cause internal conflicts with the Basin, which could easily give rise to international political friction. The only solution to the water problem, and for many other environmental problems in the Mediterranean, is for the individual countries to collaborate and manage together the resources of the region. Improved management of natural resources and of the countryside would also help tourism. More than other forms of regional collaboration, the development of tourism can also reduce the risks of conflict and of war. Even if internal competitive tourism at the regional level continues to survive, some form of integrated development would need to be considered for the whole of the Mediterranean.

Finally, the realisation of a free-exchange Euro-Mediterranean region can only take place through the principles of partnership and the sharing of responsibility. In January 1996 the European Commission presented its 'Decision of the European Parliament and Council' proposal regarding the revision of the Fifth Environmental Programme. Article 9 of this communiqué reads: 'The Community shall foster the development of practical methods to improve the system of sharing out the actions . . . so as to ensure sustainable growth. The Community shall develop improved methods of dialogue and will ensure that the appropriate range of players take part in the preparation and the carrying out of the political decisions and policies.' Such a commitment should also be applied to migration policies in the Euro-Mediterranean area.

Notes

1   The Irish case is somewhat anomalous because slow demographic growth is largely the result of continued net emigration through the 1980s and early 1990s.
2   Data in this paragraph are from the publication *Recent Demographic Development in Europe*, Strasbourg: Council of Europe, various years.
3   These Caritas dossiers have been published annually for a number of years now, and are extremely useful sources for monitoring trends on immigration to Italy and the rest of Europe. For the most recent issue see Caritas di Roma (1999).
4   In addition to personal visits to Eurostat, the basic sources for these statistics are the Council of Europe's *Recent Demographic Development in Europe* (see note 3) and the *Annual Reports* issued by SOPEMI, the OECD unit responsible for monitoring migration trends in the OECD countries.
5   Based on the latest Council of Europe data; see note 3.
6   The Greek figure is more complicated to estimate because of the dominant role of un-

registered migration, especially from Albania. The real proportion in Greece may in fact be as high as 4 or 5 per cent (see Fakiolas and King 1996).

7   NAFTA, created in 1989, includes Canada, Mexico and the United States. MERCOSUR, created in 1991, includes Argentina, Brazil, Paraguay and Uruguay. CIS, created in 1991, includes Armenia, Azerbaijan, Byelorussia, Georgia, Kazakhstan, Kyrgyzistan, Moldavia, Russia, Tadzikistan, Turkmenistan, the Ukraine, and Uzbekistan. APEC, created in 1989, includes Australia, Brunei, Canada, Chile, China, South Korea, the Philippines, Japan, Hong Kong, Indonesia, Malaysia, Mexico, New Zealand, Papua New Guinea, Singapore, the United States, Taiwan, Thailand. With the Bogor meeting, in Indonesia, an agreement was reached for the creation of a free-exchange area and technological integration by the year 2020, and to be brought forward to the year 2010 for the more economically advanced countries in the Asia–Pacific grouping.

8   Reported in *Osservatore Romano*, 7 June 1996.

References

Blotevogel, H. H. and King, R. (1996) European economic restructuring: demographic responses and feedbacks, *European Urban and Regional Studies*, 2(3), pp. 133–59.

Blotevogel, H. H., Muller-ter Jung, U. and Wood, G. (1993) From itinerant worker to immigrant? The geography of guestworkers in Germany, in King, R. (ed.) *Mass Migrations in Europe: The Legacy and the Future*. London: Belhaven, pp. 83–100.

Camilleri, R. (1997) South–North migration policies: recent international achievements, *Studi Emigrazione*, 126, pp. 195–211.

Caputo, A. and Putignano, C. (1995) Immigrazione e aspetti giudiziari, in Caritas di Roma, *Immigrazione Dossier Statistico '95*. Rome: Anterem, pp. 205–14.

Caritas di Roma (1999) *Immigrazione Dossier Statistico '99*. Rome: Anterem.

Collinson, S. (1993) *Europe and International Migration*. London: Pinter.

Commission of the European Communities (1995) *Strengthening the Mediterranean Policy of the European Union: Proposals for Implementing a Euro-Mediterranean Partnership*. Brussels: COM (95) 72 Final.

Di Liegro, L. (1995) Il Dossier Statistico: uno strumento contro il pressapochismo, in Caritas di Roma, *Immigrazione Dossier Statistico '95*. Rome: Anterem, pp. 7–12.

El Maaroufi, A. (1996) Environmental problems and perspectives in the Middle East and North African regions, *MEDIT*, 7(4), pp. 25–8.

Fakiolas, R. and King, R. (1996) Emigration, return, immigration: a review and evaluation of Greece's postwar experience of international migration, *International Journal of Population Geography*, 2(2), pp. 171–90.

Garcin, T. (1994) L'Europa affonda nel suo mare, *Limes*, 2, pp. 21–30.

Kemper, F.-J. (1993) New trends in mass migration in Germany, in King, R. (ed.) *Mass Migrations in Europe: the Legacy and the Future*. London: Belhaven, pp. 257–74.

Khader, B. (1995) *Le Grand Maghreb et l'Europe: Enjeux et Perspectives*. Paris: L'Harmattan.

Khader, B. (1997) *Le Partenariat Euro-Méditerranéen après la Conférence de Barcelone*. Paris: L'Harmattan.

Knights, M. (1996) Bangladeshi immigrants in Italy: from geopolitics to micropolitics, *Transactions of the Institute of British Geographers*, 21(1), pp. 105–23.

Liauzu, P. (1996) *Histoire des Migrations en Méditerranée Occidentale*. Brussels: Editions Complexe.

Millman, J. (1997) *The Other Americans: How Immigrants Renew Our Country, Our Economy and Our Values*. New York: Viking Penguin.

Ministerio de Trabajo y Asuntos Sociales (1996) *Los Inmigrantes Económicos en España*. Madrid: Observatorio Permanente de la Inmigración.

Montanari, A. (1993a) Migrazioni Sud-Nord: la situazione italiana nel contesto della regione Mediterranea, *Bollettino della Società Geografica Italiana*, Ser. XI, 10(1), pp. 11–34.

Montanari, A. (1993b) Food consumption, culture, quality of life and economic development in Mediterranean countries: mediation, integration and conflict, in Montanari, A. (ed.) *Food Policy and Economic Development in Mediterranean-Arab Countries*. Naples: Edizioni Scientifiche Italiane, pp. 11–47.

Montanari, A. (1998) Flows of goods and people in the Euro-Mediterranean region, in Conti, S. and Segre, A. (eds) *Mediterranean Geographies*. Rome: Società Geografica Italiana and CNR Italian Committee for the International Geographical Union, pp. 159–70.

Montanari, A. (1999) Il mercato unico e l'area Euromediterranea di libero scambio: flussi di mercati e flussi di persone, in Celant, A. (ed.) *Commercio Estero e Competitività Internazionale: imprese e squilibri territoriali in Italia*. Rome: Società Geografica Italiana, pp. 161–83.

Montanari, A. and Cortese, A. (1993) South to North migration in a Mediterranean perspective, in King, R. (ed.) *Mass Migration in Europe: the Legacy and the Future*. London: Belhaven, pp. 212–33.

Philip Morris Institute (1995) *Mediterranean Partnerships. Conference Proceedings, Real Academia de Bellas Artes, Madrid 5–6 October 1995*. Brussels: Philip Morris Institute.

Pugliese, E. (1995) New international migrations and the 'European Fortress', in Hadjimichalis, C. and Sadler, D. (eds) *Europe at the Margins: New Mosaics of Inequality*. Chichester: John Wiley, pp. 51–68.

Rampini, F. (1994) Paura dei 'Barbari' e difficoltà ad essere italiani, *Limes*, 2, pp. 191–8.

Todisco, E. (1995) Classe, ceto e cittadinanza negli immigrati e i fattori di integrazione, in Todisco, E. (ed.) *Immigrazione: dai Bisogni ai Diritti, dall'Emarginazione all'Integrazione*. Latina: Università degli Studi 'La Sapienza', pp. 9–21.

UNDP (1995) *Human Development Report*. New York: United Nations.

Vandermotten, C. and Vanlaer, M. (1993) Immigrants and the extreme-right vote in Europe and in Belgium, in King, R. (ed.) *Mass Migrations in Europe: the Legacy and the Future*. London: Belhaven, pp. 136–55.

World Bank (1997) *World Development Report 1996*. Washington DC: World Bank.

7

The Strait of Gibraltar: Europe's Río Grande?

Jan Mansvelt Beck and Paolo De Mas

The Strait of Gibraltar and the Río Grande are both natural barriers crossed by migrants who go from the poor South to the wealthy North. The fact that the Río Grande and the Strait of Gibraltar do not perfectly protect the wealthier North from 'unwanted foreigners' has led to the popular idea that migration across these barriers has similar features. In this sense, the Río Grande and the Strait of Gibraltar provide an interesting dual case study. In this chapter, attention will first be paid to the role of these barriers as obstacles to migration and as dividing lines between two worlds. Secondly, we will analyse the migration processes on both sides of the borders in a comparative way. The nature of these barriers will be questioned. Are they really a dividing line between two worlds or are they just mere border segments between states? Migration on both sides of the barrier is a much older and more massive phenomenon in Mexico and the USA than it is in Morocco and Spain. Therefore we will examine to what extent the migration history of the former two countries is nowadays repeated in the case of Morocco and Spain.

Río Grande and Strait of Gibraltar, real or symbolic barriers?

In studying the barrier role of borders a physical, a symbolic and a legal dimension can be distinguished. The case of crossing these barriers will be discussed along each of these dimensions. The question that will be answered is to what extent these natural borders form real or just symbolic barriers.

Physical barriers
The Río Grande, for Mexicans Río Bravo del Norte, forms a natural border between Mexico and the USA for almost 1,500 km, from the border cities El Paso/Ciudad Juárez to the Gulf of Mexico. For the first 350 km eastward, where the Mexican Río Conchos flows into the Río Grande, the river carries water from six to nine months per year. In the lower course the water level is comparatively

stable. At many sites it is easy to cross by walking, swimming or by lorry tyre inner tubes. While it is not very difficult to protect the border through wire fences and electronic warning systems, the river has an enormous number of bridges connecting the two countries, enabling undocumented migrants to enter into the USA as tourists or under other unofficial non-migrant titles. However, the border separation by the Río Grande only forms half of the total border length between the USA and Mexico. The Río Grande is only a border in the east and is therefore not so relevant to California, the USA's most important destination for Mexican migrants.

The Strait of Gibraltar, in medieval times also known as Bahr Al Maghreb (the Sea of the West) or the Canal of Andalusia, is a more difficult place to cross. The smallest distance to cover is about 15 km, and is a dangerous trip due to treacherous currents and high winds. Tarifa, Iberia's southernmost town, is nowadays a wind-surfers' paradise thanks to frequent strong winds. The illegal and perilous crossing in *pateras*, small vessels, is however more difficult to patrol than the Río Grande because of the enormous length of Spain's fragmented coastline, and its many beaches suitable for landing clandestine newcomers. Clandestine landings of contraband near La Linea de la Concepción, a town bordering the Rock of Gibraltar, have a long-standing reputation. Smuggled cigarettes and liquor, and at present drugs, are commonly associated with this town. Given the risky *patera* crossing it seems more likely that the majority of illegal migrants coming from Morocco to Spain use one of the big ferry boats (most of them from Ceuta or Tangier to Algeciras) in order to cross the border under a false pretext (some 800,000 ferry passengers cross to southern Spain each summer), or come to Spain stowed away aboard ships leaving Morocco's main port of Casablanca (Bodega *et al*. 1995, pp. 807–8; Cornelius 1995, pp. 337, 350). Thus, the crossing of the Strait's barrier costs generally more time and money than passing across the Río Grande.[1]

Symbolic barriers

Perhaps more than real barriers between states, the Río Grande and the Strait of Gibraltar are symbolic barriers. They are used in the media and by politicians and sometimes by scholars to give weight to value judgements, political statements and ideological discourse (Montanari and Cortese 1993). Both barriers symbolise the alarming nearness of the Third World, with its enormous potential of poor masses pressing to cross these barriers, bringing many dangers to the developed North, such as increased unemployment, the undercutting of wages, the erosion of the social security system and the threat to the country's dominant culture. Spain's former Premier, Felipe González, has been trying to make fellow politicians of Europe more conscious of the nearness and potential dangers of the Third World to Europe by giving them large photographs of Morocco taken from the Spanish coast (Cornelius 1995, p. 334). When Pete Wilson was running for the California Governorship with anti-immigrant issues, images were used in the media showing 'night-time pandemonium at the Tijuana–San Diego checkpoint with un-documented immigrants running across the border' (Smith and Tarallo 1995, p. 666). Though this event was not situated at the Río Grande, the symbolism is similar to the 'wet-back' migrant metaphor associated with the Río.

Borders as economic fault lines

The symbolic barrier consists of a threefold separation: one between two types of economy, another with two types of state, and another with two contrasting cultures. The first separates two types of economy: a productive one with well-skilled and adequately-paid labour, and an unproductive one with the reverse characteristics. As we saw in the previous chapter, demographic forces tend to aggravate these characteristics of poverty, low wages and unemployment. The two types of state refer to a liberal democratic welfare state on the one hand and, on the other, a not-so-democratic state in which welfare is mainly informally produced through personal networks based on kinship. Finally the barrier is the divide between two cultures which significantly differ according to their languages and, in the case of Spain and Morocco, also according to their religions. The state and culture divides still reflect reality. Economically, however, the fault-line idea needs revision.

At first sight the borders separate developed and underdeveloped economies, with the USA scoring seven times higher than Mexico on GNP per capita and Spain having a score 11 times higher than Morocco. However, globalisation is now changing the picture. Both Mexico and Morocco are confronted with the creation of border-free economic systems. In this respect the role of the newly-formed regional trading blocks such as the North American Free Trade Agreement (NAFTA) between the USA, Canada and Mexico, and the European Union (the Single European Market), is crucial – again as we saw in the previous chapter.

NAFTA has led to increasing economic integration between Mexico and the USA, and to a loosening of government control over the flow of goods, services, information and capital. All forms of cross-border exchange have recently intensified. Trucking provides the most concrete illustration of this trend. In 1993, on the eve of NAFTA, 1.9 million lorries crossed over from Mexico; in 1994, 2.8 million. Due to a major improvement programme of the road network in the US South-West, the number of lorries crossing the border will potentially double (Andreas 1996, p. 58). Under NAFTA, Mexican truckers will soon be allowed to operate in the border states of Arizona, California, New Mexico and Texas and, in the long run, anywhere in the Unites States and Canada. While Mexico has become a full member of NAFTA, creating a common interaction context with North America, Morocco – not being an EU member – has only reduced access to the European Market. Notwithstanding this, Morocco has tried to intensify its links with the EU. In June 1996 an agreement was signed with the EU annually increasing the trade quotas of fruits, flowers and vegetables by 3.5 per cent.

Increased trans-border interaction is now reinforcing the development of Mexican border areas by spreading the development from the old free industrial zones of *maquiladoras* more inland, through US investment in labour-intensive industries.[2] This has particularly occurred where the Río Grande forms the border and where Mexican and American twin cities like Ciudad Juárez–El Paso and Nuevo Laredo–Laredo have liberalised mutual access to health care and business activities, a phenomenon also occurring along the Arizona border. The development of Mexico's border areas therefore has mitigated economic inequality between the two sides of the frontier.

Despite its limited access to the EU, Morocco's trade is mainly with Europe, as in 1995 Europe took 62 per cent of Moroccan exports and accounted for 56 per cent of the country's imports (see van den Bremen's account in Chapter 3). Morocco's main trade partners are the major Mediterranean countries of the EU – with France being the largest partner, accounting for nearly a quarter of Moroccan imports and almost one third of exports. Spain has gradually become Morocco's second trading partner with 9 per cent of its imports and exports in 1995.[3]

There is an increase in the cross-border flows of economic goods between Morocco and Europe. But contrary to the Mexican case, there is no increase in trucking between the Mediterranean ports and Europe. Instead a decrease in the number of lorries ferried across the Strait can be observed. The Moroccan company COMANAV shipped 527 lorries to Spain and France in 1990 while in 1994 only 320 were shipped. The decrease contrasts with the increase in the shipment of containers. This shipment is mainly concentrated in the Atlantic port of Casablanca, leaving aside the Mediterranean ports of Tangier and Nador. The new container terminal in Casablanca, a new 400-metre quay, the construction of a 'roll-on roll-off' bridge and associated infrastructure have reinforced Casablanca's dominance. Casablanca handled 42 per cent of the total cargo of Morocco in 1994, and 44 per cent went through other ports of the Atlantic coast such as Jorf Lasfar, Safi and Mohammedia.[4] The orientation towards the ports of the Atlantic and container shipment have limited the possibilities of illegal migration by cross-border trucking to the north, as happens in Mexico.

Again, in contrast to Mexico, where much American capital is invested in the border regions, European capital investment in Morocco is concentrated in the cities and the irrigated plains of the Atlantic region. Nearness to Spain is even detrimental to regional development as is illustrated by the role of the Spanish enclaves Melilla and Ceuta on Morocco's north coast. The large-scale smuggling of industrial products through these enclaves was estimated to have a US $ 625 million turnover in 1995. These informal import activities frustrated industrial development in northern Morocco, as illustrated by the special development areas of Tangier and Nador which are far from dynamic (Naciri 1987, p.143; Zaïm 1992, p. 84). Instead, European aid programmes, including the Spanish ones, focus on Morocco's northern Rif provinces. The Rif is the country's main reserve of potential migrants to Europe and the biggest producer of hashish exported to Europe (Observatoire Géopolitique des Drogues 1994, pp. 54–5). By concentrating aid on the Rif, the EU hopes to reduce migration and drug trafficking which so far has generated a cosmetic urban development without any local economic base. In 1992 the King of Morocco declared a 'war on drugs and illegal migration' to persuade the EU to increase aid. Some 5,000 soldiers were posted on the Mediterranean coast to stop drugs and illegal migrants. In 1995 this action was extended to control the large-scale smuggling of manufactured goods through Melilla and Ceuta. This policing has curbed illegal migration across the Strait of Gibraltar and smuggling into Morocco but not the hashish production and trafficking. Apart from aid, infrastructure is increasing the links between Morocco and Spain. In 1992 the countries signed a 25-year agreement to build a second trans-Mediterranean gas pipeline. In the future Moroccan and Spanish power grids will be connected through a cable, while a bridge or tunnel link across the Strait of Gibraltar are also intermittently discussed.

Sassen (1996) stresses that in all highly developed countries the creation of border-free economic spaces is combined with renewed border control to keep migrants out. However this juxtaposition of border-free flows of goods and the border control of movements of people proves difficult in practice. Undocumented migration shows that real barriers for migrants do not consist of stretches of water dividing states, but are defined by law. These barriers are the legal barriers.

Legal barriers

Legal obstacles for would-be migrants have an official side – what is written in the Law – and a practical side, that is, the real sanctions after crossing the frontier. Migration laws since the beginning of Mexican and Moroccan mass migration to respectively the USA and Spain will now be discussed.

For the past century and a half, migrants from Mexico have been entering the USA. In periods of labour shortage immigration from the south was often encouraged by the US government. From 1942 to 1964 the *bracero* programme regulated the contracting of labour, mainly employed in agriculture. Young adult males were recruited in order to work seasonally during the peak periods, mainly on Californian farms. During the period 1964–86 *bracero* migration continued, though mainly undocumented. The legal quota of 20,000 migrants per year for Mexico (Klaver 1997, p. 83) could not meet the US demand for cheap agricultural labour. Though it became legally impossible for many Mexicans to find temporary employment as farm workers, in practice huge numbers of undocumented migrants were allowed to enter into the USA. During the 1979–86 period an estimated number of 115,000 undocumented Mexican migrants per year crossed the border (quoted in Jones 1995, p. 717).

The 1986 Immigration Reform and Control Act (IRCA) tried to cut the inflow of undocumented foreigners. It was partially successful regarding illegal Mexicans up to approximately 1990, more than halving the estimated annual number (Jones 1995, p. 717). Through the IRCA, illegality was able to be transformed into legal residence for about 3 million Mexican workers during the late 1980s (Cornelius 1992, p. 166). During the 1990s clandestine migration from Mexico to the USA increased to an extent that may now equal pre-IRCA numbers (Jones 1995, p. 717). An unintended effect of IRCA was the bridgehead function of legalised immigrants for relatives who wanted to join their families and hence create the next generations of clandestine incomers. IRCA put legal penalties on employers providing work to undocumented immigrants. However, the first real heavy sanction was imposed only eight years after its passing (Smith and Tarallo 1995, pp. 665–7).

In Spain, regulation of labour immigration was hardly important up to 1985, the year before Spain's integration into the European Community. Most of the legislation by that time concerned emigration and return migration, the latter mainly designed for guestworkers returning from the countries north of the Pyrenees. On a far smaller scale than Mexico, Spain had and still has its own *bracero* agreement with France. In 1973 Spain sent 135,000 seasonal workers to the French wine harvest and horticulture, a number that diminished to 25,000 in 1991. From the

1960s to the present more and more Europeans from countries like the UK and (West) Germany settled as retirement migrants in Spain. During the 1970s the foreign presence increased by a modest inflow of political refugees from Latin America, mainly from Argentina and Chile. A more significant inflow of labour migrants has only existed since the 1980s. Therefore the first 'comprehensive immigration law' of 1985 looks like an accurate reaction to increased inflow. However, in reality this law probably served to harmonise with the member states of the European Community of which Spain became a member on 1 January 1986. Like the USA, the legal difficulties of entry have not impeded an increased inflow of illegal immigrants. Spanish borders, like the US ones, are porous. Policing and patrolling are not very strict and public opinion is still relatively tolerant towards immigration (Cornelius 1995). As in the case of post-IRCA USA and the guest-worker-importing countries of Western Europe during the 1960s and early 1970s, Spain has legalised some of the clandestine foreigners, including 79,000 successful applications by Moroccans (Izquierdo Escribano 1991). Legal sanctions exist for employers who contract illegal foreigners but there is hardly any check on clandestine employment (Cornelius 1995). Policing targets are rather to be found in well-organised smuggling rings like those around the Chinese restaurants (Cornelius 1995, pp. 349–50) or in illegal open-air markets in stolen goods run by Poles in Madrid. In Spain, as in many European countries, there is no right for illegal immigrants to access public education and health facilities. In practice, however, it is possible to send children who are illegally resident to school. Unfortunately for them, they cannot obtain diplomas.

From this brief review it can be concluded that both the USA and Spain have legally restricted immigration from their southern neighbours. During the 1990s Spain had a limited scheme enabling seasonal workers to come over, a programme resembling the USA *bracero* arrangement of the 1942–64 period. Neither receiving country is able to close its border to undocumented migrants from the south. Although many legal sanctions exist, illegal immigrants can still find their way to employment in both the formal and informal sectors. A remarkable parallel of both countries is that the legal closing of their borders in the mid-1980s was immediately followed by legalisations of undocumented migrants from Mexico or Morocco.

Some features of Mexico–USA migration

Mass migration from Mexico to the USA has a century-long history. During the first 40 years of the twentieth century 820,000 Mexicans were admitted to the USA as legal migrants, most of them during the 1920s (Vernez and Ronfeldt 1991). The 1942–64 period was characterised by a predominance of seasonal migration of mainly young males to work in agriculture: a total of 4.6 million *braceros* crossing the border eight times on average and outnumbering the number of permanent legal immigrants (Vernez and Ronfeldt 1991). After the abolition of the *bracero* programme, long-term migration to the USA increased and led to the presence of more than 12 million persons having Mexican origin. The acceleration of Mexican immigration was due to an increase in both legal and illegal immigrants, the great majority now settling in urban areas instead of sojourning in the countryside. Settling was accompanied by a more balanced sex ratio of migrants and the formation of an age distribution reflecting family formation through both family reunification and female immigration. During recent decades the geographical patterns in the origin of Mexican migrants have been subject to change. In California, most migrants have their origin in the states of Jalisco and Michoacán in west-central Mexico, a journey of some 2,000 km to the main areas of settlement. By 1973 these regions accounted for approximately half the illegal immigrants into California (Dagodag 1975, p. 504). Most of the mainly unskilled migrants came directly from poverty-stricken rural areas with a *mestizo* population. During the 1980s, however, Cornelius (1992, pp. 161–3) observes a continuity of these old flows and the emergence of two new source areas: Mexico's urban agglomerations (including skilled urban-born workers); and the poorest areas inhabited by the native Indian population. Contrarily to the early 1970s, one out of five recent migrants to California have been step-migrants. According to Cornelius, the increasing educational level and urban origin of Mexican migrants reflect the reduction of the labour absorption capacity of Mexico's urban labour markets. In South Texas Jones (1995, pp. 722–6) observes a distance decay, with higher rates of Texas-bound migration in the border regions. However, after IRCA, migration from more distant regions has been increasing at the expense of the more border-prone areas. The weakening of the distance decay is due to a 'neo-employment frontier' (Jones 1995, p. 726) in north-eastern Mexico where trans-border development attracts cheap labour. Adding Cornelius' observations (1992) of a persistent and increasing poverty in the south-west, the rise of more remote areas as a 'migrant expulsion zone' may be explained.

The settlement process in the USA shows differences regarding the period of immigration. Though all border states have significant communities of Mexican origin, a distinction can be made between states where long-established communities dominate – e.g. Arizona and Texas – and California with recently-arrived migrants, reflected in a higher share of undocumented immigrants (Smith and Tarallo 1995). In Arizona and Texas relatively more second- and third-generation migrants reflect the more mature character of Mexican settlement. Though migrants' offspring show a remarkable increase in educational levels compared to their relatively low-skilled parents, they continue to lag behind the native population and other immigrant groups (Vernez and Ronfeldt 1991, p. 1191). Given the importance of formal education for upward social mobility, the widening gap in education, combined with a lagging-behind in school performance of the next generations do lead to a problematic structural disadvantage for the Mexican-origin populations of the USA. In California, where most Mexican immigrants live, the proliferation of Mexican urban neighbourhoods, the migrants' low status on the (increasingly informal) labour market, their low incomes and a marginal political participation are indicators for minority formation (Smith and Tarallo 1995, p. 668). According to these authors, the situation is less dramatic for the Mexicans of Arizona and Texas, particularly with respect to politics.

Summarising, Mexican immigration into the USA has a long history with, at present, an enormous quantitative impact on the demography of the USA border states. During the most recent decades, sojourning in agriculture shifted to settlement in metropolitan areas. The shift towards long-term migration is reflected in more balanced sex ratios and the formation of two-generation families. Integration into the American labour market takes place at the margins and will probably remain lagging behind the mean because of weaker school performance of the second and third generations. Residential segregation and persistent, low educational levels will be obstacles to joining the mainstream of American society.

From Morocco to Spain: the start of a mass movement?

The northern part of present-day Morocco was a Spanish protectorate from 1912 to 1956. Despite these colonial links and geographical proximity it was very recently that Spain became a destination for Moroccan migrants, and even then only a part came from the former Spanish protectorate. From the 1960s to the early 1980s many Moroccans passed through Spain during the holiday periods on their way to their home country and back to the countries of north-western Europe where they had migrated as guestworkers during the 1960s and early 1970s. Before the early 1980s, Spain was a transit country for Moroccan migrants; only sporadically could Moroccan street-hawkers be observed selling carpets in crowded places in Spain's cities.

Mass immigration of Moroccans to Spain was preceded by the return migration of former Spanish guestworkers who returned after the 1973 oil crisis and by the arrival of elderly and relatively wealthy persons who retired to the sunbelts. In fact the British were the most important group of registered foreign residents during the 1970s and 1980s. The closing of the borders of EC countries and Switzerland to Moroccan migrants after 1973 put Spain on the map of potential destinations. Men from the Rif and to a lesser degree the Jebala area of north-west Morocco started to migrate, most of them clandestinely. Up to 1985 immigration from Morocco was not hindered by visa requirements.

Similar to undocumented migration into the USA, numerous estimates of the number of Moroccans and other undocumented aliens in Spain have been made. Estimation techniques have been based on police registration (Bodega *et al.* 1995), key-person interviews with informants working in local organisations for or with migrants (Colectivo Ioé 1987) and consular rolls (López García 1993). Based on Colectivo Ioé (1995a, pp. 127–33), the following phases of contemporary immigration can be distinguished:

1 'Postcolonial migration' (1959–72).
   This minor flow consisted of family migration of Jews from the cities of Tangier and Tetuan (over 40 per cent from 1959 to 1964) and young Muslim males, most of them from Tetuan and Nador.
2 'European closure migration' (1973–85).
   Moroccans on their way to the European labour market saw their journey stopped at the French border in the Pyrenees. This probably caused an

early settlement in the Catalan border area with its important informal labour market (Colectivo Ioé 1995a, p. 129). The northern provinces of the Berber-speaking region were the main sending area (70 per cent from the provinces of Al Hoceïma, Nador and Tetuan). While the Rif exclusively sent males, half of the migrants from the Arab-speaking Jebala area and the big cities of Morocco were women (Ramírez Fernández 1997).

3 'Legalisation and European integration migration' (1986-present).

The most recent legalisation (1996) indicates that this phase is still continuing. The increasing immigration has been favoured by the possibilities to legalise clandestine labour situations. Spain's adhesion to the EU resulted in new legislation for migrants from Third countries, the so-called *Ley de Extranjería*. The Rif and the cities of northern Morocco remain important outmigration areas totalling more than half the immigrants to Spain. New areas of origin are Morocco's big cities of Casablanca, Rabat and Kenitra and the rural drought-stricken communities of the Atlantic plains and plateaux (Chaouia, Doukouala and Tadla).

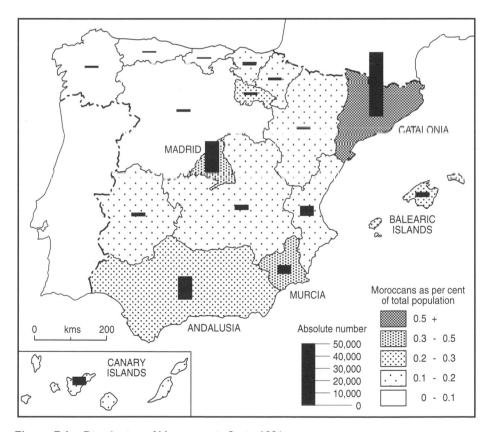

Figure 7.1 Distribution of Moroccans in Spain, 1996

Table 7.1 Moroccan residents in Spain by region, 1994 and 1996

Region	Number of Moroccans		% Population	
	1994	1996	1994	1996
Catalonia	20,431	29,459	0.34	0.48
Madrid	11,098	14,194	0.22	0.28
Andalusia	9,647	10,548	0.14	0.15
Valencia	3,700	4,313	0.09	0.11
Murcia	2,978	3,763	0.28	0.34
Canary Islands	2,402	2,525	0.16	0.16
Extremadura	1,836	1,505	0.18	0.14
Other Regions	8,375	9,760	0.06	0.07
Not recorded	3,472	1,122	–	–
Spain	63,939	77,189	0.16	0.19

Source: Ministerio de Asuntos Sociales (1995, p. 236; 1998, p. 242).

The present era is the key period in the establishment of Moroccan migrant groups in Spain because by far the greatest numbers of Moroccan immigrants have entered in recent years, and an important share have become legalised and thus increased the security of their residence. The legalisation of 1991 in particular had an enormous effect on the number of Moroccan residents in Spain, especially in the main concentrations in Catalonia and Madrid (figure 7.1). In one year their number almost tripled from 16,655 to 49,155 persons. By 1996 the Moroccan population had surpassed the British as the largest registered nationality of Spain with 77,000 against 68,000. Many British and other EU citizens are retired persons. By contrast, most Moroccans are young labour migrants. By 1996 there were 56,291 holding a work permit, followed at considerable distance by Peruvians with 9,650 permits. Though EU nationals are no longer subject to work permits, the 1991 figures indicate the importance of the Moroccan working population: at this date there were 24,481 EU citizens with permits against 38,389 Moroccans with a legal labour status (Ministerio de Asuntos Sociales 1995, p. 112; 1996, p. 163).

Whereas in France, Germany and the Benelux countries Moroccans have settled since the 1960s, in Spain they are still newcomers. Their recent arrival is reflected in the age distribution of Moroccan residents (figure 7.2). Compared to Spain's population, the huge over-representation of Moroccans in the age brackets of 20 to 39 years is noteworthy, as is the under-representation of the age group 45 years and older. The bias in migration is towards males – three out of four migrants (Actis 1995, p. 125). However, the share of women is gradually increasing, not only because of family reunification or marriage but also through the labour migration of adult women from the big cities of Morocco (Ramírez Fernández 1995, pp. 144–5). Many Moroccans are unmarried youngsters, reflected in the high share of 62 per cent of singles (Gozalvez Pérez *et al.* 1994, p. 99). The Moroccans subject to the 1991 legalisation were younger, more male and more single than the other legalised nationalities. The occurrence of so many male youngsters is not surprising

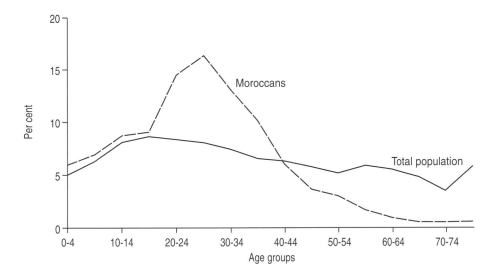

Figure 7.2   Age distribution of Moroccan residents and the total population in Spain, 1991

knowing that six out of ten Moroccans started their Spanish residence as illegal migrants between 1989 and 1992 (Gozalvez Pérez *et al.* 1994, p. 95). Spain is experiencing the start of real mass migration from Morocco. Some Spanish regions, however, receive more migrants than others (table 7.1).

Half the Moroccan residents live in the highly urbanised regions of Catalonia and Madrid, which is not surprising given their large informal labour markets. Perhaps more surprisingly a considerable proportion of Moroccans (19 per cent) live in Spain's poorest regions of Andalusia and Extremadura (figure 7.1). Murcia, Valencia and Spain's island regions also contain significant groups of Moroccans. Chain migration is important amongst Moroccans; it links the geography of destination and the geography of origin (López García 1993). Therefore the main areas of Moroccan immigration in Spain will now be studied separately in order to describe the place-bound processes of migration and integration.

The geography of Moroccans in Spain

Catalonia
One third of all Moroccans in Spain live in Catalonia. They are the most important Moroccan concentration in Spain with almost 30,000 registered migrants and 0.5 per cent of the regional population; yet this is a tiny proportion compared to California's 10 per cent Mexican-born (Klaver 1997, p. 78). When Barcelona was preparing for the Olympic Games of 1992, construction workers from Morocco and other parts of the Maghreb could easily be recognised wearing their woollen caps. By that time they were already a common phenomenon working in

the market-gardening areas on the fringe of the Barcelona agglomeration.[5]

An important study on Moroccan migrants in Catalonia has been conducted by Colectivo Ioé (1995a). According to this study, the number of immigrants increased particularly during the 1980s, strangely at a time when there were at least 300,000 unemployed and 500,000 persons working in Catalonia's informal sector. This situation is representative for all of Spain during the 1980s and 1990s, namely the curious case of high unemployment rates paradoxically accompanied by immigration of labour. Up to the early 1980s, demand for casual labour in agriculture or work in construction was matched by native migrant labour from poor areas of Spain. The improvement of the Spanish social security system opened up opportunities to benefit from public welfare instead of doing hard jobs. Welfare discouraged the native workers from filling the gaps in the labour market; these gaps were then filled by irregular foreigners like the Moroccans.

In contrast to the working population of Catalonia, where more than half are occupied in the service sector and one third in manufacturing, the Moroccans are relatively strongly represented in construction (36 per cent) and agriculture (20 per cent), and weakly in services (30 per cent) and manufacturing (13 per cent) (Colectivo Ioé 1995a, p. 140). More than 60 per cent of the Moroccan working population are in the province of Barcelona where 89 per cent are males, against 95 per cent outside this province. Nearly all the work requires no skills. Most Moroccans work in branches vulnerable to business cycles such as construction, or with seasonal fluctuations in labour demand like tourism and agriculture. Cyclical and seasonal fluctuations in labour demand imply occupational instability, which hampers structural integration into the labour market. The most fluid part of the Moroccan labour force are agricultural workers who travel from harvest to harvest, usually starting from Murcia and Almería in the south and coming to the Lleida district of Catalonia in June in order to leave for La Rioja area in early Autumn. During the Summer months in Lleida they are involved in picking peaches, pears and apples. Their numbers depend on the volume of the harvest. For 1993 a total of 3,000 migrant fruit-pickers were estimated, of whom the majority were Moroccans (Santana Afonso 1993, p. 72). The same source indicates an estimated 7 to 8 per cent of illegal migrants amongst these foreigners. The first Moroccan harvesters appeared in 1989 in the Maresme area near Barcelona. In the Lleida countryside, an old fruit production area in decline, Moroccan seasonal labour helps family farms to survive.

Labour instability is aggravated by the legal situation in which most Moroccans in Catalonia have to work. According to Colectivo Ioé (1995b, p. 147), 92 per cent of Catalonia's Moroccans have a work permit for one year. This insecure legal labour status is probably one of the reasons why the number of registered Moroccan workers initially decreased after the 1991 legalisation. Moroccans who lost their job and work permits disappeared from the statistics and started working in the informal sector. It is hard to find a regular job for young Moroccans who have to face more discrimination from employers in Barcelona than in other big cities of Spain (Colectivo Ioé 1995c, p. 54).

Housing in Spain is mainly private. The majority of Spanish workers live in cheap apartments owned by themselves and paid through savings and bank loans. For

Moroccans their saving capacity to pay the first instalment is generally too limited because of income instability and their recent arrival giving them not enough time to save. Therefore many young Moroccans live near to their workplaces and move to other locations if new income opportunities require such (Colectivo Ioé 1995a, p. 133). Some case studies suggest that in urban areas Moroccans usually rent apartments with fellow countrymen and that in rural areas accommodation varies from specially prepared hostels and sheds to sleeping in the open air (Colectivo Ioé 1995a; Santana Afonso 1993).

Table 7.2 Provinces of origin of Moroccans in Catalonia, 1991

Province	Catalonia	Spain
Nador	33.1	17.5
Oujda	7.5	14.9
Al Hoceïma	5.9	11.6
Rif and North East	46.5	44.0
Larache	16.8	9.9
Tangier	8.6	7.8
Other	18.1	28.3

Source: López García (1993, pp. 84–5), taken from legislation rolls.

During the 1995–6 school year there were 5,176 Moroccan children going to school in Catalonia, 70 per cent at primary level, 22 per cent to kindergarten and only 4 per cent in vocational training.[6] A male bias in school attendance is observed with increasing age and reflects the situation in Morocco. Catalonia's autonomy in educational affairs within the Spanish state since 1979 implies that the instruction at school is in Catalan instead of Spanish, yet Moroccan children mainly live in Spanish-speaking neighbourhoods of Barcelona. Their handicap is that integration into the neighbourhood demands command of Spanish, whereas at school proficiency in Catalan is required. The educational problems for the so-called 'second generation' are even worse for those who are brought over from Morocco when their schooling has already started. This tends to be the pattern for those migrants originating from the Rif area (especially Nador province), who make up 46.5 per cent of the Moroccan immigrants in Catalonia (table 7.2). As Obdeijn *et al.* (1995) have shown for Rif migrants in the Netherlands, family reunion in the destination tends to take place when the children are already 10–15 years of age, too old to really benefit from, or easily adapt to, the educational system in the host country.

Though step-migration and direct migration from the cities are increasing and thus bringing relatively more skilled and more female labour from Morocco to Catalonia, the majority still have a rural background. Specific zones and even individual villages of Nador are extremely over-represented as source areas for Catalonia, where their people cluster in specific districts of the Barcelona agglomeration like Viladecans (Colectivo Ioé 1995a, pp. 163–8).

Madrid

Regarding Moroccans living in Madrid, important studies have been made by Pumares (1993) about the process and patterns of Moroccan migration, and by Colectivo Ioé (1995d, pp. 205–94) on Moroccan workers in the Madrid construction sector. Since the mid-1980s Moroccan migration to Madrid has been accelerating. Three out of ten migrants are from the Rif province of Al Hoceïma, while the urbanised regions of Larache, Tangier and Casablanca are also relatively well represented with one third of the total in Madrid, compared to one quarter nationally (table 7.3).

Table 7.3 Provinces of origin of Moroccans in the Autonomous Community of Madrid, 1991

Province	Madrid	Spain
Al Hoceïma	29.6	11.6
Nador	7.9	17.5
Oujda	2.1	14.9
Rif and North East	*39.6*	*44.0*
Larache	13.7	9.9
Tangier	11.7	7.8
Casablanca	7.9	7.9
Jebala, Atlantic Coast	*32.3*	*25.6*
Other Provinces	*27.1*	*31.4*

Source: López Garcia (1993, pp. 84–5), taken from legislation rolls.

Almost half of the Moroccan workers of Madrid, more than 4,000, were employed in construction in 1992 (Colectivo Ioé 1995d, p. 20). According to the same source they accounted for 3.8 per cent of total employment in construction and 7.7 per cent of all work created during the economic expansion of the 1985–91 period. Eight out of ten construction workers come from the Al Hoceïma province. They share their rural origin with many Spanish construction workers. The strong Al Hoceïma link is part of a complex and multi-local migration chain linking that area with cities in the Netherlands, Belgium and Germany. In Al Hoceïma there is a deeply rooted subsistence behaviour in which 'the plan to migrate appears to be a "normalised" way of social reproduction' (Colectivo Ioé 1995d, p. 217).

The Moroccans are by far the most numerous group of foreign construction workers in Madrid, although they are fewer in number compared to the Spanish, as well as being significantly younger (91 per cent are in the 18–39 years bracket against 59 per cent of the Spaniards) and more single (57 per cent against 39 per cent), while like the Spaniards their general educational level is low (these and the following data on the Madrid construction sector are from Colectivo Ioé 1995d, pp. 220–43). In contrast to the low-skilled Spanish construction workers, Moroccans have a bipolar distribution with relatively more illiterates (44 per cent versus 31 per cent) and more holding a secondary school diploma (35 per cent versus 14 per cent). Those in the higher-skilled category, however, occupy the same low-category jobs as their countrymen. The Colectivo Ioé survey reveals that 26 per cent of the Moroccan construction workers came to Madrid as step-migrants, having first tried their fortune in other regions of Morocco.

Chain migration surprisingly did not lead to employment through the personal networks of fellow migrants; eight out of ten present themselves directly at the gate of the firms or the construction sites. Instead, half the Spanish workers use personal networks to be contracted. A serious problem for Moroccan construction workers is their limited access to the Spanish welfare system, particularly to unemployment benefits. Generally, they are living closer to the construction sites than their Spanish counterparts. Many Spanish teams come from villages more than one hour's drive from the sites. Labour insecurity and concomitant risks of illegal residence, as observed for Moroccans in Barcelona, also exist for the Madrid-based construction workers. The legal ones (57 per cent) have work permits which last only one year, the semi-legal ones (34 per cent) are in the process of renewing the permit or have just a resident permit and no work authorisation, while 7 per cent are illegal. But being an illegal or semi-legal worker in building activities in Madrid is even more common (20 per cent) amongst the Spanish workers. Protection by trade-unions against abuse by employers is all but absent. Moroccan construction workers in Madrid thus occupy insecure, unstable jobs and are ill-protected against bad working conditions and the loss of their employment.

Regarding female immigration, Madrid is different from Barcelona for its high proportion of working women who arrived independently as single or divorced persons (Ramírez Fernández 1997). In contrast to most males, they are mainly from the bigger cities of the Jebala and Atlantic Coast regions (Actis 1995, pp. 126–7). During the 1991 legalisation one in every five Moroccan labour migrants in Madrid was a woman. Characteristically, such women in domestic service were Arab-speaking and higher skilled than most male Moroccans. The majority work and live as maid-servants in the same houses as their 'mistresses' in the north-western and northern newly-built middle-class residential areas of Majadahonda, Pozuelo, Aravaca, Mirasierra and Moraleja. Moroccan maid-servants taking a walk with small Spanish children can be seen in the new town of Tres Cantos, with its well-paid and well-qualified population working in the surrounding firms or at the Autonomous University. Their wages were below the official level and in 1990 ranged between 50,000 and 80,000 pesetas (US $500 to $800) according to Colectivo Ioé (1991, p. 40). Legal working hours were not respected; they were roughly between 8 a.m. and 11 p.m. Many of these young women entered Spain as tourists or transit passengers. Employment was often previously arranged by Moroccan agencies. Often Moroccan women do not have any health care insurance. In the case of illness the insurance of a friend is fraudulently used, a practice not uncommon amongst Spaniards themselves.

The service branch of immigrant employment in Madrid also includes so-called informal activities, of which the Moroccans working as street-hawkers in the underground stations are the most visible example. They are all Arab-speaking from the

Jebala region; many of them have lost their previous jobs in Spain and now take refuge in petty trade.

Destinations connected with agriculture and tourism

Andalusia, Valencia, Murcia, the Canary Islands and Extremadura all have important groups of Moroccans (table 7.1). With the exception of Extremadura, with its sole agricultural orientation, these regions combine a considerable tourist industry with export-oriented agriculture. Moreover, these regions are Morocco's nearest neighbours. The Moroccans are predominantly males coming from the Rif and Oujda, employed or semi-employed in agriculture. Particularly since the early 1990s migration to rural areas has increased (Santana Afonso 1993, p. 26). Working conditions are hard, while the seasonal labour demand in agriculture makes employment almost automatically unstable. Although regional variations in agricultural seasons are at the origin of harvest circulation, Moroccan agrarian labourers do tend to settle in determined areas such as the municipality of El Ejido, located in the export-oriented area of Campo de Dalías in the Andalusian province of Almería. In El Ejido, as in the irrigated plains of Murcia, the agricultural calendar runs from September to May, providing comparatively more work than agriculture in other regions, where shorter growing seasons prevail.[7]

Agricultural technology is innovative as it is based on irrigation, cultivation on sand layers, plastic greenhouses and modern marketing. The small family farms dominating the rural scene use Moroccan temporary labour in addition to family labour. Fluctuations in labour demand at the farm level oblige many casual labourers to switch frequently from one family farm to another (Santana Afonso 1993, p. 32). This instability, combined with the enormous heat in the greenhouses and the easy access to unemployment benefits, has made local agriculture more and more unattractive to native casual labour. Spanish seasonal migrants are increasingly joining the local temporary workers in the processing plants, where labour conditions are better.

From 1991 there has been a limited quota of temporary foreign workers in agriculture in Spain. In 1995, 4,500 Moroccans entered in this quota which totalled 5,500 (Ministerio de Asuntos Sociales 1996, p. 282). The figures for Spanish seasonal labourers annually sent to France show the reverse picture as they diminished from 66,026 in 1985 to 3,045 in 1994. The increasing quotas for Moroccans and the shrinking of labour circulation from Spain to France reflect the substitution process of Spanish by Moroccan temporary labour in agriculture.

In El Ejido legal migrants are generally working one hour more per day than they should do according to labour legislation. They are paid per day and not per hour and it is not uncommon for them to be obliged to work on Sundays. The legal stock of Moroccan workers is accompanied by an unknown number of illegal ones. The labour conditions of the latter are much worse than the former because employers usually do not pay social security dues and do not respect minimum wages. Generally the workers live outside the nuclear villages in isolated and often abandoned farmsteads, called *cortijos*, where many amenities like showers are missing. Moroccan agricultural workers not only work in booming agri-businesses, as in El Ejido, or in the strawberry harvest of Lepe near the Andalusian Atlantic coast; by circulating through Spain they help the survival of more traditional forms of agriculture like olive growing in Andalusia and tobacco farming in Extremadura. In this sense many Moroccans are the new *jornaleros* (day labourers) of southern Spain, the old ones living on the dole or having migrated to the big cities.

Labour instability and the recency of migration are reflected in the near total absence of women and children. Therefore it is still questionable whether the fluid potential of Moroccans working as seasonal labourers will finally result in the establishment of a stable group of migrants in the rural areas. Whereas in the USA Mexican migrants first came as seasonal workers in agriculture and were followed by new urbanites at later stages, in Spain Moroccans started to establish themselves in the cities, and only recently became involved as a casual labour force in agriculture.

The remaining Moroccans in these regions work in the tertiary sector. Surprisingly, the important tourist industry hardly employs Moroccans, as they usually work either as self-employed traders or in domestic service. Traders are mainly from the Arab-speaking Jebala region.

Conclusion

The Strait of Gibraltar and the Río Grande are both symbolic barriers and gates. They are used as metaphors in politics, the media and social sciences. Regarding the Strait of Gibraltar, the barrier metaphor better reflects present-day reality than in the case of the Río Grande, which only partly separates the US from Mexico. Economic integration between Mexico and the US is now moving the development frontier southward. In northern Morocco cosmetic urbanisation based on migrants' earnings in Europe cannot hide the lack of a solid economic base. Spatial discontinuity between Morocco and Europe is therefore greater than on the US–Mexican border. The twin cities of US–Mexico foster Mexican border development whereas, on Morocco's northern coast, the two Spanish enclaves frustrate industrial development, a backwash effect of smuggling.

Geographical discontinuity is also reflected in the migration to northern Europe, a spatial jumping in which Moroccan migrants only see Spain as a transit country. In contrast to the US–Mexican border, which sees its economic interaction increase, the Strait of Gibraltar border suffers from the leap-frogging of Moroccan exports which are directly shipped from the Atlantic coast to north-west Europe.

With modern transport links, the real barriers for potential migrants are not the physical but the legal ones. In this sense the integration of Spain into the EC in 1986 shifted the European border to the Strait, and to the two enclaves of Ceuta and Melilla on the Moroccan side. In the near future the US Río Grande will possibly shift to the borders between Mexico and its neighbouring states of Guatemala and Belize.

In comparing Spain and the USA as migration targets for Moroccans and Mexicans respectively there are remarkable differences. Spain's immigration from the south is very recent in comparison to American immigration with its long history. The volume of Mexican migration to the USA by far outnumbers

Moroccan migration to Spain. Recent migration history reveals that the first Moroccan labour migrants settled in big cities like Barcelona and Madrid to be followed by others seeking jobs as temporary labourers in agriculture. In contrast in California, the first phase of circulating seasonal labourers in agriculture was followed by a second one of poorly-qualified workers who settled in urban areas.

In their respective host societies both Moroccan and Mexican migrants are at the edge of the labour and housing market and the educational system. The Mexicans tend to form better organised communities in the USA, particularly in Arizona and Texas. The Moroccan groups are too recently settled to form strong associations and pressure groups. Moreover, as research in older migrant communities of the Netherlands reveals, Moroccans, and Riffians in particular, do not organise themselves to the extent other migrant groups do (Obdeijn *et al.* 1995). Madrid and Barcelona are no 'new Los Angeles', where jobs are created due to rapid economic growth, nor are the rural areas of southern Spain really equivalent to their American counterparts. Some Spanish agricultural areas do perhaps seem to be 'new Californias' as pointed out in Chapter 10 of this book, but their small-scale family farms hardly resemble California's enormous estates. The back-breaking work is gradually handed over to legal and clandestine migrants from Morocco. In other Spanish regions traditional agriculture is now continuing through Moroccan casual labour. It is at first sight surprising that Spain, having high unemployment rates, offers so many job opportunities to foreigners. This is because in Spain jobs are left as a consequence of the expansion of the welfare state, and in particular the introduction of unemployment benefits, which is in sharp contrast to real job creation by the market in the USA.

Though there is a considerable legal barrier at the Strait of Gibraltar, the challenge for potential migrants from North Africa to cross it will even be greater. Stagnating economic development and high natural population increase will remain the main incentive to move abroad, to *Al Kharisch*, as Europe is called.

Notes

1   The ferry prices are quite moderate, but there are risks of refusal of entry; passage on a *patera* can cost from $700 to $1200 per person (Cornelius 1995, p. 337; Santana Alfonso 1993, p. 29).
2   *Maquiladoras* are factories which are exempt from Mexican law, located just south of the US/Mexican border, but owned by US capital. A *maquiladora* utilises cheap Mexican labour in assembly, processing or some other labour-intensive form of manufacturing, whilst the legal and fiscal regime favours US owners.
3   Data from *Revue d'Information de la BMCE*, 230, May 1996, pp. 3, 7.
4   Data from *Annuaire Statistique du Maroc*, 1995, pp. 235–6.
5   During a 1987 field excursion, Amsterdam geography students were struck by the near-absence of Spaniards and the omnipresence of Maghrebians tilling the land.
6   Statistics from Ministerio de Educación y Cultura, Oficina de Planificación Estadistica.
7   In February 2000 a major conflict arose in El Ejido; several days of race riots occurred following the killing of a Spanish woman by a mentally unstable Moroccan. See *The Guardian*, 9 February 2000, p. 15, for a report on the situation.

References

Actis, W. (1995) Mujeres marroquíes en España, in Martín Muñoz, G. (ed.) *Mujeres, Democracia y Desarrollo en el Magreb*. Madrid: Editorial Pablo Iglesias, pp. 125–9.

Andreas, P. (1996) US–Mexico: open markets, closed border, *Foreign Policy*, 103, pp. 51–70.

Bodega, I., Cebrián, J. A., Franchini, T., Lora-Tamayo, G. and Martín-Lou, A. (1995) Recent migrations from Morocco to Spain, *International Migration Review*, 29(3), pp. 800–19.

Colectivo Ioé (1987) *Los Inmigrantes en España*. Madrid: Caritas, Documentación Social 66.

Colectivo Ioé (1991) *Trabajadoras Extranjeras de Servicio Doméstico en Madrid, España*, Geneva: ILO, International Migration for Employment Working Paper 51.

Colectivo Ioé (1995a) *Presencia del Sur, Marroquíes en Cataluña*. Madrid: Fundamentos.

Colectivo Ioé (1995b) *La Discriminación Laboral a los Trabajadores Inmigrantes en España*. Geneva: ILO, Department of Employment Working Paper 9.

Colectivo Ioé (1995c) La inmigración marroquí en Cataluña, in López García, B. (ed.) *Atlas de la Inmigración Magrebí en España*. Madrid: Ministerio de Asuntos Sociales, pp. 146–51.

Colectivo Ioé (1995d) *Inmigrantes y Mercados de Trabajo en España*. Madrid: Colectivo Ioé.

Cornelius, W. A. (1992) From sojourners to settlers: the changing profile of Mexican immigration to the United States, in Bustamente, J. A., Reynolds, W. A. and Hinojosa Ojeda, R. A. (eds) *US Mexico Relations, Labor Market Interdependence*. Stanford: Stanford University Press, pp. 155–95.

Cornelius, W. A. (1995) Spain: the uneasy transition from a labor exporter to labor importer, in Cornelius, W. A., Martin, P. L. and Hollifield, J. F. (eds) *Controlling Immigration, a Global Perspective*. Stanford: Stanford University Press, pp. 331–69.

Dagodag, W. T. (1975) Source regions and composition of illegal Mexican immigration to California, *International Migration Review*, 9(3), pp. 499–511.

Gozalvez Pérez, V. and Team (1994) La inmigración marroquí en España, un flujo reciente, clandestino, de crecimiento rapido y con dificultades para su integración sociolaboral, *Cuadernos de Geografía Valencia*, 55, pp. 91–107.

Izquierdo Escribano, A. (1991) La inmigración ilegal en España, *Revista de Economía y Sociología del Trabajo*, 11, pp. 18–38.

Jones, R. C. (1995) Immigration reform and migrant flows: compositional and spatial changes in Mexican migration after the Immigration Reform Act of 1986, *Annals of the Association of American Geographers*, 85(4), pp. 715–30.

Klaver, J. (1997) *From the Land of the Sun to the City of the Angels. The Migration Process of Zapotec Indians from Oaxaca, Mexico to Los Angeles, California*. Amsterdam: Netherlands Geographical Studies 228.

López García, B. (1993) La inmigración marroquí en España: la relación entre las geografías de origen y destino, *Política y Sociedad*, 12, pp. 79–88.

Ministerio de Asuntos Sociales (1995, 1996, 1998) *Anuario de Migraciones*. Madrid: Dirección General de Migraciones.

Montanari, A. and Cortese, A. (1993) South to North migration in a Mediterranean perspective, in King, R. (ed.) *Mass Migration in Europe: the Legacy and the Future*. London: Belhaven Press, pp. 212–33.

Naciri, M. (1987) Les villes méditerranéennes du Maroc: entre frontières et périphéries, *Hérodote*, 45, pp. 121–44.

Obdeijn, H., De Mas, P. et De Ruiter, J. J., eds (1995) *L'enseignement aux élèves marocains de la deuxième génération dans les pays d'acceuil. Actes du Colloque Maroco-Européen, Leiden 15–17 juin 1995*. Amsterdam: IMES, Universiteit van Amsterdam/Rijks Universiteit Leiden, paper.

Observatoire Géopolitique des Drogues (1994) *Etat des Drogues, Drogue des Etats*. Paris: Hachette.

Pumares, P. (1993) La inmigración marroquí, in Giménez Romero, C. (ed.) *Inmigrantes Extranjeros en Madrid, Estudios Monográficos de Colectivos de Inmigrantes*. Madrid: Comunidad Autónoma de Madrid, Consejería de Integración Social, pp. 119–221.

Ramírez Fernández, A. (1995) Las inmigrantes marroquíes en España. Emigración y emancipación, in Martín Muñoz, G. (ed.), *Mujeres, Democracia y Desarrollo en el Magreb.* Madrid: Editorial Pablo Iglesias, pp. 143–58.

Ramírez Fernández, A. (1997) *Migraciones, Género e Islam: Mujeres Marroquíes en España.* Madrid: Universidad Autónoma de Madrid, doctoral thesis.

Santana Afonso, I. (1993) *La Mano de Obra Marroquí en el Sector Agrícola.* Madrid: Dirección General de Migraciones.

Sassen, S. (1996) *Transnational Economies and National Migration Policies.* Amsterdam: Institute for Migration and Ethnic Studies, Universiteit van Amsterdam.

Smith, P. and Tarallo, B. (1995) Proposition 187: global trend or local narrative? Explaining anti-immigrant politics in California, Arizona and Texas, *International Journal of Urban and Regional Research*, 19(4), pp. 664–76.

Vernez, G. and Ronfeldt, D. (1991) The current situation in Mexican immigration, *Science*, 251(4998), pp. 1189–93.

Zaïm, F. (1992) Les enclaves espagnoles et l'économie du Maroc méditerranéen. Effets et étendue d'une domination commerciale, in El Malki, H. (ed.) *Le Maroc Méditerranéen, la Troisième Dimension.* Casablanca: Editions Le Fennec, pp. 37–85.

8

The geography of recent immigration to Portugal

Maria Lucinda Fonseca

Portugal in the European international migration system

Economic restructuring and the demographic, social and political changes which
have taken place in Europe in the last few decades have brought about profound
shifts in the patterns of European international migration. As Chapter 6 showed,
Southern Europe has played a pivotal role in the reshaping of these migration flows,
and the 'migration turnaround' of Southern Europe – from region of mass emigration to one of mass immigration – has been much discussed in recent years by
geographers and migration specialists (see, for example, King and Black 1997; King
and Rybaczuk 1993; Misiti *et al.* 1995). As both of the previous chapters have
demonstrated, Spain and Italy are subject to strong migration pressures from across
the Mediterranean, especially from countries such as Morocco and Tunisia. Greece,
too, is the recipient of trans-Mediterranean migration flows, notably from Egypt,
although the major migrant presence in Greece comes from its northern neighbour
Albania. At the other end of Southern Europe, on the fringes of the Mediterranean,
Portugal has both similarities with, but also important differences from, these new
Southern European migration patterns.

The similarities involve macro-structural features at a European scale. The
economic crisis of the 1970s and the restrictions on the entry of immigrants imposed
by the migrant-receiving industrialised nations of Western Europe drastically
reduced the flow of labour migrations from Southern Europe and caused a mass

This study is part of a larger research project undertaken by the Centre of Geographical Studies at Lisbon
University and is financed by JNICT (the National Council for Scientific and Technological Research)
and DGOT (Directorate-General for Land Management), Project 34/95.

return to the regions of origin. From the mid-1970s on, substantial return flows took place to Portugal, Spain, Southern Italy and Greece (King 1984). These flows continued through the 1980s and even the 1990s, although at a much reduced rate. At the same time Italy, Spain, Greece and Portugal became poles of attraction for migrants coming from developing countries in Africa, Asia and Latin America. As we shall see in more detail later in the chapter, what makes Portugal rather distinct from the other South European countries is the specific set of countries supplying these immigrants.

Several factors help to explain the development of these migratory flows into Southern Europe. Besides economic reasons such as the differences in income, wages and employment opportunities, geographic, cultural and linguistic ties between the origin and receiving countries must be taken into account. Ease of entry is another important consideration. Border controls are not as strict as in the countries of Northern Europe, and it is relatively easy to obtain a tourist or student visa. As has been pointed out by several migration scholars, however, the migratory networks that are formed with the support and solidarity of the already-settled immigrant communities tend to assume an important role (Castles and Miller 1998, pp. 24–7; Faist 1997), especially in a situation where the destination countries have increased restrictions on entry of foreigners.

Immigration flows into Portugal and other Southern European countries are made up of three major types. First, as already noted, Southern Europe has become, since the 1980s, a 'new door for entry to the European Union' for many African, Asian and South American migrants (Robin 1994). Some of this immigration has been classed as illegal. Second, there is an increase in the number of foreigners from developed countries, chiefly Western Europe. These migrants fill highly qualified positions, mainly in multinational enterprises. This migration process is closely linked to the internationalisation and expansion of all European economies, but especially those of the Southern EU. Thirdly, the coastal regions of southern Portugal and Spain also attract many retired foreigners, above all British and German, who are drawn by the warm and sunny climate and a lower cost of living (Williams *et al.* 1997). Yet Portugal distinguishes itself from the other three Southern EU countries in that these immigration flows co-exist with the continuing emigration of Portuguese workers to other European countries such as Switzerland, Germany and France, and to North America. This paradox will be explained later in the chapter.

Despite the fact that foreign immigration into Portugal started to experience significant growth as early as the mid-1970s, research on the issue only started to be published in the 1990s, some of the most important studies being those by Baganha (1998), Eaton (1996, 1998, 1999), França (1992), Guibentif (1996), Machado (1993, 1997), Malheiros (1995, 1996a, 1996b, 1997), Rocha-Trindade (1995) and Saint-Maurice (1997). The present chapter presents an analysis of the geography of immigration into Portugal. The phenomenon of immigration is framed within the dialectical relationship between, on the one hand, processes of Portuguese economic and social restructuring and, on the other, the country's new role in the European international migration system. The account starts by looking at the question of how the recent development of immigration to Portugal is affected by the restructuring of the economy, changes in employment conditions and Portugal's position within the international division of labour. Next, attention will be focused on the geographical pattern and character of the stocks and flows of migrants and on their main social and economic impacts on Portuguese regions. Finally, the chapter examines the legal context and the reception and settlement processes of the immigrants in Portugal.

Economic restructuring and international migration in Portugal

In this section of the chapter I outline the main features of the recent evolution of the Portuguese economy, laying particular stress on those elements which are important in contextualising recent migration trends.

Following the period of economic recession and political instability that characterised much of the 1970s, the Portuguese economy entered a new phase of growth starting in the mid-1980s. The country benefited from the combined effects of a particularly favourable international situation, the consolidation of democracy, and the re-establishment of the political stability needed to recover the confidence of economic agents and stimulate the growth of both national and international investment. The aid received from European structural funds following Portugal's entry into the European Union on 1 January 1986 was crucial in the improvement of basic infrastructures such as water and electricity supply, sanitation, and transport and communications networks, as well as enhancing social capital in terms of professional training and incentives for the development of endogenous potential in the different regions of the country (Jacinto 1993; Reis 1996). As a result, as table 8.1 shows, Portuguese GDP grew at a greater rate than the EU average, the rates of unemployment and inflation went down, and both national and foreign investment rose.

The Portuguese economy entered a new phase of recession at the beginning of the 1990s, triggered largely by the changing European and international market context: the slump in the international economy; the process of restructuring of Portuguese firms and of the public sector imposed by the setting up of the Single European Market; reform of the Common Agricultural Policy; increase in competition due to the general globalisation of the economy and the transition of the Eastern European countries to market economies; and the construction of the European Economic and Monetary Union (Fonseca 1994). Between 1993 and 1995 the annual growth rate of Portuguese GDP was lower than the EU average, and unemployment rose sharply (table 8.1). At the same time, there was also a substantial drop in the growth of fixed capital formation. After the middle of the decade, there have been good signs of a recovery of the Portuguese economy – for instance GDP growth was more than twice the EU rate in 1996 – but unemployment remains high by Portuguese standards.

Economic restructuring has had important repercussions on the structure of the labour market, which in turn provides some of the context for understanding the significance of recent immigration into Portugal. The strategy of more flexible work practices adopted by Portuguese firms in an effort to reduce labour costs and

Table 8.1 Evolution of the Portuguese economy, 1985–96

Year	GDP real growth (%)		Unemployment (%)	
	Portugal	EU	Portugal	EU
1985	2.8	2.5	8.5	10.8
1986	4.3	2.6	8.4	10.7
1987	5.1	2.7	7.1	10.4
1988	4.0	4.1	5.7	9.8
1989	5.5	3.4	5.0	9.0
1990	4.2	2.9	4.7	8.4
1991	2.2	1.5	4.1	8.7
1992	1.5	1.1	4.1	9.4
1993	−1.2	−0.4	5.5	10.5
1994	0.5	2.9	6.8	11.1
1995	2.3	2.4	7.2	10.8
1996	3.3	1.6	7.3	10.9

Sources: Various reports and documents from the Instituto Nacional de Estatística, Bank of Portugal, European Commission.

Notes: EU 12 up to 1993; EU 15 after 1994. The criterion for recording Portuguese unemployment changed in 1992.

increase competitiveness and productivity gave rise to an increase in the numbers of short-term contracts and of part-time workers. Between 1993 and 1996 part-time workers increased from 7.2 per cent to 8.7 per cent of total workers, and the proportion of workers with temporary contracts grew from 10.9 to 12.4 per cent. At the same time the structure of unemployment changed, with the long-term unemployed increasing from 29.3 to 42.0 per cent of the total unemployed.[1] Other key changes in the labour market have been the rise in the number of women engaged in paid work and the increase in the proportion of jobs and gross value added generated by the service sector. Inter-censal comparisons (1981–91) show how dramatic has been the rise of the service sector at the expense above all of agricultural employment. Between 1981 and 1991 the number employed in service activities rose by more than 31 per cent, the share of the service sector increasing to account for half of total employment in 1991. This trend was strongly correlated with the rise in the number of working women – from 34 to 40 per cent of the labour force. Agriculture lost more than 280,000 workers during 1981–91, a reduction of some 40 per cent. Finally, manufacturing industry maintained a constant employment share over the period under discussion, despite showing an increase of around 76,000 workers; this reflected the overall expansion of the labour market during the decade.

If we were to pick out one key feature from the set of labour market changes just described, it would be the rise of the service sector. The changing size and configuration of the tertiary sector also have profound implications for different types of employment available to immigrants, as we shall see later. The expansion of services started in the 1960s with the first modest growth of tourism, then accelerated more rapidly in the late 1970s and 1980s. There are two essential factors behind this growth: in the first place, the launching of the welfare state following the revolution of 25 April 1974 and the widening of local power from 1979. Secondly, in the 1980s, the ever-increasing internationalisation of the Portuguese economy, especially after EU entry, the growth of national and foreign investment, and the re-privatisation of the banks and insurance companies all created conditions for the sustained growth of service sector activities (Ferrão and Domingues 1994).

Tertiary sector growth and technological development, combined with profound social changes in Portuguese society (rural-urban migration, higher levels of education amongst young people etc.), produced a period of intense occupational mobility, including the rise of a new elite of highly qualified management staff. At the same time, new dualisms in society and the labour market emerged, based on class, gender, race and ethnic origins. The progressive segmentation of the labour market can be seen in the distinction between stable jobs and precarious employment, between highly-paid and low-wage activities, and between the formal and the informal economy (Almeida *et al.* 1992). Once again migration trends are closely interwoven with the overall picture.

Despite the limited data available, there is no doubt that Portuguese emigration to Western Europe fell off sharply during the 1970s; that which remained consisted mainly of family reunification. Instead, the dominant flow was of emigrants returning to Portugal. At the same time, the country was faced with the problem of integrating half a million refugees from its ex-colonies during 1975–76. Portuguese emigration gained momentum again after the mid 1980s, a somewhat paradoxical situation given the euphoria about rapid economic growth and accession to the European Community. Numbers were much lower than the mass emigration of the 1960s, however, and most of the movement – to Switzerland, France and Germany – was of a temporary character.[2] Also during the 1980s, the immigration of workers from the Portuguese-speaking countries of Africa (known as PALOP) increased rapidly. These immigrants took up the jobs which were increasingly rejected by the Portuguese – jobs which were unskilled, under-paid and frequently illegal. Construction was – and remains – the main sector involved.

Although it seems a paradox, the co-existence of emigration and immigration in Portugal derives from the increased segmentation of both the national and international labour markets and the semi-peripheral situation of Portuguese society in the context of the world economy. In other words, the international hierarchy of employment and wages is the structural factor which explains why many Portuguese continue to emigrate to Switzerland, France etc., occupying in the employment system and social pyramid of those countries the same positions that Cape Verdeans, Angolans, Mozambicans and Guineans do in Portugal. At the same time, the development of industry and services and the internationalisation of the Portuguese economy attract managerial and skilled technical staff from other European countries and from North America, while the difference in wages between Portugal and the more developed countries of Western Europe and the increase in unemployment in Portugal are an incentive for semi-skilled and unskilled Portuguese workers to emigrate. On the other side of the development spectrum, Portugal receives unskilled labour from the PALOPs and provides them with Portuguese skilled personnel (although relatively few in number) who have a socio-professional profile similar to the Europeans and North Americans that it receives.

Evolution and composition of the foreign population in Portugal

There is no statistical information on the number of foreigners resident in Portugal prior to 1960. According to the 1960 census, 29,428 foreigners lived in the country, including 19,794 Europeans (11,713 Spanish) and 6,357 Brazilians. The foreign contingent remained relatively stable during the 1960s but started to increase at the end of the decade. This immigration had a dualistic nature. On the one hand, links with Europe following Portugal's membership of the European Free Trade Association (EFTA) in 1960, the opening to foreign investment and the growth of tourism along the Algarve coast brought to Portugal company directors, managers and qualified technical personnel from advanced European countries, including some foreign retirees in the Algarve (Williams and Patterson 1998). On the other hand, the spurt of industrial growth and urban expansion, within the context of the colonial wars and the legacy of mass emigration of young workers to Western Europe, created job openings in construction and unskilled services for the first

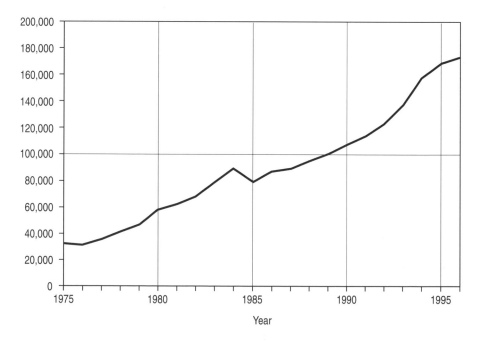

Figure 8.1 Evolution of legally resident foreign population in Portugal, 1975–96

wave of African immigrants from Cape Verde (Esteves 1991). There were also some Africans from the PALOPs and some ethnic Indians from Mozambique who settled in Portugal as refugees in 1974 and 1975 as a result of the decolonisation process. Although these refugees were not counted as foreigners since they opted for Portuguese nationality, they played an important role in the later development of migration flows from their countries of origin, creating family, social and ethnic community networks. Migrants from Macao and Timor also started to arrive after 1975, the former concerned about the absorption of the territory into China in December 1999, and the latter fleeing the climate of insecurity that was felt in East Timor after the start of the civil war in 1975 and the subsequent occupation of the territory by Indonesia (Rocha-Trindade 1995). Recent events in East Timor following the overwhelming vote for independence in September 1999 will change the status of the Timorese in Portugal: instead of refugees they will become immigrants with a similar status to those from the other Portuguese-speaking countries, and in the near future some student and labour migration will probably develop.

Figure 8.1 shows the profile of increase of resident foreigners in Portugal since 1975. Africans quickly replaced Europeans as the main group. By 1981 for instance, their respective numbers were 27,948 and 18,931. By 1986 Cape Verdeans constituted 30.2 per cent of all legal immigrants and made up the largest foreign community in Portugal. While the incoming flow from Cape Verde continued in the late 1980s and 1990s, it was rivalled by the influx from other PALOPs and from Brazil. Indeed by 1996 the Cape Verde total (39,546) was roughly matched by the sum of immigrants from other PALOPs, whilst the Brazilian total exceeded 20,000. Table 8.2 provides a detailed statistical breakdown of the evolution of the various foreign populations in Portugal between 1986 and 1996, when the overall numbers practically doubled from 87,000 to 173,000. The greatest proportional

Table 8.2 Evolution of the legally-resident foreign population in Portugal by region and country of origin, 1986–96

Region and country of origin	1986 no.	1986 %	1991 no.	1991 %	1996 no.	1996 %	growth (%) 1986–91	growth (%) 1991–96
Europe	24,040	27.6	33,011	29.0	47,315	27.4	37.3	43.3
Africa	37,829	43.5	47,998	42.1	81,176	46.9	26.9	69.1
Cape Verde	26,301	30.2	29,743	26.1	39,546	22.9	13.1	33.0
Angola	3,966	4.6	5,738	5.0	16,282	9.4	44.7	183.8
Guinea-Bissau	2,494	2.9	4,770	4.2	12,639	7.3	91.3	165.0
Mozambique	2,475	2.8	3,361	2.9	4,413	2.6	35.8	31.3
S. Tomé	1,563	1.8	2,183	1.9	4,234	2.4	39.7	94.0
Americas	21,676	24.9	27,902	24.5	36,516	21.1	28.7	30.9
USA	6,326	7.3	7,210	6.3	8,503	4.9	14.0	17.9
Brazil	7,470	8.6	12,678	11.1	20,082	11.6	69.7	58.4
Asia	2,958	3.4	4,458	3.9	7,140	4.1	50.7	60.2
Total	86,982	100.0	113,978	100.0	172,912	100.0	31.0	51.7

Source: Portuguese Ministry of Home Affairs.

increases were of migrants from Angola and Guinea. Migrants from Asia, mainly Chinese, Indians and Pakistanis, also increased more rapidly than the aggregate of immigrants.[3]

Besides factors of an economic nature, the rise in the number of immigrants coming from the PALOPs was undoubtedly influenced by the cultural relations established during the colonial period and by the fact that PALOP citizens spoke Portuguese. In addition, there was no political control on the entry of foreigners into Portugal until the start of the 1990s. The ease with which people entered and settled in Portugal, at a time when virtually all other West European countries maintained a tight control over their borders and raised barriers against the entry of non-EU citizens, was also reflected in the growth of 'illegal' immigration, which in turn has led to a series of social problems, especially in the Lisbon Metropolitan Area, the region with the main concentration of immigrants.

Despite the fact that the Portuguese economy suffered a slump in the early-mid 1990s, the number of foreigners with residence permits rose by 51.7 per cent between 1991 and 1996, a faster rate of increase than the 31.0 per cent registered for the previous five-year period (table 8.2). Although a part of this increase was due to a campaign carried out in 1993 to legalise immigrants who were in the country illegally, the high number of requests for legalisation (some 36,000) made during the second phase of the campaign between March and December 1996 demonstrates the size of the migratory influx into Portugal and the numbers who had entered illegally during the 1990s.

Immigrants and the labour market

Most immigrants to Portugal are economic migrants. This can be indicated by their low average age, the large proportion of males, and their higher than average rates of labour force participation (53.5 per cent as against 48.9 per cent). After the economically active share, the largest groups are made up of housewives and students. Retired persons and those who live off accumulated incomes only account for 5.5 per cent of the total; these mainly come from North European countries (especially the UK) and the USA.

As far as the structure of the economically active immigrant population is concerned, figure 8.2 shows that foreign workers are basically employed in two segments of the labour market, and that participation in these segments is heavily conditioned by national origin.[4] The largest group comprises unskilled workers who fill the most unpleasant and lowest-paid jobs in construction, transport, cleaning and domestic service. According to the classification employed in figure 8.2, production workers, labourers etc. make up half of foreign workers' employment, although the figure is much higher, nearly 80 per cent, for PALOP workers. At the other extreme of the socio-professional hierarchy are directors, management staff and self-employed professionals who come mainly from the developed countries of Western Europe and North America, and also from Brazil. The Asians, especially those of Indian origin, are distinct from the other immigrant groups in that most of them work in commerce. According to Malheiros (1996a, 1997), they form two

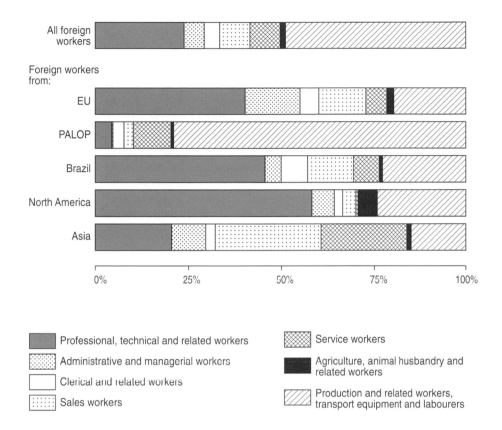

Figure 8.2   Employment structure of the main foreign worker groups in Portugal, 1996

commercial enclaves in Lisbon: one based on furniture shops belonging to the Ismailians; the other based on trading chinaware, electrical products and toys imported from the Far East, controlled by Hindus and Muslims. More recently, a large number of Chinese restaurants belonging to immigrants from China and Macao have opened, again concentrated mainly in Lisbon.[5]

The differences in the socio-occupational structures of Portugal's various immigrant communities are closely related to the migrants' motivations, expectations and their distinct migratory circuits. African and Asian immigrants tend to follow a similar sociological path to that of the Portuguese who emigrated to France and Germany in the 1960s and early 1970s. The migration is long term, leading to the establishment of settled ethnic communities, but also with cycles of return and to-and-fro migration; it is very dependent on information and personal contacts with relatives, friends and co-nationals. Both the Portuguese in Europe and the PALOP migrants in Portugal occupy the lowest rungs of the occupational hierarchy. The immigrants have tended to cluster in certain places and districts, where they form rather closed communities which reproduce many of the characteristics of their countries of origin.

On the other hand, the highly qualified professional groups of immigrants mostly migrate on a short-term basis. The tendency is to work on a temporary basis for an international firm with the aim of seeking professional promotion in the near future. The case of the Brazilian immigrants is a little more complex. Besides the high-status immigrants who are motivated by professional opportunities (for instance as doctors or dentists) and by the growth of foreign investment and business openings in Portugal, the serious economic situation and the social instability in Brazil have led to an increasing number of unskilled and semi-skilled immigrant workers, most of whom work in restaurants, hotels, commerce and other services.

The increase in the number of migrants coming to Portugal since the mid-1980s has not brought about any significant change in the immigrants' demographic and social characteristics; indeed the socio-occupational profile of the foreign population resident in Portugal was almost the same in 1996 as it had been in 1986. This fact, combined with relative stability in the recruitment areas for labour migration, mainly limited to Portuguese-speaking countries, gives Portugal a special position in the context of Southern European migration. It suggests that Portugal, unlike Spain and Italy, is not on the migratory routes that make their way to the Northern EU countries and are travelled by migrants arriving from North Africa, the Middle East and beyond. Rather, Portugal is regarded as merely a niche in the global migration system and is targeted mainly by people from Brazil and the PALOPs. This rather singular behaviour shows that Portugal is not an attractive destination in comparison with other European countries. For PALOP migrants, however, the weak economic attraction (compared to other European destinations) is compensated by the greater ease of entry and by cultural and linguistic factors which, at least in theory, make for a faster integration. It is, however, now becoming more common for some African migrants with Portuguese residence permits to try to migrate onward to other countries of the European Union.

Spatial distribution of the foreign population in Portugal

The absolute geographic distribution of foreign residents in Portugal (figure 8.3) is closely tied to the level of economic and urban development of the Portuguese regions. In 1996, 64.4 per cent of legally registered immigrants lived in the districts of Lisbon and Setúbal. Outside the Lisbon Metropolitan Area, the largest immigrant numbers are to be found in the Algarve (roughly co-terminous with the *distrito* of Faro), the Oporto Metropolitan Area, and the urban-industrial areas of the northern and central coastal strips. The number of immigrants in the northern and central interior, in the Alentejo (Beja and Evora), and in the Autonomous Regions of Madeira and the Azores, is minimal. The high concentration of immigrants in the Lisbon area reflects the clustering of job opportunities there. Besides being the national capital and the dominant economic centre of the country, it is also where public works and their associated construction activities registered the greatest increase during the 1980s and 1990s, reflecting a strong dynamism generated both by private agents and by several mega-projects such as Expo '98, the new Tagus Bridge, the expansion of the metro network and the new rail link between Lisbon

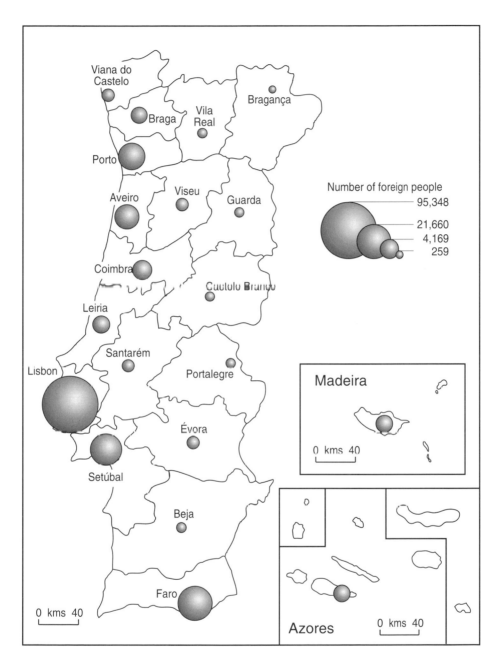

Figure 8.3   Regional distribution of foreigners legally resident in Portugal, 1996

and the south bank. In addition, there is the simple fact that most migrants arrive at Lisbon and it is easier for them to remain there, where ethnic community networks are already highly developed, than to move elsewhere in the country.

A disaggregated analysis of regions of origin shows that the main immigrant groups have distinct patterns of regional settlement in Portugal. The African and Asian communities are those that present the greatest geographical concentration. The two districts of Lisbon and Setúbal, which had 28 per cent of the national population in 1991, accounted for 82 per cent of the African immigrants and 74 per cent of the Asians in 1996. This trend towards spatial clustering is favoured by the density of social networking mechanisms amongst low-status immigrant groups in the metropolitan area of the capital, which also contains a greater diversity of immigrant nationalities than any other part of Portugal. By contrast, a large share of the British immigrants, many of whom are retired people, live in the Algarve. Germans are also found in quite large numbers in the Algarve, and also in Lisbon. The Brazilians are the group which shows the least concentrated spatial distribution. Although 47 per cent of them live in the districts of Lisbon and Setúbal, many are scattered across the northern and interior areas that were strongly affected by Portuguese emigration to Brazil in the past, thus giving the impression of an ebb tide of returnees and their Brazilian-nationality descendants.

Figure 8.4 maps the location quotients (LQs) of the main foreigner groups and confirms their very different spatial patterns.[6] Africans and Asians are over-represented in Lisbon and Setúbal, Brazilians have their strongest over-representation in northern districts such as Oporto, Aveiro and Braga, North Americans also are over-represented in northern and central regions (again reflecting counter-stream migrations), and EU citizens are over-represented in the Algarve and Alentejo.

The spatial distribution of the residential neighbourhoods of immigrants and ethnic minorities within the Lisbon Metropolitan Area shows a generally more scattered pattern, when compared with North American and North European cities, with a lesser concentration in the inner- and central-city districts and a wider spread amongst the suburban areas. However, in a more detailed spatial analysis recently carried out by Malheiros (2000), a distinct tendency towards the clustering of immigrant communities according to their geographical origin and socio-cultural characteristics is visible. Malheiros finds that spatial self-segregation mechanisms are well developed, both amongst Europeans and North Americans, and among the African immigrants. The Europeans and North Americans are over-represented in the higher status and more prestigious areas of Lisbon, in Cascais and along the railway line to Estoril. The Africans, besides having a more scattered spatial distribution, displaying a lower spatial segregation level, show relative concentrations in the municipalities located on the immediate periphery of Lisbon – in particular along the borders with Amadora and Oeiras municipalities, and in Seixal on the south bank of the Tagus.

Figure 8.4 (opposite) Location Quotients for the main foreigner groups in Portugal, 1996

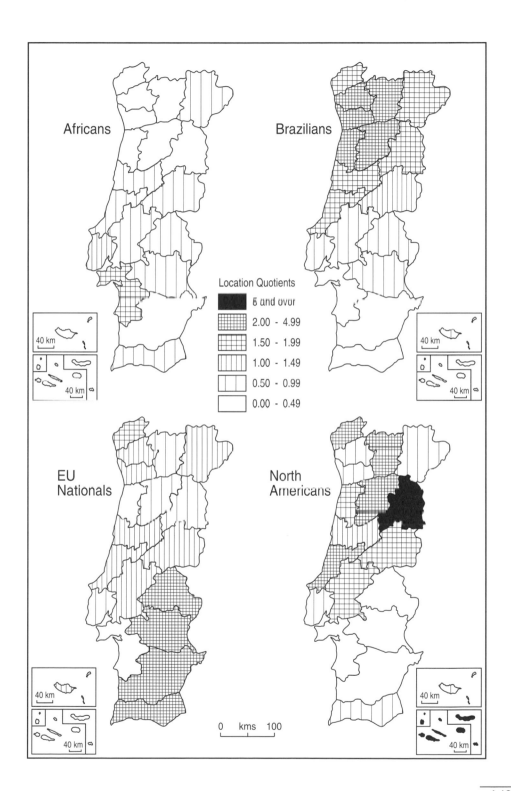

Africans

Brazilians

Location Quotients

5 and over

2.00 - 4.99

1.50 - 1.99

1.00 - 1.49

0.50 - 0.99

0.00 - 0.49

EU
Nationals

North
Americans

0   kms   100

Table 8.3   Dissimilarity indices for the major immigrant groups in Portugal, 1996

	Africans	North Americans	EU citizens	Brazilians	Asians
Africans					
North Americans	36.6				
EU citizens	34.4	38.5			
Brazilians	38.6	24.9	26.3		
Asians	10.6	34.1	26.5	33.2	

Note: See explanation in note 7.

Table 8.4   Dissimiliarity indices for African groups in Portugal, 1996

	Angolans	Cape Verdeans	Guineans	Mozambicans	Santomese
Angolans					
Cape Verdeans	11.0				
Guineans	14.9	16.2			
Mozambicans	7.1	9.2	18.4		
Santomese	9.9	11.7	20.4	5.1	

Note: See explanation in note 7.

Table 8.5   Dissimilarity indices for the major European nationalities in Portugal, 1996

	UK nationals	Spaniards	Germans	French
UK nationals				
Spaniards	49.7			
Germans	23.9	32.1		
French	42.7	19.6	23.1	

Note: See explanation in note 7.

Further exploration of the geographical distribution of migrant groups can be made by calculating dissimilarity indices for paired migrant groups. The index measures the degree of (dis)similarity between the spatial patterns of two immigrant communities. The higher the index, the greater the degree of dissimilarity or non-correspondence between the spatial distribution of the two groups in question; the lower the index, the closer the distributions are to each other.[7] A sequence of tables (8.3–8.5) presents the key results, based on the districts (*distritos*) of mainland Portugal (mapped in figures 8.3 and 8.4) and the Autonomous Regions of the Madeira and Azores Islands. Table 8.3 sets out the dissimilarity indices for the major immigrant groupings: Africans, Asians, North Americans, Brazilians and non-Portuguese EU citizens. The Africans are quite similar to the Asians (reflecting the concentration of both groups in Lisbon) but highly dissimilar to the other groups. North Americans have a geographical pattern which is closer to Brazilians than to Europeans because the former two groups are associated with counterwaves from former Portuguese emigrants to Brazil, the USA and Canada. Table 8.4 shows, as one would expect, that the various African nationalities have low dissimilarity indices amongst themselves. Finally, among the European nationalities (table 8.5), it is interesting to note that the British and the Germans have a low dissimilarity, as do the French and Spanish; some common social and cultural elements are indicated as possible explanatory factors here.

Immigration policies and inter-ethnic relations

Just as the scholarly literature on immigration only started to appear at the beginning of the 1990s, so the perceived political relevance of the immigration issue also dates from this time. Various factors contributed to the sudden interest of the government and politicians at this time: the phenomenon of immigration had become very visible, especially in the Lisbon Metropolitan Area, due to the rapid growth in the size of ethnic minority populations; and this was being paralleled by an increase in inter-ethnic tension, directed above all towards the Africans who were the victims of social ostracism, racism and xenophobia, much of which was widely disseminated by the media. On the other side one could witness the ever-increasing intervention of immigrant associations and non-governmental organisations in defence of immigrant communities and ethnic minorities. Finally there were at this time the international obligations assumed by the Portuguese government, namely the clause in the Schengen agreement relating to migration control, and the special statute granted to the citizens of Portuguese-speaking African countries and Brazil (Rocha Trindade 1995).[8]

The first operation to legalise illegal immigrants (October 1992 to March 1993), during which residence permits were granted to 39,166 people, marked the beginning of a change of attitude on the part of Portuguese government with regard to immigration. In an effort to respond to the new responsibilities that Portugal had to face as a new country of immigration within the EU, a series of decrees was issued which changed the regulations regarding the entry and stay of foreigners in Portugal. These were Decree-Law 59/93, regulating the entry, residence and expulsion of foreigners, Law 70/93 (Right of Asylum and Statute of Refugee) and Law 25/94 (Alteration of Nationality Law). Some of these laws, especially the one on asylum, were strongly criticised, caused an enormous controversy among the political parties and in public opinion, and reached the point where they created problems between the majority Social Democrat Party and the President of the Republic Mário Soares. The changes embodied in these decrees clearly demonstrated an increase in the control of the external borders of Europe under Portuguese jurisdiction and a reduction of entry and settlement possibilities in Portugal (Malheiros 1995).

The victory of the Socialist Party in the parliamentary elections of 1995 did not result in any profound change in the immigration policies which had been delineated by the previous government. The new government recognised the need to strengthen the mechanisms to combat illegal immigrations and its obligation to control the entry into Portugal of non-EU citizens. At the same time it has recognised both the problem of increased crime and the insecurity felt by Portuguese people on the one hand and, on the other hand, it has tried to emphasise the social and economic ostracism to which many PALOP immigrants are subject, and has made great efforts to recognise the rights of the immigrant populations and to create conditions that facilitate their integration into Portuguese society. The following measures have been taken:

- The creation, in February 1996, of a High Commissioner for Immigration and Ethnic Minorities, directly responsible to the Cabinet (Decree-Law 3A/96). The High Commissioner is responsible for promoting the improvement of the living conditions of immigrants and ethnic minorities and their inclusion in Portuguese society, while respecting their identity and culture. The Office of the High Commissioner also works in collaboration with immigrants' associations, members of ethnic minorities and NGOs, in order to promote actions and design policies to eradicate the main causes of social exclusion, and to offer advice dealing with legal and bureaucratic problems.
- A second campaign to legalise the status of so-called illegal migrants took place between March and December 1996, this time with input from immigrant associations; 36,000 candidates applied for this regularisation. This legalisation specifically targeted migrants from Portuguese-speaking countries, and the rules included some positive discrimination in favour of those from Brazil and the PALOPs. Those who were successful received a residence permit valid for one year, renewable up to a three-year period, after which it can be converted into a permanent residence permit.
- Revision of the law regulating the entry, residence and expulsion of foreigners (Decree-Law 59/93), replaced by Decree-Law 244/98. The new legal framework expressly recognises the right to family reunification in Portugal, provided that the petitioner is residing legally in Portugal and can furnish proof of housing and sufficient economic means to support the family. Moreover Law 244/98 also enables children born in Portugal to parents who are legal residents, to be covered by the same residence card granted to the parents, provided the parents present a request within six months of the child's birth.
- The access of immigrant families living in shanty towns in the Lisbon and Oporto Metropolitan Areas to special housing schemes (known as PER and PER Families) on an equal footing with Portuguese families. PER (*Programa Especial de Relojamento*) is a scheme based on a contract between the government and the metropolitan municipalities which aims to eradicate shanty housing by 2001. PER Families, set up by Decree-Law 79/96, allows people to buy a house for permanent habitation that is suit-

able for the size of the family or to restore a house they own elsewhere on condition that it will be used as the family's permanent residence.

- Immigrant families legally residing in Portugal can also benefit under the same conditions as Portuguese families from the Guaranteed Minimum Income (Law 19A/96) to combat poverty and exclusion. Apart from the financial support, the policy includes housing and health initiatives, educational programmes, employment services, and opportunities for professional training. Furthermore, Law 20/98, regulating the working conditions of foreigners in Portugal, states that legally-resident foreigners are entitled to the same working conditions as nationals.

At a local level, the concern to integrate immigrants and ethnic minorities can mainly be noted in the municipalities of the Lisbon Metropolitan Area, where such communities are both diverse and numerous. Special mention must be made of the efforts to improve housing conditions and provide schooling and professional training. The work of non-governmental organisations such as Obra Católica das Migrações (Catholic Aid for Migrants), SOS Racismo and the Portuguese Council for Refugees is important in helping to combat racism, promote intercultural exchange and defend the rights of immigrants and ethnic minorities.

Yet despite the efforts recently made by central and local authorities, and by immigrant associations, unions and NGOs, the political participation of immigrant communities still leaves much to be desired. Confirming this fact is the low number of non-EU citizens who appear on the list of electors following the electoral enrolment that took place in May 1996 (only 9,686 non-Europeans, of whom 9,432 were Cape Verdeans). It must be said that the politicisation of ethnic minorities is a very recent phenomenon in Portugal; perhaps by the same token, the issue of migration has yet to become overtly politicised. There are no political parties or organised ultra-nationalist movements with xenophobic policies in Portugal. However, the existence of nazi-like skinheads, and the sporadic conflicts which erupt between groups of Portuguese and ethnic minority youths, especially in the Lisbon area, are clear signs of an increase in racial tension and are a reason for increasing concern for both politicians and the public at large.

Notes

1 These data are from the Employment Surveys published by INE, the Instituto Nacional de Estatística. Long-term unemployed are those who have been without work for at least one year.

2 As Baganha (1993, p. 826) explains, temporary emigration to these European countries often works in the same way as permanent emigration, since many of the temporary emigrants remain illegally in the host countries or continually renew their work contracts.

3 The number of immigrants of Indian origin is far greater than those with Indian nationality with residence permits in Portugal. Many who came as a result of decolonisation had Portuguese nationality. According to Malheiros (1995) there are about 35,000 Indians living in Portugal.

4 It should be mentioned here that the data on employment structure are somewhat incomplete, and hence potentially biased, as they do not include illegal workers.

5   The strong orientation of the Chinese and Indians to the commercial sector has to do with the experience they already possessed in this field before migration to Portugal, and also reflects the good business opportunities provided by the growth of the economy and the rise in the purchasing power of Portuguese families in the last 20 years.

6   Location quotients are statistical measures of the degree of over- or under-representation of one immigrant group compared to the total immigrant population. If a particular group of immigrants is geographically distributed in equal proportion to the total immigrant population, all regions have an LQ of 1.0. If there are twice as many, proportionally, in a region compared to the distribution of the total immigrant population, LQ = 2.0; if there are only half as many, LQ = 0.5.

7   Put another way, the index of dissimilarity (ID) measures the proportion of one group which would have to shift its area of residence in order to have the same distribution as the group with which it is being compared. For an accessible description of the method see Peach and Rossiter (1996, pp. 111–12).

8   In 1994, the Law on Nationality was revised, introducing some positive discriminatory rules for foreigners from Portuguese-speaking countries. First, regarding nationality by birth, those who are born on Portuguese national territory may be considered Portuguese if their parents are foreigners who have been legally residing in Portugal for at least six years if they came from a Portuguese-speaking country, or ten years if from another country. Second, regarding acquisition of Portuguese nationality through the process of naturalisation, petitioners have to have maintained legal residence in Portugal for at least six years if they are from a Portuguese-speaking country; ten years in other cases.

References

Almeida, J. F., Costa, A. F., Machado, F. L., Nicolau, I. and Reis, E. (1992) *Exclusão Social – factores e tipos de pobreza em Portugal.* Oeiras: Celta Editora.

Baganha, M. I. (1993) Principais características e tendências da emigração portuguesa, in *Estruturas Sociais e Desenvolvimento – Actas do 2° Congresso Português de Sociologia.* Lisbon: APS/Editorial Fragmentos, Vol.2, pp. 819–35.

Baganha, M. I. (1998) Immigrant involvement in the Portuguese economy: the Portuguese case, *Journal of Ethnic and Migration Studies,* 24(2), pp. 367–85.

Castles, S. and Miller, M. J. (1998) *The Age of Migration.* London: Macmillan.

Eaton, M. (1996) Résidents étrangers et immigrés en situation irrégulière au Portugal, *Revue Européenne des Migrations Internationales,* 12(1), pp. 203–12.

Eaton, M. (1998) Multicultural insertions in a small economy: Portugal's immigrant communities, *South European Society and Politics,* 3(3), pp. 149–68.

Eaton, M. (1999) Immigration in the 1990s: a study of the Portuguese labour market, *European Urban and Regional Studies,* 6(4), pp. 364–70.

Esteves, M. C., ed. (1991) *Portugal, País de Imigração.* Lisbon: Instituto de Estudos para o Desenvolvimento, Caderno 22.

Faist, T. (1997) The crucial meso-level, in Hammar, T., Brochmann, G., Tamas, K. and Faist, T. (eds) *International Migration, Immobility and Development: Multidisciplinary Perspectives.* Oxford: Berg, pp. 187–217.

Ferrão, J. and Domingues, A. (1994) Portugal: as condições terrítoriais de um processo de terciarização vulnerável, *Finisterra,* 29(57), pp. 5–42.

Fonseca, M. L. (1994) Portuguese labour market – challenge and change, in *Regional Conference of the International Geographical Union, Prague, 22–26 August 1994: Papers Presented by the Human Geography Unit of CEG.* Lisbon: Centro de Estudos Geográficos (EPRU, no. 41), pp. 9–22.

França, L., ed. (1992) *A Comunidade Caboverdiana em Portugal.* Lisbon: Instituto de Estudos para o Desenvolvimento.

Guibentif, P. (1996) Le Portugal face à l'immigration, *Revue Européenne des Migrations Internationales,* 12(1), pp. 121–40.

Jacinto, R. (1993) As regiões portuguesas, a política regional e a reestruturação do território, *Cadernos de Geografia*, 12, pp. 25–39.

King, R. (1984) Population mobility: emigration, return migration and internal migration, in Williams, A. M. (ed.) *Southern Europe Transformed*. London: Harper and Row, pp. 143–78.

King, R. and Black, R., eds (1997) *Southern Europe and the New Immigrations*. Brighton: Sussex Academic Press.

King, R. and Rybaczuk, K. (1993) Southern Europe and the international division of labour: from emigration to immigration, in King, R. (ed.) *The New Geography of European Migrations*. London: Belhaven, pp. 175–206.

Machado, F. L. (1993) Etnicidade em Portugal – o grau zero de polítização, in *Emigração/Imigração em Portugal: Actas do Colóquio Internacional sobre Emigração e Imigração em Portugal (Séculos XIX e XX)*. Lisbon: Editorial Fragmentos, pp. 407–14.

Machado, F. L. (1997) Contornos e especificidades da imigração em Portugal, *Sociologia – Problemas e Práticas*, 24, pp. 9–44.

Malheiros, J. (1995) Refugees in Portugal and Spain: a preliminary approach on reception policies and integration prospects, in Delle Donne, M. (ed.) *Avenues to Integration: Refugees in Contemporary Europe*. Naples: Ipermedium, pp. 145–76.

Malheiros, J. (1996a) Communautés indiennes à Lisbonne, *Revue Européenne des Migrations Internationales*, 12(1), pp. 141–58.

Malheiros, J. (1996b) *Imigrantes na Região de Lisboa: os anos da mundança*. Lisbon: Edições Colibri.

Malheiros, J. (1997) Indians in Lisbon: ethnic entrepreneurship and the migration process, in King, R. and Black, R. (eds) *Southern Europe and the New Immigrations*. Brighton: Sussex Academic Press, pp. 93–112.

Malheiros, J. (2000) Urban restructuring, immigration and the generation of marginalised spaces in the Lisbon region, in King, R., Lazaridis, G. and Tsardanidis, C. (eds) *Eldorado or Fortress? Migration in Southern Europe*. London: Macmillan, pp. 207–32.

Misiti, M., Muscarà, C., Pumares, P., Rodríguez, V. and White, P. (1995) Future migration into Southern Europe, in Hall, R. and White, P. (eds) *Europe's Population: Towards the Next Century*. London: UCL Press, pp. 161–87.

Peach, C. and Rossiter, D. (1996) Level and nature of spatial concentration and segregation of minority ethnic populations in Great Britain, 1991, in Ratcliffe, P. (ed.) *Ethnicity in the 1991 Census, Vol.3: Social Geography and Ethnicity in Britain*. London: HMSO, pp. 111–34.

Reis, R. (1996) Os fundos estruturais em Portugal: impactes no desenvolvimento do território, in Oliveira, C. (ed.) *História dos Municípios e do Poder Local*. Lisbon: Círculo de Leitores, pp. 402–33.

Robin, N. (1994) Une nouvelle géographie entre concurrences et redéploiement spatial. Les migrations ouest-africaines au sein de la CEE, *Revue Européenne des Migrations Internationales*, 10(3), pp. 17–31.

Rocha-Trindade, M. B. (1995) *Sociologia das Migrações*. Lisbon: Universidade Aberta.

Saint-Maurice, A. (1997) *Identidades Reconstruídas – Capoverdianos em Portugal*. Oeiras: Celta Editora.

Williams, A. M., King, R. and Warnes, A. M. (1997) A place in the sun: international retirement migration from Northern to Southern Europe, *European Urban and Regional Studies*, 4(2), pp. 115–34.

Williams, A. M. and Patterson, G. (1998) An empire lost but a province gained: a cohort analysis of British international retirement migration in the Algarve, *International Journal of Population Geography*, 4(2), pp. 135–55.

9

Tourism and development in the Mediterranean Basin: evolution and differentiation on the 'fabled shore'

Allan M. Williams

Internationalisation has been the dominant feature of late-twentieth-century tourism. Globally, the number of international tourist arrivals rose from 25 million in 1950 to 69 million in 1960, and then to 166 million in 1970, 288 million in 1980, 358 million in 1990 and 594 million in 1996. These are the bare statistical facts behind the massive tourist movements which have played a major role in the shaping of international economic relationships and in the economic development of particular countries and localities. International tourism growth has outstripped international trade growth and, in the process, has locked particular economies into global flows of people, currencies and company transactions (Williams 1995). Localities have not been passive recipients in these processes. Instead, the particularities of place and of tourism development have helped to shape global tourism. This is especially true of the Mediterranean Basin which has become a byword for summer coastal mass tourism, even if this is only one of its tourist products and the Mediterranean is only one of the world's 'pleasure playgrounds' (Shaw and Williams 1994).

The character of the Mediterranean region as a tourism destination has changed over time, as has the importance of individual destinations within the region. During the nineteenth century, the socially exclusive Grand Tour began to be appropriated by Northern Europe's middle classes. The role of Thomas Cook in this popularisation provided a harbinger of the decisive influence that the international tour companies would play in shaping the production and consumption of Mediterranean tourism after the 1950s (Brendon 1992). In the late nineteenth and early twentieth century, there was a social reconstruction of Mediterranean tourism when it attracted seasonal winter visits by Europe's wealthy elites. Reynolds-Ball (1914) saw the Mediterranean shores as the world's greatest winter playground, in other words as a 'pleasure periphery' dependent on the economic dynamism provided by the growth of a wealthy upper middle class in Northern Europe.

This tourism model survived the inter-war years little changed. But from the 1950s Mediterranean tourism was transformed by international mass holiday-making, which was fundamentally different in terms of the social construction and consumption of the 'tourist gaze' (Urry 1990). Marchena Gómez and Vera Rebollo (1995, p. 111) have written that Mediterranean coastal tourism after the 1950s differed from that of the late nineteenth and early twentieth century in terms of its mass social character, its wider areal transformation of the coastal regions, and an increasing preference for hotter southerly latitudes within the region. It is socially constructed primarily as a sand, sea and sunshine product, with secondary features related to low prices, particular types of entertainment, and the scope for modified individual behaviour (Shaw and Williams 1994, pp. 180–92). Norman Lewis (1984, p. 153), writing about the arrival of mass tourism in Farol in Catalonia in the 1950s, emphasises the standardisation inherent in this model of tourism which, in this case, was brought about by a local entrepreneur, Muga:

> The effect of Muga's tidying up was a deadening one. The ancient handsome litter of the sea-front had possessed its own significance, its vivacity and its charm . . . Muga, high priest of the standardisation and the monotony that lay in wait, was consumed with a passion for make-believe. Walls with castellations and arrow slits for the archers had been ruled out, but before the villagers knew what was happening to them he had made a start on a Moorish-style cafe with a dome and horse-shoe arches . . . . Muga wondered what all the fuss was about. The tourists were coming and Moorish cafes were what was expected of Spain.

This powerful quote illustrates both the compelling anticipation of economic gains from mass tourism, and how the social construction of the Mediterranean tourism experience could rapidly change the built and natural environments of localities, even if the 'Moorish' architecture symbolised a new *reconquista*! There is the suggestion of an overarching popular image of the Mediterranean region which bridges its northern and southern shores. It is important, however, not to overemphasise standardisation for, while it is true that (unlike some other forms of tourism) mass tourism played down place differences, local and national differences remained important in the economic organisation of tourism and in the way that tourism impacted upon particular districts. There were also differences in the way and the timing of the insertion of particular localities, and even countries, into the Mediterranean mass tourism market. They were not simply passive recipients of the 'pleasure periphery' status, but were able to actively seek, modify or reject such a role.

There were a number of reasons why some localities and countries actively sought to promote Mediterranean mass tourism. Tourism has a powerful instrumental role which encompasses cultural change and modernisation, political legitimisation, and economic development (Shaw and Williams 1994, pp. 9–16). This chapter focuses mainly on the economic role of tourism, but that cannot fully be understood without reference to its cultural and political dimensions. The notion of tourism as an economic instrument also has to be disaggregated into several components including: its contribution to the current account and to overall strategies of economic development, particularly industrialisation; its role in direct employment generation, an increasingly important goal in the face of economic restructuring and persistently high unemployment rates; and finally its deployment as a component of local and regional development strategies (Williams and Shaw 1988).

In the 1960s and 1970s, the principal economic interest of tourism for the state was as a source of international receipts which could help counterbalance the current account deficit and help to pay for capital goods imports. However, there was also appreciation that the employment-generating capacity of tourism could facilitate inter-sectoral transfers of labour, especially from low-productivity, labour-intensive agriculture. In the face of rising and persistent unemployment after the mid-1970s, tourism had particular appeal as one of the few economic sectors with the capacity to generate employment. Measuring tourism employment is problematic, but Jenner and Smith (1993, p. 33), for example, estimate that the sector provides approximately 3 million direct jobs and 2 million indirect jobs in the Mediterranean. The employment impact of tourism is anyway only hinted at by statistical data, for it is conditional upon the organisation of labour in the sector and how this relates to the division of labour in other industrial sectors and households, including the role of labour migration (King 1995; Williams and Montanari 1995).

Reference to the overall economic impact of international mass tourism on the Mediterranean is necessarily an over-simplification. In reality, there have been a series of impacts which are highly contingent on the form of the mass tourism product, local economic structures (the availability of local capital, organisation of labour markets etc.), the role of domestic tourism, and company structures (particularly the degree of external control exercised by tour companies). It is not possible to examine all of these aspects within the confines of this chapter, and some of these themes are anyway considered elsewhere (Montanari 1995a, 1995b; Williams 1995, 1997). Instead, this contribution has two main concerns. The first is to examine the evolution of tourism in recent decades, and in particular since the 1960s. The main emphasis is on international mass tourism, because of the leading role of this sector.[1] My particular emphasis will be on the *differentiated* evolution of Mediterranean tourism. In the second part, I investigate the economic role of tourism, again emphasising national and regional differentiation.

Before proceeding to the analysis of tourism flows and economic impacts, it is first necessary to visit briefly some definitional issues. The first is how the Mediterranean region is to be defined, an issue already discussed at some length in Chapter 1. The simplest approach is to take the countries bordering the Mediterranean Sea, but this excludes Portugal whilst including large parts of countries such as France and Algeria which have non-Mediterranean characteristics. Alternatively, national-level data can be disaggregated so as to include only the specifically 'Mediterranean' regions (which still begs the question of definition): this is the approach of Jenner and Smith (1993). These definitions produce very different estimates of the scale of Mediterranean tourism. In 1991 the Mediterranean Sea countries attracted 187 million foreign tourists, while the Mediterranean regions attracted 113 million (Jenner and Smith 1993, p. 7). This chapter considers all the countries bordering the Mediterranean, plus Portugal, as broadly encapsulating the 'socially-constructed Mediterranean mass tourism product'.

The second point is to emphasise that the analysis in this chapter will be at the level of the individual states. This is, of course, problematic in that the impact of tourism is highly polarised inter- and intra-regionally (Williams and Shaw 1991), and there is also the problem, noted already, that many of the countries bordering the Mediterranean contain regions which are not Mediterranean in character. This does not usually constitute a major problem as the vast majority of international tourists tend to be concentrated in and around the Mediterranean littoral. However, it is problematic in a few cases – such as France – where the majority of tourists are to be found in non-Mediterranean regions. Nevertheless, country-level data are employed in this analysis on the grounds that the regional data in most countries are inadequate for anything other than the most simplistic of disaggregations, and because the emphasis here is on differentiation rather than on producing global estimates for the Mediterranean region.

Most of the data presented are derived from the World Tourism Organisation's *Yearbooks of Tourism Statistics* which, in turn, do little more than collate the statistics produced by national governments. There have been recent attempts to try and standardise estimates of the economic impacts of tourism, notably by the OECD (1995), but because these include only a few of the (northern) Mediterranean countries they cannot be used for a comparative analysis of the region as a whole. Reliance on World Tourism Organisation data does require certain qualifications about the analysis. First, the coverage is not complete, and data are lacking for Albania and Lebanon, and are very limited for Libya; all three countries are relatively insignificant as international tourism destinations. Difficulties also arise because of differences in the definitions of tourists and in the means of collecting statistics (Williams and Shaw 1991). In particular, whilst 'tourist arrivals' are used wherever possible, some countries only produce statistics on 'visitors' which include day and over-night visitors. There are also important differences in the methodologies used to calculate tourism expenditures. In most countries, there are further problems related to the relatively large informal sector in tourism, which means that a substantial part of tourism accommodation, overnights and expenditures are unrecorded.

Key factors behind the evolution of Mediterranean mass tourism

Mediterranean mass tourism expansion was underpinned by a number of socio-economic shifts in Northern Europe in the postwar period. On the demand side it was linked to the emergence of mass consumption in the 1950s and 1960s, which contributed to and was a product of aggregate economic growth in Northern Europe. There was a once-off redistribution of income, expanded public expenditure, and the success of organised labour in winning entitlements to paid annual leave (Shaw and Williams 1994, pp. 75–6). Changes in the production and delivery of tourism services also helped to open up the Mediterranean as an arena of mass tourist consumption. These included technology-led reductions in transport costs, and the evolution of a highly competitive package tour industry. The first air inclusive package holiday was from the UK to Corsica in 1950, but by 1970 the volume of passengers carried by charter flights had already surpassed that on scheduled flights (Pearce 1987). The international tour companies benefited from scale economies in purchasing, marketing and (later) the use of computer technology which – together with oligopsonistic relationships with sub-contracting enterprises in the resorts – allowed them to reduce costs and prices (Williams 1995, 1996). The opening up of the Mediterranean region to mass tourism was also reinforced by state investment, particularly in respect of key infrastructure such as airports.

International mass tourism originated in a few European Mediterranean countries, regions and resorts but subsequently diffused to most other countries in the Mediterranean Basin. The French and Italian Rivieras were the leaders in the era of elite tourism, but Spain came to assume the mantle of market leader in the era of mass tourism. Whereas it only attracted 1.3 million foreign visitors in 1951, this had surged to 14.1 million in 1964 (Manrique 1989, p. 341). By the late 1960s, international mass tourism had also been implanted in parts of Italy, Greece, Portugal and the former Yugoslavia. In the 1970s there was further 'diffusion' of mass tourism away from the northern shores of the Mediterranean; Morocco and Tunisia recorded significant increases in international arrivals in this period, while Turkey, Israel and Egypt also bid for tourism expansion in the 1980s, although with mixed results. Finally, a small group of countries has not developed mass tourism, either because of ideological choice or constraints such as wars and civil conflicts. Amongst these are Albania, Libya, Algeria, Syria, the Lebanon and – in the 1990s – the former Yugoslavia.

Over time, then, there have been significant changes in the geography of Mediterranean mass tourism, and these have been shaped by three principal influences: price levels, product cycles and the role of tourism as a positional good, and military/political crises. Each of these is considered in turn.

Mediterranean mass tourism, as we have seen, was socially constructed from the 1950s as a relatively standardised product where competition was predominantly based on prices (Williams 1996), and the resulting depression of real prices stimulated a strong expansion of demand. One measure of this is provided by the analyses of the prices of package tours from Northern Europe to the principal Mediterranean destinations undertaken by Spain's Dirección General de Política Turística (1992). There is considerable variation in prices for particular tourism products, with the former Yugoslavia at 63 and Corsica at 158 representing two extremes around a mean of 100 for prices in all destinations. There is some semblance of an inverse relationship between price and growth rates, with France, the south of Italy and some Spanish regions having price levels above average and relatively low rates of demand growth in recent years. However Cyprus, Portugal and the Mediterranean coast of Morocco – all of which have had high rates of expansion of demand – also have above average prices. At the other extreme, Malta, Greece and Egypt (as well as some Spanish regions) have prices below the mean, but only have relatively modest growth rates. Therefore while prices are important, they do not constitute a sufficient explanation of the changing geography of mass tourism in the Mediterranean.

The diffusion of mass tourism can partly be explained in terms of the product cycle, which is reinforced by the fact that tourism can be regarded as a positional good (Urry 1995). As products age and fashions change so particular tourism destinations may become more or less in demand – provided of course that they remain broadly price-competitive. The importance of the product cycle is evident in the slow growth in more 'mature' destinations such as France, Spain, Italy, Malta and, more recently, Greece. In contrast, other countries – such as Turkey, Tunisia and Morocco – have relatively 'young' tourism industries and are still in the growth phase of the product cycle. This is an appealing explanation, but it is also an over-simplification, for resorts (let alone regions or countries) are not a single product but an amalgam of tourism products at different points in their cycles. Individual resorts also have the capacity to reinvent themselves through promotional activities and product investment.

The third major influence on the spread of mass tourism is constituted by the discontinuities associated with domestic or international crises. These can be generalised, such as the effects of the 1991 Gulf Crisis which depressed tourism growth throughout the eastern and southern Mediterranean countries. Alternatively they may be country-specific, such as the effects of military coups and the Cyprus conflict on the Greek and Cypriot tourist industries (Buckley and Papadopoulos 1986, p. 87). Contested territorial sovereignty, when it spills over into violence, can also have a negative impact on tourism; this is evident in the case of ETA campaigns in Spain and PPK activities in south-eastern Turkey, especially as both have targeted tourists at particular times.

The above discussion only indicates some of the broader influences on the changing geography of Mediterranean mass tourism, and the changing positions of particular countries or resorts require detailed attention to the contingencies of place and time. However, these brief remarks do provide a context for the following statistical overview of the evolution of Mediterranean tourism.

Mediterranean tourism: a statistical overview

The Mediterranean is the world's single most important tourism region, accounting for 31 per cent of global tourist arrivals in 1994; in terms of individual countries, it also held three of the top four places and six of the top 25 positions in that year. Over time the relative importance of the Mediterranean region has been in decline and since 1973 world growth rates have exceeded those in the region in most years (figure 9.1). This reflects global shifts in the distribution of production and consumption capacity towards the Pacific Rim (at least prior to the regional economic crisis of 1997–8), as well as the globalisation of the tourist consumption sphere of Europeans.

The three dominant destinations are France (61.3 million tourist arrivals), Spain (43.2 million) and Italy (27.5 million) (table 9.1). These are followed in rank order by Greece, Portugal and Turkey. In absolute terms, then, tourism remains very

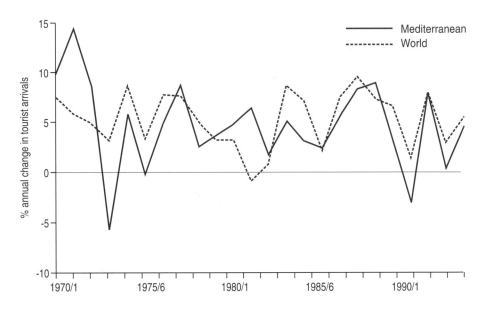

Figure 9.1 Evolution of global and Mediterranean international tourism, 1970–95

much a northern Mediterranean product, even if large proportions of the arrivals in France, in particular, are outside of the Mediterranean regions. While there has been strong recovery since the late 1990s in Croatia and Slovenia, tourism remains weak in the other components of the former Yugoslavia. On the southern shores of the Mediterranean, tourism is strongest in Tunisia and Morocco, followed by Egypt. However, arrivals in Egypt have been devastated by outbreaks of violence; and Algeria's nascent tourism industry has been still-born for similar reasons. On the eastern edge of the Mediterranean, Israel is an important destination. A modest recent recovery is evident in Lebanon.

It should be noted, however, that the relative 'weight' of tourism, compared to the resident populations, produces a somewhat different picture (see table 9.1). The top-ranked are Cyprus and Malta, with approximately three tourists per resident. This reflects their strong tourism industries and their relatively small populations. Thereafter, the Southern European dominance is reasserted; France, Spain and Greece are the only other countries to receive more tourists than they have residents (an index value greater than 1), while Portugal only narrowly falls below this threshold. No other country exceeds the level of more than one tourist per two residents, although Italy and Tunisia narrowly miss this cut-off point. At the other extreme, Algeria, Syria and Egypt – although having vastly different absolute numbers of arrivals – have index values of less than 0.05. Tourism impacts, therefore, remain highly polarised, even if differences in populations are taken into account.

The overall pattern, then, is of fairly strong and consistent growth across most of the Mediterranean Basin. This is reflected in the change rates for the long period

Table 9.1 Tourist arrivals, 1970–94 ('000)

	1970	1980	1987	1994	Tourist arrivals per head of population 1994
Albania	–	–	–	28	0.01
Algeria	236	920	778	805	0.03
Croatia	–	–	–	2,293	0.55
Cyprus	127	353	949	2,069	2.85
Egypt	–	1,253	1,311	2,356	0.04
France		30,100	–	61,312	1.07
Greece	1,407	4,796	7,564	10,713	1.03
Israel	419	1,116	1,379	1,839	0.35
Italy	14,188	22,087	25,749	27,480	0.48
Lebanon				228	0.07
Macedonia (FYROM)	–	–	–	185	–
Malta	171	729	746	1,176	3.17
Morocco	747	1,425	2,248	3,465	0.13
Portugal	–	2,730	6,102	9,132	0.93
Slovenia	–	–	–	748	0.38
Spain	15,320	23,403	32,900	43,232	1.10
Syria	–	347	618	237	0.02
Tunisia	411	1,602	1,875	3,856	0.45
Turkey	446	921	2,468	6,034	0.10
Yugoslavia	–	5,955	8,907	91	0.01

Sources: Grenon and Batisse (1989); World Tourism Organisation (1987, 1996).
Notes: 'Yugoslavia' means former Yugoslavia in 1980 and 1987, the new Federal Republic of Yugoslavia in 1994. Some years are different: Algeria, Syria and Yugoslavia are 1982 not 1980, and Egypt and Syria are 1986 not 1987.

1970–94 (table 9.1), with virtually all countries for which data are available having growth in tourist arrivals in excess of 100 per cent. The exceptions are Syria, Libya and Yugoslavia, related to military/political reasons. In Italy the tourism industry would appear to be in the mature stage: numbers just failed to double over the 25-year period, with marked slowing down since 1980. Also noteworthy is the case of Spain. Although it continued to experience formidable absolute growth in tourist numbers over this period, its growth rate had already peaked in the 1960s. At the other extreme are Cyprus and Turkey with overall growth rates in excess of 1000 per cent.

These aggregate growth rates are, however, deceptive in some ways. For example, Cyprus has been experiencing an overall deceleration of growth – in context of

Table 9.2 Patterns of growth in tourism of Mediterranean countries, 1970–72, 1980–82, 1990–92

| Overall growth | Evolution of growth across three time periods | | |
	Rising	Declining	Inconsistent
Strong	Turkey	Greece	Malta
		Cyprus	Tunisia
			Morocco
			Egypt
Weak		France	Algeria
		Spain	Libya
		Italy	Syria
		Yugoslavia	Israel

strong overall demand – while Turkey has been on a rising curve. This can be illustrated by considering data for three time periods: 1970–2, the end of the 'golden era' of European economic expansion; 1980–2, a relatively troubled period for tourism expansion; and the early 1990s. The Mediterranean countries can be grouped according to whether they experienced consistently rising or falling growth rates across the three time periods and whether they experienced stronger overall growth (being ranked 1–7 for 1970–92) or weaker growth rates (ranked 8–15). This simple classification produces the broad groups set out in table 9.2. Only one country has strong and rising growth: Turkey, starting from a low base, has gradually but consistently been experiencing increasing demand. At the other extreme there are a number of countries which have experienced diminishing growth rates across the three periods; these are all located in Southern Europe, and are countries with more mature tourism products – Cyprus, Greece, France, Spain, Italy, and the special case of the former Yugoslavia. They are differentiated by the fact that the first two had stronger overall growth rates, that is growth rates were slowing down but in the context of strong overall expansion in the period as a whole. In contrast, Spain and Italy had weaker growth, and France occupied an intermediate position. These differences are broadly in line with the stages of their tourism product cycles, although France is particularly difficult to interpret given its varied and complex tourism industry, and the fact that most tourism activity is located outside its Mediterranean regions. Finally there is a group of southern and eastern shore countries which have experienced inconsistent growth trends. In most of these cases the key would seem to be real or perceived political/military instabilities. Those countries which experienced strong overall growth have all had governments which have pursued active tourism development policies.

Tourism and economic development

The contribution of international tourism to the Mediterranean countries as a whole in 1990 has been quantified as an estimated $67 billion, whilst in the Mediterranean regions it is an estimated $40 billion (Jenner and Smith 1993, pp. 7, 28). This represents an estimated 2 per cent of GNP, which rises to 5 per cent if domestic tourism and indirect multiplier effects are included. There are, however, significant national differences, largely reflecting levels of demand but also influenced by the tourism product (length of stay, price levels) and industrial structures (especially the degree of external ownership). In 1994 the largest international tourism receipts were in France, Italy and Spain, all within the range of $21 billion to $25 billion (table 9.3). They were followed at some distance by Turkey, Portugal and Greece. At the other extreme are several countries where tourism receipts do not exceed $1 billion.

As would be expected, the picture is greatly modified if the ratio of tourism receipts to population size is considered. Malta and Cyprus as small, tourism-dependent island economies are special cases and have per capita receipts of $1,770

Table 9.3 International tourism receipts, 1982 and 1994 ($ million)

	1982	1994	Receipts per capita 1994 ($)
Algeria	197	49	2
Croatia		1,427	299
Cyprus	292	1,700	2,341
Egypt	886	1,384	25
France	6,991	24,700	430
Greece	1,527	3,905	377
Israel	894	2,300	441
Italy	8,339	23,927	419
Lebanon		672	174
Malta		639	1,770
Morocco	354	1,265	49
Portugal	896	3,828	389
Slovenia		932	468
Spain	7,126	21,853	554
Syria	224	800	58
Tunisia	545	1,302	150
Turkey	370	4,321	73

Source: World Tourism Organisation (1987, 1996).

and $2,341 respectively. They are followed by a group of prosperous Southern European countries which have receipts in the range of $350–550: France, Greece, Italy, Portugal and Spain. Interestingly, Israel also belongs to this group, reflecting a relatively high expenditure per visitor, as well as its relatively small population.[3] In contrast, all of the North African countries have receipts of less than $60 per capita, with the exception of Tunisia ($150). Turkey's impressive growth in tourism arrivals and absolute receipts is not yet reflected in the per capita rankings, due to the country's relatively large population, now approaching 60 million.

While in absolute terms tourism is of greatest importance to the economies of the islands and of Southern Europe, its contribution needs to be assessed against the relative sizes of their economies. Table 9.4 provides a perspective on this in relation to exports and GNP. The detailed raw data show that, while tourism numbers were greatest in Southern Europe (cf. table 9.1), the pattern of economic impact of tourism tended to be far more complex. In general, the relative importance of tourism during the 1980s increased most in Egypt, Morocco, Turkey, Cyprus and Malta. On the basis of these data, the Mediterranean countries can be classified into five groups in terms of the economic role of tourism (table 9.4):

- The island economies of Malta and Cyprus, where tourism accounts for more than one fifth of GNP, and a very large proportion of export earnings. The role of tourism in the larger economy of Cyprus is particularly noteworthy, especially its enormous contribution to the international

Table 9.4 Classification of Mediterranean countries according to the economic weight of tourism, 1992

Tourism as per cent of GNP	Tourism as a percentage of export earnings				
	>88	26–88	20–35	10–13	<13
>20	Cyprus Malta				
4–8		Egypt Morocco	Tunisia		
3–5			Portugal Spain Greece Turkey (Israel)		
1.5–2				France Italy	
<1.5					Algeria Libya Syria (Albania) (Lebanon)

balance of payments. In addition, the economic significance of tourism has increased markedly in both countries, particularly in Cyprus.

- Tourism accounts for between 4 per cent and 8 per cent of GNP in the leading North African destinations of Egypt, Morocco and Tunisia, and for between 26 per cent and 88 per cent of their export earnings. Although all three countries have faced difficult conditions in the 1990s, due to the Gulf War and adverse publicity regarding the rise of Islamic fundamentalism, tourism continues to be a major component of their economies, if not the leading export sector. The economic importance of tourism increased markedly in Morocco and Egypt in the period under review, but its contribution to exports declined in Tunisia. Subsequently, Egypt's international tourism industry has been severely blighted by fundamentalist violence against tourists.

- In Portugal, Spain, Greece and Turkey, receipts are equivalent to 3–5 per cent of GNP, and at least one fifth of exports. Spain's membership of this group is in spite of the fact that the relative importance of tourism has been declining due to the growth and increasing diversity of the national economy. In these countries, the share of tourism in GNP has been rising, but the share of exports has been static, reflecting the increasing internationalisation of other sectors of the economy. Israel potentially could also join this group, even though its tourism market is quite distinctive and highly vulnerable to changing political conditions.

- Although France and Italy are, respectively, the first and fourth ranked countries in the world in terms of tourism arrivals, their enormous absolute tourism receipts account for less than 2 per cent of GNP, reflecting the overall size, strength and diversity of these economies. Nevertheless tourism does account for at least a tenth of their exports.

- The final group contains those countries where tourism is of minimal importance, and includes Algeria, Libya and Syria (as well as Albania and Lebanon for which data are not available). In this group tourism accounts for a minimal percentage of export earnings, except Syria (12.7 per cent) which also has a higher tourism/GNP proportion (2.5 per cent).

Tourism, then, makes a varied contribution to the economies of the Mediterranean countries, reflecting differences in both the industry itself and in national development trajectories.

Further perspectives on the economic role of tourism

The simple picture which has been painted so far of the role of tourism in the Mediterranean economies needs to be modified to take into account other features of both the tourism industry and the individual economies. Here we focus on three of these: market segmentation and the role of the international tour companies, the Mediterranean countries as *sources* of international tourism within the region, and domestic tourism.

Market segmentation and the international tour companies

There is strong segmentation, in terms of the origins of the tourists, in the international markets for Mediterranean tourism. There are two principal markets for Mediterranean tourism: Northern Europe, and the Arab countries (for the southern Mediterranean destinations). The evolution of a strong Northern European market is based on the social construction of Mediterranean tourism, historical links and the leisure consumption potential of prosperous societies. According to Jenner and Smith (1993, p. 46), Europe accounts for more than half of the total Mediterranean market, with Germany, France and the UK alone accounting for 39 per cent of arrivals in the region. The leading international flows are strongly concentrated on Italy, Spain and Greece, underlining the fact that Mediterranean tourism continues to be essentially an intra-European phenomenon (see Williams 1997). Table 9.5 summarises the degree of market segmentation for most of the major receiving countries, although the data should be treated cautiously as they combine information on visitors and tourists. The data confirm that not only is there heavy dependence on Europe, but often the reliance is on only one or two countries. Europe accounts for at least 80 per cent and usually more than 90 per cent of the tourists arriving in most of the Southern European countries and the islands. Cyprus and Malta, reflecting their colonial ties, are particularly dependent on the UK market.

Table 9.5 Market shares: arrivals at frontiers, 1994

	UK	Germany	% dependency on: France	Europe	Middle East and Africa
Algeria	0.2	0.3	4.8	7.5	33.5
Cyprus	46.9	8.4	1.9	91.0	5.8
Egypt	9.0	9.4	3.0	48.2	36.7
France	18.9	17.5	–	86.3	2.3
Greece	22.8	22.7	5.8	93.9	0.6
Israel	10.7	10.6	8.7	61.0	7.0
Italy	3.5	16.0	15.6	91.7	0.6
Lebanon	3.3	3.9	10.4	35.1	39.0
Malta	45.1	17.0	5.8	92.7	3.8
Morocco	3.3	6.2	12.7	37.1	23.6
Portugal	14.1	8.8	6.6	93.6	–
Spain	14.9	15.8	22.3	83.7	4.2
Tunisia	3.7	12.2	6.5	33.7	64.6
Turkey	8.5	14.9	3.5	83.4	4.0

Source: World Tourism Organisation (1996).
Notes: Data for Tunisia are 1993; for Algeria and Morocco, significant shares of arrivals are from non-specified countries.

The second most important market for the Mediterranean region is the Arab countries. The Arab countries are the largest sources of arrivals, if not always of overnights, for Algeria and Tunisia, and major sources for Morocco and Egypt. The data on Morocco must be approached with caution: while the 1994 data show more tourist arrivals from Europe than from Africa and the Middle East, the country of origin of more than one third of arrivals was not known.[4] There are two distinct streams of Arab tourism: arrivals from neighbouring Arab countries, often of a short duration, and longer-stay tourists from Saudi Arabia and the Gulf states. The latter flow has tended to be confined to a wealthy elite, but after the oil price rises of the mid-1970s, there was a growing stream of middle-class clients from these countries.

The dependence of most of the destinations on a small number of countries of origin means that these economies are relatively vulnerable to external changes. This dependency is reinforced by the role of the international tour companies, with individual destination resorts and regions being highly dependent not only on particular countries but also on a small number of tour companies (Pearce 1987). This is illustrated by the regional differences in the gross operating profits of four-star hotels in Portugal: a gross profit of 39.6 per cent of turnover in Lisbon compares to 30.8 per cent in the Algarve which is dependent on mass tourism, and to only 20.1 per cent in Madeira which is almost exclusively reliant on air travel and the international tour companies (Economist Intelligence Unit 1993, p. 38).

The degree of dependence on tour companies is a function of three things: accessibility by individual transport, familiarity, and reliance on the European market (where the tour companies are most significant). As a result, the role of the tour companies, and the concomitant degree of external control, is greatest in Cyprus and Malta, followed by Greece, Turkey, and Portugal; in these instances more than 60 per cent of tourist arrivals are channelled through tour companies (data from Grenon and Batisse 1989). Spain shows some evidence of becoming a more familiar (to Northern Europeans) and mature destination, with increasing numbers of second homes, and this is reflected in a lower level of dependence (48 per cent) on the tour companies. Tour companies have more limited roles in most North African countries, the exception being Tunisia (56 per cent), which has a relatively well-developed package holiday industry. Therefore, although Southern Europe has a far larger mass tourism industry than North Africa, it is also generally more dependent both on tour companies and on relatively small numbers of markets.

A quiet revolution: the Mediterranean as a source of international tourists

Economic growth in the Mediterranean Basin has meant that many of the individual countries have become important countries of origin for international tourism, as well as destinations. France, Italy and Spain are, respectively, the third, fourth and fifth largest sources of visitors to the Mediterranean region and have significant levels of international tourism expenditures.

Table 9.6 provides a perspective on this by presenting data on absolute expenditures, and the ratio to receipts, in 1994. Tourism expenditure represents more than 50 per cent of receipts in the two most developed countries, Italy and France, as well as in the special case of Israel. In general, expenditures are of lower relative

Table 9.6 International tourism expenditure, 1994

	Expenditure $m	Expenditure as % of receipts
Albania	4	80.0
Cyprus	176	10.4
Croatia	552	38.7
Egypt	1,067	77.1
France	13,773	55.8
Greece	1,125	28.8
Israel	2,600	−1.1
Italy	12,181	50.9
Malta	176	27.5
Morocco	302	23.9
Portugal	1,698	44.4
Slovenia	316	33.9
Spain	4,188	19.2
Syria	400	50.0
Tunisia	216	16.6
Turkey	866	20.0

Source: World Tourism Organisation (1996).

importance in the less developed economies (Turkey, Tunisia and Morocco) and the islands. There are two main exceptions to this pattern; the still relatively low level of expenditure in the case of Spain, reflecting the enormous scale of its tourism receipts, and the high level of expenditure compared to receipts in Egypt, as a result of the limited scale of its international tourism industry and the growing number of wealthy Egyptians travelling abroad.

Another perspective on intra-Mediterranean tourism is given by figure 9.2 which shows the main flows (exceeding 100,000 in 1994); these are displayed separately in two diagrams, in recognition of the different data bases available – visitors and tourists. While many of the flows represent short-term, cross-border movements of visitors, shoppers and those in transit, there is also clear evidence of the importance of flows within the region. As would be expected, the two main sources are France and Italy, followed by Spain, but there are also other tourism sub-systems, including those based on Morocco/Algeria, and on Turkey. With increasing incomes in many parts of the region, these movements are likely to increase further in future, dependent of course on political conditions.

Domestic tourism

While the precise form of market segmentation may vary throughout the Mediterranean region, a high degree of reliance on foreign markets and on a small

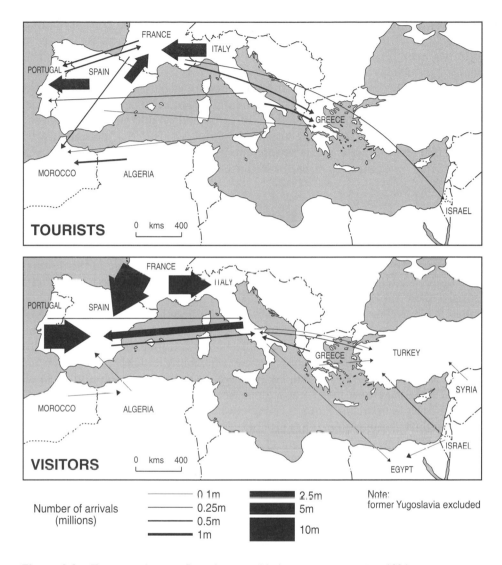

Figure 9.2 Tourist and visitor flows between Mediterranean countries, 1994

number of countries of origin is common to most of the main destinations. As was emphasised earlier, this means that they are particularly vulnerable to external shocks and to cyclical fluctuations. The rapid economic development in parts of the Mediterranean Basin does, however, mean that domestic markets for tourism have also been expanding; not all the increase in disposable incomes spills over into international outbound tourism. Reliance on foreign markets is least in the more prosperous northern Mediterranean countries which have relatively large domestic markets. The domestic market share (as a proportion of all tourist overnights) is particularly significant in France (63 per cent in 1992) and Italy (68 per cent), where there is a strong preference for domestic destinations despite these populations having some of the highest incomes in Europe. The other Southern European countries are more strongly dependent on foreign markets both because of their smaller populations and their lower per capita incomes – although this position is changing rapidly. In Spain and Portugal, the domestic share is 45 and 40 per cent respectively, whilst in Turkey it is 35 per cent and in Greece 25 per cent. In contrast, those countries with small populations (Cyprus and Malta) or weak domestic markets (much of North Africa) display the strongest degree of dependence on foreign markets. Domestic tourism represents only 2 per cent of total overnights in Cyprus, and 7 per cent in Tunisia, for instance.

Conclusions

The economic and geographic impact of tourism in the Mediterranean Basin has been changing over time. At its simplest, this has involved change in the destinations of international tourists. In the process of evolving from elite tourism to mass tourism, there has been a geographical diffusion of tourism from the French and Italian Rivieras, first to Spain, then to most of Southern Europe and the islands of Cyprus and Malta, and eventually to selected countries within the eastern and southern Mediterranean. There have also been, and continue to be, changes in the tourism product with self-provisioning, greater flexibility and more activities being grafted onto the basic coastal holiday package, as well as the evolution of rural, cultural and urban tourism in some regions (see Montanari and Williams 1995).

While mass tourism has brought foreign exchange earnings and higher incomes to the Mediterranean region, it has also brought dependence. This dependence has been reinforced by the interlocking of a reliance on a few markets and a small number of major international tour companies. However, this chapter has also demonstrated that simply to focus on the external dependence of the Mediterranean region on Europe is over-simplistic. In the course of the postwar period, and in part supported by the growth of mass tourism, the Mediterranean region has become an important centre of production and wealth (Dunford 1997). Domestic tourism has co-existed with international mass tourism in all these countries and this, too, has become a mass consumer product in France, Italy and, more recently, Spain. These three countries have also become important dynamic nodes of tourism origin, generating flows of tourism within (and beyond) the Mediterranean region.

Tourism development has also been accompanied by tourism differentiation within the region. There are clear differences not only between the northern and southern shores, and between the politically more sensitive eastern and more stable western sub-regions, but also between those countries which do and do not have tourism development strategies. This poses the question of whether tourism has been acting as a force for overall economic convergence or divergence in recent years. This issue has been considered elsewhere (Williams 1997), and it is sufficient here to note that there is no clear relationship between either growth in receipts (measured in dollars) and levels of GNP per capita, or between tourism's contribution to the GNP and GNP per capita.

This conclusion may be modified if we take into account the growth of incomes and consumption within the Mediterranean region. Although some of this is siphoned off into domestic tourism, thereby reinforcing the structures of inequality, it is also probable that internationalisation of demand is leading to even more rapid growth in outbound tourism, some of which is directed at other Mediterranean countries. As a result there may be considerable potential for the *net* contribution of international tourism (taking account of receipts and expenditure) to lead to greater economic convergence within the region. That convergence, however, will only apply to those countries which have been successful in pursuing tourism strategies, and there are several countries such as Albania and Libya which largely remain outside the international tourism system.

Tourism is likely to expand in future in the Mediterranean region, even if its share of the global market declines due to distance from the emerging new centres of consumption in the world economy, increased competition from destinations in the earlier stages of the product cycle, and growing concerns about the environmental quality of the Mediterranean. Any estimate of future growth is problematic, as the experiences of Yugoslavia and Egypt demonstrate. Nevertheless, the Blue Plan estimate that there will be upwards of 380 million tourist arrivals in the Mediterranean region by 2025 does not seem unrealistic (Grenon and Batisse 1989). It is far more problematic to try and predict the distribution of tourism in the year 2025 across the Mediterranean region. There are still relatively 'immature' destinations such as Morocco, Egypt and Turkey which have considerable potential for further development. Whether they will do so depends on the strength of both their Arab and European markets, as well as domestic and international political relations. There is also the potential to build, or rebuild, new tourist industries in Albania, the Lebanon and parts of the former Yugoslavia, although this is dependent on the achievement of greater political stability. Even if they fail to attract foreign tourists, most of these countries stand to benefit from the expansion of their domestic tourism markets following anticipated economic growth. The success of the region will also depend on the ability of the established European Mediterranean countries to maintain or enhance their market shares. That, in turn, is dependent on the latter's ability to respond to significant shifts in holiday spending, including the demand for more individualised holidays (Marchena Gómez 1995).

Even more problematic are the distributional questions related to tourism, and this has to be seen in terms not only of economic costs and benefits, but also of environmental and cultural ones. The key issues centre on the distribution of revenues between the international tour companies and the host economies and societies, the spatial distribution between the coastal and interior zones, and the distribution of tourism income between the state, fractions of tourism capital and the tourism labour force. While the outcomes of each of these are to some degree constrained by the globalisation of tourism, they are not inevitable but can, depending on local political and economic structures, be contested. The current debate about sustainable tourism is in fact one attempt to contest the current distributions of costs and benefits, even if the discussion is largely framed in terms of the language of environmental management (Frangialli 1993).

Notes

1   This is not to deny the existence of other forms of tourism in the Mediterranean region such as rural tourism, cultural tourism, business visits etc. (see Claval 1995), nor of the importance of domestic tourism (but statistical sources are limited on this).
2   Jenner and Smith (1993) thereby estimate that, for example, 80 per cent of tourists to Italy visit Mediterranean regions of that country; the proportions are 50 per cent in Morocco and only 15 per cent in France. In Malta and Cyprus they are 100 per cent.
3   Israel receives large numbers of religious tourists, as well as visitors from North America and Europe who are members of the widespread Jewish diaspora.
4   The year 1994 was an unusual one for tourism to Morocco. Morocco closed the border with Algeria, following an attack on an hotel in Marrakech; as a result Algerian 'shopping tourists' were reduced dramatically. The annual data show a more or less constant number of Europeans arriving over the four years 1992–5 (about 1.3 million per year); arrivals from Algeria slumped from 1.66 million in 1992 and 1.28 million in 1993 to 70,000 in 1994 and 13,000 in 1995.

References

Brendon, P. (1992) *Thomas Cook: 150 Years of Popular Tourism.* London: Secker and Warburg.
Buckley, P. J. and Papadopoulos, S. I. (1986) Marketing Greek tourism – the planning process, *Tourism Management,* 7(1), pp. 86–100.
Claval, P. (1995) The impact of tourism on the restructuration of European space, in Montanari, A. and Williams, A. M. (eds) *European Tourism: Regions, Spaces and Restructuring.* Chichester: Wiley, pp. 247–62.
Dirección General de Política Turística (1992) Los precios de los packages turísticos temporada verano 1992, *Estudios Turísticos,* 115, pp. 55–86.
Dunford, M. (1997) Mediterranean economies: the dynamics of uneven development, in King, R., Proudfoot L. and Smith, B. (eds) *The Mediterranean: Environment and Society.* London: Arnold, pp. 126–54.
Economist Intelligence Unit (1993) Portugal, *EIU International Tourism Reports,* 1, pp. 23–42.
Frangialli, F. (1993) El turismo en el Mediterraneo: la apuesta del desarrollo sostenible para un gran destino frágil, *Estudios Turísticos,* 119–120, pp. 5–21.
Grenon, M. and Batisse, M. (1989) *Futures for the Mediterranean Basin: The Blue Plan.* Oxford: Oxford University Press.
Jenner, P. and Smith, C. (1993) *Tourism in the Mediterranean.* London: Economist Intelligence Unit.
King, R. (1995) Tourism, labour and international migration, in Montanari, A. and Williams, A. M. (eds) *European Tourism: Regions, Spaces and Restructuring.* Chichester: Wiley, pp. 177–90.
Lewis, N. (1984) *Voices of the Old Sea.* London: Hamish Hamilton.
Manrique, E. G. (1989) El turismo, in de Ory, V. B. (ed.) *Territorio y Sociedad en España II, Geografía Humana.* Madrid: Taurus, pp. 341–67.
Marchena Gómez, M. J. (1995) New tourism trends and the future of Mediterranean Europe, *Tijdschrift voor Economische en Sociale Geografie,* 86(1), pp. 21–31.
Marchena Gómez, M. J. and Vera Rebollo, F. (1995) Coastal areas: processes, typologies and prospects, in Montanari, A. and Williams, A. M. (eds) *European Tourism: Regions, Spaces and Restructuring.* Chichester: Wiley, pp. 111–26.
Montanari, A. (1995a) Tourism and the environment: limitations and contradictions in the EC's Mediterranean region, *Tijdschrift voor Economische en Sociale Geografie,* 86(1), pp. 32–41.
Montanari, A. (1995b) The Mediterranean region, in Montanari, A. and Williams, A. M. (eds) *European Tourism: Regions, Spaces and Restructuring.* Chichester: Wiley, pp. 41–65.

Montanari, A. and Williams, A. M., eds (1995) *European Tourism: Regions, Spaces and Restructuring*. Chichester: Wiley.

OECD (1995) *Tourism Policy and International Tourism in OECD Countries 1992–1993*. Paris: OECD.

Pearce, D. G. (1987) Spatial patterns of package tourism in Europe, *Annals of Tourism Research*, 14(2), pp. 183–201.

Reynolds-Ball, E. (1914) *Mediterranean Winter Resorts*. London: Kegan Paul, Trench, Trüber and Co.

Shaw, G. and Williams, A. M. (1994) *Critical Issues in Tourism*. Oxford: Blackwell.

Urry, J. (1990) *The Tourist Gaze*. London: Sage.

Urry, J. (1995) *Consuming Places*. London: Routledge.

Williams, A. M. (1995) Capital and the transnationalisation of tourism, in Montanari, A. and Williams, A. M. (eds) *European Tourism: Regions, Spaces and Restructuring*. Chichester: Wiley, pp. 163–76.

Williams, A. M. (1996) Mass tourism and international tour companies, in Barke, M., Towner, J. and Newton, M. T. (eds) *Tourism in Spain: Critical Issues*. Wallingford: CAB International, pp. 119–35.

Williams, A. M. (1997) Tourism and uneven development in the Mediterranean, in King, R., Proudfoot, L. and Smith, B. (eds) *The Mediterranean: Environment and Society*. London: Arnold, pp. 208–26.

Williams, A. M. and Montanari, A. (1995) Introduction: tourism and restructuring in Europe, in Montanari, A. and Williams, A. M. (eds) *European Tourism: Regions, Spaces and Restructuring*. Chichester: Wiley, pp. 1–16.

Williams, A. M. and Shaw, G. (1988) Tourism: candyfloss industry or job generator, *Town Planning Review*, 59(2), pp. 81–104.

Williams, A. M. and Shaw, G., eds (1991) *Tourism and Economic Development: Western European Experiences*. London: Belhaven.

World Tourism Organisation (various years) *Yearbook of Tourism Statistics*. Madrid: World Tourism Organisation.

10

Northern Europeans and the Mediterranean: a new California or a new Florida?

Vicente Rodríguez Rodríguez, Pere Salvà Tomàs and Allan M. Williams

There is a popular argument that Southern Europe has the potential to become the California of Europe. This is based on the simple – perhaps simplistic – argument that a combination of economic, social and environmental factors will lead to a highly productive economic base, rooted in high technology industries, where the supply of labour and entrepreneurship will be sustained by inward migration. In this chapter we argue that such an hypothesis misreads the situation in Southern Europe, even in those regions which would appear to best approximate the Californian model. Instead, large parts of Southern Europe – especially the Mediterranean coastal regions – have tended to become regions of consumption, and the focus of lifestyle-related migration rather than labour migration.

Walker (1997) provides a useful summary of the Californian experience. California has grown to be bigger than all but six national economies, and had a Gross Domestic Product in 1995 of $700 billion. During its long boom between 1975 and 1990, 5.5 million new jobs were generated as 'California took over as the principal engine of US economic growth', especially in the electronics and aerospace sectors. A risk-taking business climate and the existence of sophisticated venture capital mechanisms favoured entrepreneurship. In addition, almost 5 million migrants were attracted to California during the 1970s and the 1980s. As a result, 'California has enjoyed a virtuous circle of investment, employment and spending in a highly diversified economy, and its skilled labour and ample capital have sustained a high rate of innovation that keeps Californian products in demand far and wide' (Walker 1997, p. 346).

Various attempts have been made to reproduce the Californian success story in Europe, most notably in the form of technology parks, characterised by the attraction of inward investment and the support of indigenous companies in the high technology sectors (Peck *et al.* 1996). While the best-known European example is Cambridge, there are several instances of technopoles in Southern Europe, including Sophia Antipolis in France (Longhi and Quere 1993) and the Seville Technology Park (Peck *et al.* 1996, p. 54). The latter was linked to the Seville Expo and involved a combination of state and private capital; by 1995, 1,500 jobs had been created in this pole, but it only secures regional rather than national (let alone global) significance. A similar picture emerges in the case of the Andalusia Technology Park in Málaga, which has mainly attracted high technology production rather than research and development, and continues to rely mostly on 'external technology' (Peck *et al.* 1996, pp. 61–2).

Any comparisons to California therefore remain at best premature, and there is no clear evidence of the positive circle of investment, employment, capital, immigration and innovation described by Walker (1997). There are some parallels in terms of population immigration. King *et al.* (1997), for example, have noted the demographic turn-around in Southern Europe which, after the 1970s, became a region of net immigration rather than of net emigration. While some of these flows involve unskilled and skilled labour migrants, the migration streams from the northern to the Mediterranean regions of Europe have been largely consumption-orientated. This chapter focuses on these latter flows and, in particular, on international retirement migration to the Mediterranean regions of Europe, prompting the alternative interpretation that they constitute the Florida rather than the California of Europe. Empirical material presented in the chapter draws on survey results collected by the authors in various research projects concerned with foreign retirement migration and property ownership in Spain and other destinations.[1]

Northern Europeans in the Mediterranean region

International migration to Southern Europe has been heterogeneous in terms of social and labour market characteristics. For example, there are three main migration flows to the Spanish Mediterranean areas in terms of geographical origins, while each of these could be further sub-divided in social terms:

- Europeans, including both labour and residential immigrants, many of the latter being retired;
- Latin Americans, generally consisting of skilled workers, often with ties established by earlier Spanish emigration; and
- Africans, who are generally unskilled labour immigrants.

The last two categories have increased rapidly since the 1980s, and their immigration has become a political issue around which racist and other xenophobic politics and protests have crystallised. In contrast, Spanish and other Mediterranean societies have generally been more welcoming to the first category of immigrants.

Any attempt to analyse international migration to the Mediterranean countries, especially international retirement migration, faces considerable difficulties in defining and measuring the phenomenon (Williams *et al.* 1997). In practice, 'retirement' migration is a very variegated form of mobility constituted of a continuum of flexible situations, representing different degrees of temporal and property commitments to the destination areas. The continuum includes, *inter alia*, seasonal migrants, owners of second homes, long-term tourists, permanent legally-registered residents, and non-registered residents; the fact that these categories often shade into each other poses considerable definitional problems (O'Reilly 1995; Warnes 1991; Williams *et al.* 1997). It is by no means easy to calculate – even approximately – how many retired foreigners really live in these areas, especially in countries such as Spain and Portugal, where foreigners apparently have little interest in registering with the authorities.

Eurostat provides a broad overview of the distribution of foreign population stocks in Southern Europe (table 10.1), although the data are not specifically for the Mediterranean regions of these countries, and they are not disaggregated by age and economic activity. Nevertheless, the figures do demonstrate the importance of immigration from a small number of Northern European countries, and in particular from the UK, Germany and France. The UK is the largest source of European foreign nationals resident in Spain, Portugal and Greece, and holds second place in Italy after Germany. France holds third place in all three countries, and there are also significant numbers of immigrants from the Netherlands, Sweden and Switzerland, with the last of these being particularly important in Italy.

The use of comparative international migration data is fraught with difficulties, and while national-level analyses pose their own challenges, these do at least permit detailed comparisons of different and sometimes conflicting data sources, as is evident in the case of Spain. In Spain, general statistical information about foreign populations is provided in several official publications, especially the *Anuario de Migraciones* or Migration Yearbook, and data on work permits granted to foreigners. The only regional or municipal sources are the Census and the

Table 10.1   Northern Europeans resident in Southern European countries, 1993

Country of origin	Italy		Greece		Spain		Portugal		Total	
	'000	%	'000	%	'000	%	'000	%	'000	%
France	25.4	15.7	8.0	7.5	22.6	11.4	3.7	10.9	59.7	11.9
Germany	39.5	24.4	14.1	13.3	30.5	15.4	5.4	15.9	89.5	17.9
Netherlands	7.0	4.3	3.7	3.5	10.5	5.3	2.0	5.9	23.2	4.6
Sweden	3.2	2.0	2.3	2.2	5.3	2.7	0.7	2.1	11.5	2.3
Switzerland	18.2	11.2	3.6	3.4	5.6	2.8	0.7	2.1	28.1	5.6
UK	28.4	17.5	20.7	19.5	53.4	27.0	9.3	27.4	111.8	22.4
European Total	162.0	100.0	106.0	100.0	198.0	100.0	34.0	100.0	500.0	100.0

Source: Eurostat (1996).

Population Register (*Padrón*), which contain information about *registered* foreigners, yet even these various sources do not always coincide. In response to these difficulties, the 1996 *Padrón* for the Balearic Islands has applied a new record management model, and is considered to be a more complete register of the islands' residents. A resident is defined as someone who lives in the municipality on a regular basis on the date that the *Padrón* is renewed, and a foreign resident is a non-Spanish resident. This basic register can be complemented with other surveys and indirect sources of information (such as special registers in town halls, tenants' registers for apartments and other dwellings, and airport departure and arrival records).

In Spain, the 1991 Census is the main official source of information about foreign residents. Inevitably, there are major differences between this and other, non-official, sources relating to the legal status of the immigrants, their length of stay in Spain, and the methods of data collection. The census records almost 54,000 Northern Europeans aged over 55 years living in Spain but this is likely to be a

Table 10.2 Older North European populations (>55 years) in Spain recorded by the Census, 1991

Region Province	Legally resident	%	de facto population	%
Andalusia	14,867	31.5	17,923	33.3
Almería	689	4.6	804	4.5
Cádiz	451	3.0	630	3.5
Cordoba	21	0.1	24	0.1
Granada	820	5.5	879	4.9
Huelva	45	0.3	52	0.3
Jaén	18	0.1	18	0.1
Málaga	12,692	85.4	15,735	85.8
Seville	131	0.9	141	0.8
Balearic Islands	3,418	7.2	3,549	6.6
Canary Islands	5,246	11.1	6,204	11.5
Las Palmas	1,258	24.0	1,487	24.0
Tenerife	3,988	76.0	4,717	76.0
Catalonia	2,670	5.7	2,894	5.4
Barcelona	1,355	50.8	1,420	49.1
Girona	879	32.9	991	34.2
Lleida	17	0.6	17	0.6
Tarragona	419	15.7	466	16.6
Valencia	18,599	39.4	20,630	38.3
Alicante	17,672	95.0	19,650	95.2
Castellón	281	1.5	301	1.5
Valencia	646	3.5	679	3.3
Total, Spain	47,192		53,850	

Note: This table only lists Mediterranean coastal and island provinces. Percentage data for the regions are percentages of the national total, those for the provinces are percentages of the totals of the regions.

considerable underestimate, even though their spatial distribution is probably broadly accurate (see table 10.2). Three-quarters live in coastal and Mediterranean regions, especially Andalusia (mainly Málaga province) and the Autonomous Region of Valencia (especially Alicante province), where elderly Europeans constitute sizeable proportions of the total foreign populations, with a very large floating population (the difference between legal and *de facto* residents) in some provinces.

The data provided in this and other sources produce such different estimates that all these statistics must be used with due caution. According to consular estimates, there are more than 200,000 British living in Spain (between 50,000 and 100,000 on the Costa del Sol); yet according to the 1991 Census there are barely 15,000! There are even bigger differences between the official sources and some other calculations (Galacho 1991; Paniagua 1991). It would seem that the level of non-registration may be as high as 80 per cent (Mullan, 1992). According to the European Retired Immigrants in Andalusia (ERIA) survey, which was conducted in April and May 1996, 28 per cent of respondents stated that they had not registered with the authorities, either because they were unfamiliar with bureaucratic procedures, were convinced that it was not necessary, wished to remain anonymous or were only staying temporarily (Rodríguez *et al.* 1998).

In the Balearic Islands, foreign residents are classified in terms of how long they spend on the islands each year: permanent foreign residents are normally considered to live on the islands all year round; temporary ones stay for more than three months. The latter are particularly difficult to quantify. A nationality-based multiplier index that takes into account variables such as the figures provided by the Spanish Police, the occupation rate of dwellings, airport and port passenger figures, and opinion poll and survey data, has also been established. According to these data, at the end of 1995 there were 35,089 legally-established permanent foreign residents, 79 per cent of whom were European (table 10.3). This figure is much higher

Table 10.3 Foreigners resident in the Balearic Islands, 1995

Country of origin	Permanent	Temporary	Total
Germany	7,667	21,873	29,540
UK	8,347	10,010	18,352
France	2,149	372	2,521
Sweden	1,543	308	1,851
Italy	1,216	317	1,532
Netherlands	1,226	269	1,495
Switzerland	893	200	1,093
Other European countries	4,657	588	5,245
Europeans	27,698	33,931	61,629
Foreigners	35,089	39,167	74,256

Sources: Dirección General de la Policía; Padrón.

– 74,256 – if temporary foreign residents are added (83 per cent of whom are European). This higher figure equates to almost 10 per cent of the population of the Balearic Islands in 1996. Just over three-quarters of the total permanent foreign residents live in Majorca, with the remainder on Ibiza and Formentera (18 per cent) and Menorca (4 per cent). However, these percentages change slightly if temporary foreign residents are taken into account.

In terms of nationalities, the nearly 30,000 Germans constitute the largest group of foreigners (48 per cent of all Europeans) in the Balearics. Numbers have risen significantly since 1986. Germans tend to have a sharply defined seasonal residential pattern. In contrast, more British (who make up 30 per cent of the European total) are permanent residents, and first arrived in the islands in the 1970s. The other nationalities (French, Swedish, Italian, Dutch and Swiss) each barely account for 5 per cent of the European total. German residents are the largest group in Majorca and Ibiza, whereas the British are the largest group in Menorca.

The attraction of 'the South'

Southern Europe is attractive to Northern European migrants for a number of reasons, as can be seen in the case of Spain. Even in the context of Southern Europe, many of Spain's coasts are highly regarded by Northern European citizens for their climate (Marchena 1987): there are high annual temperatures, averaging between 16° and 22° in Palma de Majorca and the Canary Islands; in winter, the average temperature remains above 10° C; the number of hours of sunshine a year varies from 2,700 in Palma to 2,900 in Alicante; and there are very few rainy days, especially in summer. These climatic conditions facilitate a more relaxed, associational and outdoor lifestyle, together with the possibility of being able to re-create supposedly more 'traditional' ways of life (Williams *et al.* 1997). In short, these elements combine to form a Mediterranean *sunbelt* for retired Europeans from Northern Europe, similar to that which exists in parts of southern and western North America in relation to metropolitan areas in the colder north.

One of the most important features of this outdoor lifestyle is golf. The golf market is a major attraction for Northern European tourists in general, and for retired people in particular. Demand for golfing facilities in Southern Europe, especially in Spain, is well-established, as is shown not only by the number of golf courses, often associated with residential complexes, but also by the flow of 'golf tourists', two-fifths of whom come from the United Kingdom and 30 per cent from Germany and Scandinavia (Priestley 1995). Most of these tourists head for Andalusia and the Balearic Islands. The Spanish climate has a positive bearing on golf tourism, which is not subject to the seasonal nature of mass tourist demand.[2] Golf is also instrumental in encouraging retired foreigners to become permanent residents, because many real-estate companies use the lure of golf courses to sell high-class housing.

In addition, Spain is also attractive due to reductions in travel time and cost, and to improvements in access facilitated by the liberalisation of air transport (Barke and France 1996, p. 289). Most regular flights between Palma, Alicante or Málaga and the main airports of Central and Northern Europe (Hamburg, Frankfurt, Brussels, Amsterdam, Manchester, London) take two to two and a half hours; the door-to-door trip for some travellers can be as short as four hours. Charter flights may take slightly longer, but their low cost makes them more attractive. Therefore, even though the coasts of Southern Europe are on the periphery of Europe, Málaga, Alicante, the Balearic Islands and the Algarve are well connected to the major cities of Europe. Moreover, the development of tourism access in these regions has helped to reinforce their international connections for other purposes (Cattan 1995), while travel agencies and tour companies have helped to reduce costs in terms of money and time (Williams 1996).

Many Northern Europeans have also been drawn to the South by differences in the cost of living. If one considers consumer spending rates, based on purchasing power parity, it remains far cheaper to live in Spain, Greece or Portugal than in the Northern European countries today, even though the differences have been reduced by economic development and convergence (OECD 1995). Furthermore, pensions and retirement benefits in Spain and Portugal are far lower than the EU average (61 per cent and 37 per cent of the EU mean respectively), whereas in Belgium, Germany and Holland benefits are well above the EU average. Although this gap too has narrowed in recent years, it means that retired Northern Europeans are still likely to have relatively more disposable income, especially if they have occupational and personal pensions, and perhaps other sources of income.

Retired immigrants have tended to rebuild their economic and social networks, both informally and by making use of a well-developed framework of associations. Religious associations, clubs and organisations and 'grapevine' networks provide good examples of this process. This has made it easier for other potential migrants, and has helped to reinforce the mechanisms that attract new migrants (Myklebost 1989). Their experience as tourists in previous years, very often in Southern Europe, operates in a similar way to reduce the risks and disadvantages associated with international migration.

Table 10.4 Companies owned/managed by foreigners in Spain, 1997

Region Province	No. of companies	Foreign-owned as percentage all companies	No. of commercial companies	Foreign-owned as percentage all commercial companies
Andalusia	5,519	2.0	4,552	2.6
Málaga	3,500	5.5	2,882	8.0
Balearic Islands	2,805	4.9	2,167	6.8
Canary Islands	7,793	9.3	6,116	13.4
Catalonia	5,482	1.4	3,818	2.0
Barcelona	3,288	1.0	2,020	1.4
Girona	1,508	3.2	1,241	5.7
Valencia	4,483	2.0	3,612	3.0
Alicante	2,924	3.5	2,324	5.5

Source: Chambers of Commerce.

In addition to retirement migration to the Mediterranean regions, there has also been a concomitant flow of labour migrants to these areas, in large part to provide services for particular national groups of expatriates and tourists. Barke and France (1996) have documented some of the business activities associated with these labour migrations to the Costa del Sol, such as bars and restaurants, travel and car rental agencies, and shops. Establishments owned or managed by foreigners, especially commercial undertakings, are to be found throughout the coastal and island regions of Spain (table 10.4). More than 10 per cent of the shops on the Canary Islands are owned or managed by foreigners. In Málaga, the Balearic Islands and Alicante the percentages are somewhat lower – in the range 3–6 per cent – but foreign businesses remain a notable component of these regional economies, especially as the proportions would be significantly higher in particular municipalities such as Benidorm and Torremolinos. Although not all these establishments cater solely, or even largely, for the consumer needs of foreign residents, the advertisements in regional foreign-language publications reveal that there is a range of businesses run by foreigners on the Costa del Sol and that these survive and proliferate in niche markets. Examples include real-estate companies (including real-estate financing), bars and restaurants, health-related services, telecommunications companies, educational services, pet-care services, and general household services.

The various types of Northern European immigration have had a major impact on the economies of the destination regions. The Register of Foreign Real-Estate Investment provides an indicator of the scale of foreign investment in the Spanish housing market. In the three-year period 1994–6, foreigners were recorded as having invested more than 200 billion pesetas (approximately £1 billion) in real estate, although the actual total is estimated to be much higher. Nine-tenths of the registered foreign investment is accounted for by the five regions that have the highest level of tourism development: Andalusia, the Balearics, the Canaries, Catalonia and Valencia. Of particular importance are the islands, and the provinces of Barcelona, Girona, Alicante and Málaga. Within these, certain municipalities have attracted exceptional levels of investment. Calvià and Andraitx account for almost half of the Balearic total. In Málaga province, Marbella, Estepona and Mijas account for 70 per cent of foreign real-estate investment, most of the purchasers being from the United Kingdom and Germany (Subdirección General de Planificación y Prospectiva Turística 1988). The main investors in Torrevieja and Denia, in Alicante province, are from the United Kingdom, Germany and Switzerland. On the Costa del Sol, as in most other touristic regions, foreigners tend to live in distinctive types of houses (mostly semi-detached or detached houses, with an average surface area of 200m²). Generally these are grouped into residential estates called *urbanizaciones*. Associated leisure facilities, such as a swimming pool or a tennis court, are close by – in the residential development or attached to individual dwellings. Other than the initial jobs in construction, these developments have also had a more lasting impact on employment; each *urbanización* employs 20 maintenance employees on average (Asociación Provincial de Urbanizadores de Málaga 1990). In addition, the immigrant communities become the focus of international tourism. According to survey data (King *et al.* 2000), each British resident in the Costa del Sol receives six visitors a year on average, and in addition to this there are vast numbers of second homes used by both their owners and those they let or lend these properties to.

In addition, the migrant communities have also transformed the social and cultural context in the regions of destination. This has facilitated the process of integration for later migrants. One notable example is the way in which the foreign language media have established a strong foothold on the Spanish coasts. English publications are dominant, such as the weekly *Sur In English* on the Costa del Sol (50,000 copies), *The Entertainer* on the Costa del Sol and Costa Blanca (43,000), the *Majorca Daily Bulletin* on the Balearic Islands, and monthly magazines such as *The Reporter* (20,000) and *LookOut* (20,000). There are also German publications such as *Leben auf der Balearen*, *Majorca Magazin* or the Costa del Sol's *Nachrichten*, as well as other foreign language publications in, for example, Swedish and Finnish. Foreigners are also made to feel 'at home' in Spain by the availability of foreign radio stations, Spanish radio stations that include slots for foreign residents, and regional or satellite TV stations.

Religious, social, cultural and leisure associations also serve to bring together people of different nationalities who share common interests. According to informal records, there are more than 60 religious associations of different faiths and 80 other associations (catering for example for music, dancing, sports, card games, and politics) on the Costa del Sol alone. Most are run by foreigners, but there are rarely any formal nationality barriers to membership. They meet in public places (often in bars and municipal offices) and arrange a wide variety of activities, especially in the autumn and winter. Examples of the leading associations include the Royal British Legion, The International Club, Rotary Club, Oasis in Majorca, and Help in Alicante (Balao 1994; Mullan 1992). These associations provide social and health-related services, helping individuals to resolve administrative problems, obtain public health care, and raise funds to pay for their activities. In response to the needs of these significant foreign communities, there have been municipal initiatives to support retired residents, such as the establishment of Foreigners' Departments in Mijas, Calvià and Alfaz del Pi (Balao 1994). Benalmádena, Fuengirola and Estepona in the Costa del Sol have also allocated resources for integrating foreigners.

Case studies of international retirement migration to Southern Europe

The previous discussion has identified retirement migration as an important component of the population flows from Northern to Southern Europe. We have stressed that it is notoriously difficult to quantify these communities, and that there are very few substantive studies of either motivations or experiences of retirement in Southern Europe (Williams *et al.* 1997). However, some insights are available from three recent research projects engaged in by the authors. The first of these is a comparative study of British retirees in Tuscany, Malta, the Costa del Sol and the Algarve, and it provides a broader framework for considering two Spanish case studies: a comparative survey of British, German, Nordic and other national retiree groups in the Costa del Sol (the ERIA study); and a study of North European residents in the Balearic Islands. Selected findings from these research studies are now presented.

British retirees in four South European destinations

One of the first tasks of this research was to provide estimates of the older British populations in the four regions chosen for the survey research (King *et al.* 1998, 2000; Williams *et al.* 1997). On the basis of the available statistics, adjusted to take into account the opinions of key informants, this study provided the following estimates for the numbers of British people, including those of official retirement age (60 for women and 65 for men), living in the four areas:

- Tuscany: 4,000 in total, of whom half are retired;
- Malta: 4,000–5,000 in total, of whom 30–40 per cent are retired;
- Costa del Sol: 50,000 in total, of whom one third are of retirement age;
- Algarve: 10,000 in total, of whom at least one sixth are of retirement age.

These data do not include the early-retired who, in some cases, may almost double the total number of British people who have retired from full-time work; in other words, it is not unreasonable to estimate that between one third and two-thirds of these British populations are effectively retired.[3] This confirms the significance of international retirement migration to these Mediterranean regions, at least in the case of the British. It is a pattern which is broadly confirmed by the data available on the distribution of British pensions paid to recipients living in Southern Europe (Williams *et al.* 1997).

In the absence of detailed secondary data on the British living in Southern Europe, the above-mentioned research project has undertaken the largest questionnaire survey to date of the expatriate communities in these areas. The results are reported in detail elsewhere,[4] but some of the salient features of these four retiree communities are summarised here in order to underline both similarities and differences in their socio-demographic profiles (table 10.5). The main features of the four communities are as follows:

- Tuscany. This is the oldest-established community (with almost one half having arrived before 1985), and in many ways the most distinctive. It has a high-status social profile which is particularly evident in terms of educational background. Of the four, this is also the group with the highest levels of lifetime mobility, as indicated by the fact that more than 60 per cent lived abroad for at least part of the five years prior to migration to the region. Their choice of Tuscany was far less influenced by climate than was the case in the other destinations. Instead, admiration of the region (and of Italy in general) and work and family connections were major attractions. A high proportion of the group lives permanently in Tuscany; few are seasonal residents.
- Malta. Given its historical and colonial ties to the UK, Malta also has a distinctive and relatively long-established retired British population. This

Table 10.5 International retirement to Southern Europe: profiles of British expatriate communities in four destinations

	Tuscany	Malta	Costa del Sol	Algarve	Total sample
Percentage arrived post-1985	49.0	54.3	61.0	71.2	60.1
Mean age	69.1	68.4	66.3	65.6	67.1
Percentage Social Class 1 & 2	71.0	54.9	67.7	72.4	65.7
Mean age left full-time education	19.6	16.7	17.1	17.5	17.4
Percentage giving climate as main attraction	15.4	37.5	48.1	44.2	40.4
Percentage who holidayed in area before migration	75.8	77.9	95.7	93.0	87.6
Percentage who lived exclusively in UK for 5 years pre-migration	37.1	66.3	68.1	61.6	62.6
Percentage spending at least 45 weeks per year at the destination	63.2	64.0	51.2	53.9	57.0

Source: Based on King *et al.* (1998).

is the group with the lowest occupational status and the least formal education, reflecting in part the large numbers of non-commissioned personnel from the armed forces. As in Tuscany, a large proportion spend virtually all the year in the destination region.

- Costa del Sol. Three-fifths of this survey population had arrived after 1985, reflecting the more recent peaking of retirement migration here. Virtually all the migrants had been tourists in the region, almost one half were primarily attracted by the climate, and two-thirds had lived exclusively in the UK prior to emigrating to the Costa del Sol. Only one half live all year round in the destination. In contrast to Malta and Tuscany, this is a group with far less experience of living abroad, and with fewer prior connections to the region in terms of work and family links. Instead, their holiday experiences have played a leading role in their destination choice, and many are seasonal migrants (of varying duration).
- Algarve. This British population resembles that in the Costa del Sol in many ways, the main difference being a somewhat later peaking of emigration, so that 71.2 per cent had arrived since 1985. For the same reason, the average age was younger and, unlike the Costa del Sol, there was a tendency to settle inland and not right on the coast.

This brief synthesis serves to highlight the contrasts that exist between different destination countries within the overall British retirement populations, as well as some of the differences in retirees' motivations and backgrounds. While confirming the broad importance of retirement migration, it also underlines the need to study specific types of communities. This is further illustrated in the next two sections which consider two contrasting Spanish destination regions – the Costa del Sol and the Balearic Islands.

The Costa del Sol: a case study of retirement migration

The Costa del Sol has been subject to a complex process of development over the last 30 or more years, involving increases in tourism facilities, in real-estate trans-actions and in the concentration of population in the coastal areas (Barke and France 1996). Residential tourism and migration have increased the number of foreigners, especially retired Northern Europeans, but it is difficult to calculate their exact number due to the limitations of the published data sources. The following discussion draws on primary data obtained from the aforementioned ERIA survey, which was conducted among Europeans aged 55 or above living along the Costa; the sample was stratified in accordance with the distribution of European nationalities recorded in the 1991 Census (British 63 per cent; German 15 per cent; Scandinavian 13 per cent; Dutch 9 per cent).

The over-50 population is relatively young, with an average age of 66 years; 8 per cent are younger than 55 years old and only 14 per cent are aged over 75. They came to Spain after having retired young (mean age of 57) and have been living in Spain for an average of nine years, although 12 per cent have been in Spain for more than 16 years, especially the Scandinavians (29 per cent). There are few signs of local mobility: in 1996, three out of four were living in the same dwelling as when they first arrived in Spain.

The importance of the earlier-mentioned climatic, environmental and cost-of-living factors is borne out by the reasons given by the survey respondents for moving to Spain upon retirement (table 10.6). A pleasant climate, a different way of life, lower costs and previous experience of living in Spain are the main reasons given. More than 90 per cent of the interviewees – with little national variation – stated that the climate (mild temperatures, abundant sunshine and little rain) was the key

Table 10.6 International retirement migration to the Costa del Sol: reasons for emigration by national group (%)

Reasons	Nationality				Total
	British	German	Nordic	Dutch	
Mediterranean climate	90.4	86.7	97.4	100.0	91.6
Spanish way of life	52.7	37.8	50.0	39.3	48.8
Lower cost of living	28.7	24.4	31.6	39.3	29.4
Holidays in Spain	29.8	22.2	31.6	14.3	27.4
Cure for health problems	17.0	33.3	34.2	17.9	21.7
Owning a house in Spain	19.1	4.4	7.9	14.3	15.1
Number	189	45	38	28	300

Source: Rodríguez et al. (1998).
Note: The column totals exceed 100% because the question has a multi-response structure.

factor that made them decide to move to Spain. The Spanish outdoor lifestyle, and flexible hours – in the sense of not being strictly timetabled – are also important attractions, above all for the British and Scandinavians. A fifth of the interviewees, especially the Germans and Scandinavians, considered the climate to be beneficial for particular health problems, especially those related to the ageing process. In addition, almost 30 per cent of interviewees stressed cost-of-living differences, mentioning that it was cheaper to live in Spain than in their country of origin; this applied particularly to the Dutch, Belgians and Scandinavians. Previous knowledge of the area was also important, particularly that gained through having been on holiday or owning a house prior to coming to live there.

The lives of older Northern Europeans on the Costa del Sol tend to be far more consumption- than production-oriented, not least because most have moved to Spain upon retirement. According to the 1991 Census, only 6 per cent of older Northern Europeans living on the Costa del Sol had a job whereas, before coming to Spain, more than 70 per cent had worked and only 9 per cent had already retired (with quite a low retirement age of about 60). Indeed early retirement, together with relatively high levels of income especially compared to Spain, have been important in stimulating and facilitating emigration to the Costa; this is especially true of Germans.

Despite having retired from full-time employment, many retired people do work after coming to Spain. Work – which is subject to many definitional nuances – is considered to be beneficial by a majority of respondents, either because they need the income (stressed by 16 per cent of the total sample, but by a higher proportion of the British), or because they believe this will help them retain their mental and physical health, or to keep in touch with people (27 and 15 per cent respectively). Although the majority viewed work positively, there was also a body of opinion which was opposed to this, with 33 per cent stressing the importance of resting. The survey revealed that one out of every five older foreign people living on the Costa del Sol actually did work, either for a salary or on a voluntary basis. Amongst these, those who stressed the positive aspects of work – such as the Scandinavians, the Dutch and the Belgians – are more likely to be working. The majority are men, tend to live alone (37 as compared to 25 per cent of all the interviewees), and spend more than six months a year in Spain. Work is very much part-time, and almost three-quarters of those who work do so for less than 14 hours a week, averaging between two and three hours a day. In terms of national differences, the British are distinctive, being likely to work fewer hours, and also being more likely to participate in voluntary work. There are fewer national differences in the type of job undertaken, where the most striking characteristic is a reluctance to continue in the same line of work as they had before retiring to Spain. Others work in activities related to their new life and leisure time; some 40 per cent (and 60 per cent in the case of the British) do voluntary work, manage leisure activities or chair a club, organisation or property owners' association.

The ERIA survey results present an overwhelming picture of satisfaction with the experience of retirement migration to the Costa del Sol. Almost 78 per cent of respondents believed that retired people live better in Spain than in their country of origin; this is particularly true of the British and Scandinavians. The reasons for this correspond to those which originally motivated their emigration, with climate and lifestyle being especially important. The 'Spanish lifestyle' was also highly valued by all nationalities, but especially by the British. This does not, however, indicate a profound knowledge of or participation in Spanish society; rather the 'Spanish way' is constructed in a very superficial fashion, often as a counterpoint to the 'bad Britain' they have left behind (see O'Reilly 1995). The main blemish in this picture of overall satisfaction with emigration is lack of ability in the Spanish language. More than 70 per cent of respondents, but especially the British and Scandinavians, considered that linguistic inability placed limits on their lifestyles, although this can be ameliorated through involvement in a wide range of associational and commercial activities. In fact, they know that they do not *need* to speak Spanish to live on the Costa del Sol.

Foreign European residents in the Balearic Islands

The flow of European residents to the Balearic Islands is different to that to most other Mediterranean regions, because it not only includes retired and leisure-time residents but also a relatively large number of labour and business migrants in the tourism and hospitality industries, especially in those sectors providing services to foreign residents (doctors, dentists, lawyers, real-estate agents etc.). Although there is a significant increase in the number of European labour migrants, above all of professionals and skilled workers in international companies, most European migration is for retirement purposes. The majority of foreign residents are women, reflecting the feminisation of the foreign working population, especially those in jobs related to the tourism sector, and gender differences in survival rates. The age profile of the resident Europeans (figure 10.1) displays two main peaks in the 40–44

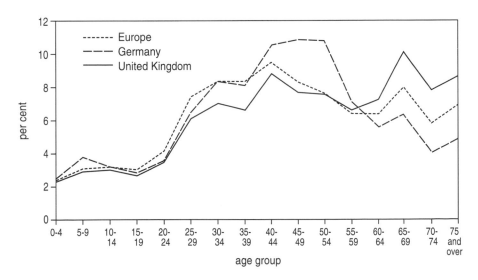

Figure 10.1   Age structure of foreign residents in the Balearic Islands, 1991

and 65–69 age groups, as well as a relatively high average age (50 years). The first peak is constituted largely of the owners and employees of small or medium-sized service businesses (typically restaurants, bars, shops and discos), who have set up businesses to meet the demands of tourists of their own nationality. There are also some professional and business people who run businesses elsewhere from the Balearics, or professionals who work at home (often using advanced forms of information technology and communications) and others who are involved in foreign investment in the Balearic Islands. There are differences between the nationalities: the British tend to be older (a large proportion are aged 60–65), having arrived during the first wave of European immigration in the 1970s; whereas most Germans are 40–54 years old, and tend to be business executives or early-retired (Salvà Tomàs 1994).

As would be expected, different types of residents have different motivations for migration. In general, an important role is played by mass tourism (the Balearics received 9 million tourists in 1997). At one level, this has generated demand for skilled workers, especially those specialising in supplementary tourist activities, such as sport and leisure instruction. Moreover, the promotion of tourism on the Balearic Islands has prompted a boom in temporary and permanent leisure residents, above all from Germany and the UK. Viewed from this residential

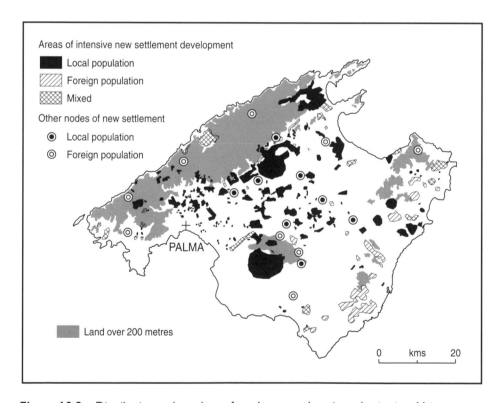

Figure 10.2   Distribution and typology of rural areas undergoing urbanisation, Majorca

perspective, the Balearic Islands are affected by the processes of counter-urbanisation emanating from the main metropolitan areas of Europe, and the islands are used for week-end, short and long holiday breaks.

Other factors influencing European migration are the islands' accessibility and reductions in the time-distance-cost relationship, the availability of large numbers of flights, well-resourced promotional campaigns, the lure of the climate and the natural rural landscape. These basic factors are complemented by the good communications infrastructure, allowing new tele-working methods afforded by technological breakthroughs (mobile phones, the Internet etc.), as well as changes in economic activities, with a boom in personal and corporate services. The result is the presence of a large group of residents who run their companies from homes on the islands, thereby combining work with a less stressful and more balanced lifestyle.

The mass-tourism boom has provoked a deep change in the previous traditional territorial model in the Balearic Islands, having a major impact on what had been essentially rural areas (Binimelis 1996). There has been intensive urbanisation in natural areas, an increase in the scattered rural population, a transformation of the morphology of the traditional rural landscape and the advent of new land uses and functions. All these changes underline the significance of external influences, for the demands generated by the indigenous population are far less dynamic than those of the immigrants, not least because of differences in purchasing power. The resultant developments are not only used for leisure activities, but increasingly as permanent homes. These foreign residents have settled in communities in both the coastal areas and in traditional inland urban areas. They have also settled, in a more dispersed manner, in the countryside (figure 10.2).

The growth in the number of European foreign residents, especially of those who come to the islands for the purpose of 'rest and tranquillity', has had a number of consequences, including the following:

- An increased demand for housing and rural land, which is either rented or bought in housing estates.
- The appearance of land speculation processes, involving a wide variety of social agents, so that the price of land is no longer based on its farming value, but on its use in real-estate operations in 'virgin' potential urban environments. This is linked to foreign demand, which is far more powerful than endogenous demand, with high levels of purchasing power and favourable exchange rates, especially in the case of the German mark. The result is that land prices in rural areas currently top $12,000 (£7,500) per hectare (Seguí 1997).
- The appearance of new non-rural owners in the rural milieu is associated with a new type of ownership, with plots being closed off to ensure privacy, as a typical feature of 'urban' behaviour. There are also new types of farming activities, such as 'hobby farming' carried out by 'urbanites' as a leisure activity and to meet their own needs.
- Finally, there is an anti-immigrant reaction on the part of the local population who feel increasingly threatened and 'colonised' by Northern

Europeans (especially Germans), who buy up all available land and even organise themselves politically to promote their own interests in the municipal elections.

International retirement migration has constituted one of the main streams of foreign immigration to the Balearic Islands.[5] Residential tourism for retired people began in the 1970s, even if the 'take-off' occurred around 1986, when Spain joined the European Community. Retired foreigners had bought land or real estate before then, but only on a limited scale. It was in the 1970s that foreigners began buying significant numbers of apartments in tourist residential estates, as well as small farms and country houses which they restored or renovated; the British were the dominant group in a number of the enclaves which evolved in the islands. Most are artists or retired people seeking a quiet, peaceful life and an attractive environment in which to live (Salvà Tomàs and Binimelis 1993). There have also been major capital investments by foreigners from oil-producing countries, especially Iranians (Formentor, Costa de Levante), and South Americans (Pollença). This interest in buying country land and houses has surged since 1986, when major new investment trends emerged.

In summary, then, a number of linked development and migration processes has led to the growth of a socially heterogeneous foreign population in the Balearic Islands. The Europeans living on the Balearic Islands can be divided into two main groups and four different types in all, depending on their labour market status, their residential status and the length of their stay. The first group is formed by temporary working residents who rent houses near where they work (especially at S'Arenal, Palma beach and city, and the different towns in the municipality of Calvià), whereas the permanent residents, many of whom are business-people, tend to own houses in coastal and inland estates, and country estates. The second group is formed by European residents whose migration was linked to leisure interests and free time. This group can be divided into three subtypes:

- Long-term holiday residents (of more than one month) tend to buy apartments and houses in coastal residential estates set apart from the main tourist areas and often located near pleasure harbours, in an attempt to combine peace, sunshine, the sea and sailing.
- Pre-retired and retired people who spend long periods on the islands tend to buy semi-detached houses in new residential estates built in rural areas, and to live alongside the indigenous residents.
- Foreigners with considerable purchasing power make major investments in buying large rural plots and dwellings in mountainous zones or areas of beautiful landscape, or which are of historical value.

Conclusions

The introduction to this chapter set out the question of whether some regions in Southern Europe may constitute a 'new California'. Starting from Walker's (1997) description of the virtuous circle of investment, migration and innovation which characterised the long boom in California, it is clear that no Southern European region has yet reproduced the same conditions, although it is possible to identify some of the key elements in southern France. Instead, focussing only on the dimension of international migration from Northern Europe, this chapter advances the argument that some regions in Southern Europe correspond more to 'new Floridas' because of the weight of retirement migration in their demographic profiles. However, even this label must be treated cautiously, for all the case studies reviewed point to considerable diversity amongst the destination areas and the different national migration streams.

It would be inaccurate to describe Southern Europe as a macro-region of consumption rather than of production just on the basis of the evidence presented here. There are of course significant regional production structures in many parts of Southern Europe (see Dunford 1997; Hudson and Lewis 1984) which have not been the subject of this chapter. However, it is equally true that the economic significance of the consumption role of particular regions should not be underestimated. First, the inter-weaving of the demands of retirement migrants with those of tourists has had implications for land and housing markets, for the landscape and environment, and for the delivery of social, health and commercial services. Consumption in Southern Europe, supported by incomes and wealth deriving from Northern Europe, has generated jobs and incomes in these destination areas, as well as individual and social costs for the indigenous communities and the local and national administrations. A significant proportion of these jobs, and of business opportunities, has been filled by Northern Europeans, but to date these are dedicated to serving niche national markets, and there is little evidence that they have become strongly embedded in the regional economies or have developed an independent capacity to expand and diversify into other markets.

The role of these regions as spheres of consumption also carries other economic meanings. Changes in work practices and in communications technologies are leading to increased locational flexibility in some occupations and industries. This is most clearly evident in the growing numbers of foreigners living in the Balearic Islands who are tele-workers or long-distance commuters, deriving their current income streams from Northern Europe but living in the south. While the scale of this phenomenon should not be exaggerated, it may point to a future for some Mediterranean regions neither as new Floridas nor new Californias, but instead as new and distinctive foci of increasingly complex working, leisure and travel careers in Europe.

Notes

1   For contextual and more detailed accounts of this body of research see King *et al.* (1998, 2000), Salvà Tomàs (1994), Salvà Tomàs and Binimelis (1993), Rodríguez *et al.* (1998), Williams *et al.* (1997). Vicente Rodríguez acknowledges that this chapter is a partial output of the ERIA research project on 'European Retired Immigrants in Andalusia' (Inmigrantes Europeos Jubilados en Andalucía) financed by the Spanish government (CICYT) under grant SEC95-0120 and by the Instituto de Estadística de Andalucía. Pere Salvà Tomàs acknowledges the financing of his project on 'Rurbanisation, urban planning and foreign-owned property on the island of Majorca' from the European

Commission, DGI CYP, project no. PS5-0104. Allan Williams acknowledges research undertaken jointly with Russell King and Tony Warnes on the project 'International Retirement Migration from the UK to Southern Europe' funded by the Economic and Social Research Council, project no. R000235688.

2   It is worth mentioning here that golf has also become a popular middle- and upper-class activity for Spaniards, as the success of several top Spanish golfers attests.

3   Younger-retired British seem to be particularly numerous in the Algarve. It is also important to acknowledge that these early-retired migrants (many in their 50s and some still in their 40s) have different lifestyles and activity patterns to the older-retired in their 70s or older. Different again are the activities of young adult visitors and residents in places such as Marbella or Ibiza.

4   See King and Patterson (1998), King *et al.* (1998, 2000), Warnes and Patterson (1998), Williams and Patterson (1998); other papers from this project are still being written.

5   Other immigrations have been of low-skilled labour migrants working in the construction and tourist industries, who mainly come from African countries, especially Morocco.

References

Asociación Provincial de Urbanizadores de Málaga (1990) *Guía de Urbanizaciones.* Málaga: Patronato Provincial de Turismo de Málaga y Costa del Sol.

Balao, P. (1994) *Ciudadanos Europeos Mayores en España. Aproximación a la Situación Actual.* Madrid: Ministerio de Asuntos Sociales, 3 vols.

Barke, M. and France, L. A. (1996) The Costa del Sol, in Barke, M., Towner, J. and Newton, M. T. (eds) *Tourism in Spain. Critical Issues.* Wallingford: CAB International, pp. 265–308.

Binimelis, J. (1996) *Caracterització, Tipificació i Pautes de Localització de les Arees Rururbanes a l'Illa de Mallorca.* Majorca: Departament de Ciències de la Terra, Universitat de les Illes Balears, doctoral thesis, 3 vols.

Cattan, N. (1995) Attractivity and internationalisation of major European cities: the example of air traffic, *Urban Studies,* 32(2), pp. 303–12.

Dunford, M. (1997) Mediterranean economies: the dynamics of uneven development, in King, R., Proudfoot, L. and Smith, B. (eds) *The Mediterranean: Environment and Society.* London: Arnold, pp. 126–53.

Eurostat (1996) *Statistics in Focus: Population and Social Conditions.* Luxembourg: Eurostat.

Galacho, F.B. (1991) Problemas de cuantificación del turismo residencial en la Costa del Sol malagueña. Una propuesta de método de medición, *III Jornadas de Población Española,* Torremolinos: Asociación Geografos Españoles, pp. 57–70.

Hudson, R. and Lewis, J. R. (1984) Capital accumulation: the industrialisation of southern Europe? in Williams, A. M. (ed.) *Southern Europe Transformed: Political and Economic Change in Greece, Italy, Portugal and Spain.* London: Harper and Row, pp. 179–207.

King, R., Fielding, A. J. and Black, R. (1997) The international migration turnaround in Southern Europe, in King, R. and Black, R. (eds) *Southern Europe and the New Immigrations.* Brighton: Sussex Academic Press, pp. 1–25.

King, R. and Patterson, G. (1998) Diverse paths: the elderly British in Tuscany, *International Journal of Population Geography*, 4(2), pp. 157–82.

King, R., Warnes, A. and Williams, A. M. (1998) International retirement migration in Europe, *International Journal of Population Geography,* 4(2), pp. 91–111.

King, R., Warnes, A. and Williams, A. M. (2000) *Sunset Lives: British Retirement Migration to the Mediterranean.* Oxford: Berg.

Longhi, C. and Quere, M. (1993) Innovative networks and the technopolis phenomenon: the case of Sophia Antipolis, *Environment and Planning C: Government and Policy,* 11(3), pp. 317–30.

Marchena, M. (1987) *Territorio y Turismo en Andalucía.* Seville: Junta de Andalucía.

Mullan, C. (1992) *A Report on the Problems of the Elderly British Expatriate Community in Spain.* London: Help the Aged.

Myklebost, H. (1989) Migration of elderly Norwegians, *Norsk Geografisk Tijdsskrift*, 43(3), pp. 191–213.

OECD (1995) *National Accounts. Vol. 1 Main Aggregates.* Paris: OECD.

O'Reilly, P. (1995) Constructing and managing identities: 'residential tourists' or an expatriate community in Fuengirola, Southern Spain, *Essex Graduate Journal of Sociology*, 1, pp. 25–38.

Paniagua, A. (1991) Migración de noreuropeos retirados a España: el caso británico, *Revista Española de Geriatría y Gerontología,* 26(4), pp. 255–66.

Peck, F., Stone, I. and Esteban, M. (1996) Technology parks and regional development in the southern European periphery: the Andalucía case, *European Urban and Regional Studies*, 3(1), pp. 53–65.

Priestley, G. K. (1995) Sports tourism: the case of golf, in Ashworth, G. J. and Dietvorst, A.G. J. (eds) *Tourism and Spatial Transformations.* Wallingford: CAB International, pp. 205–23.

Rodríguez, V., Fernández-Mayoralas, G. and Rojo, F. (1998) European retirees on the Costa del Sol: a cross-national comparison, *International Journal of Population Geography,* 4(2), pp. 183–200.

Salvà Tomàs, P. A. (1994) Los nuevos flujos de inmigración extranjera en las Islas Baleares en la década de los noventa, in *Inmigración Extranjera y Planificación Demográfica en España.* La Laguna: Universidad de La Laguna, pp. 517–23.

Salvà Tomàs, P. A. and Binimelis, J. (1993) Las residencias secundarias en la isla de Mallorca: tipos y procesos de crecimiento, *Mediterranée*, 79(1–2), pp. 73–6.

Seguí, J. J. (1997) *El Mercat Inmobiliari a les Balears.* Majorca: Departament de Economía, Universitat de les Illes Balears, doctoral thesis, 2 vols.

Subdirección General de Planificación y Prospectiva Turística (1988) Inversiones extranjeras en inmuebles. Provincias de Alicante y Málaga, *Estudios Turísticos,* 99, pp. 45–112.

Walker, R. (1997) California rages: regional capitalism and the politics of renewal, in Lee, R. and Wills J. (eds) *Geographies of Economies.* London: Arnold, pp. 345–56.

Warnes, A. M. (1991) Migration to and seasonal residence in Spain of Northern European elderly people, *European Journal of Gerontology,* 1(1), pp. 53–60.

Warnes, A. M. and Patterson, G. (1998) British retirees in Malta: components of the cross-national relationship, *International Journal of Population Geography*, 4(2), pp. 113–33.

Williams, A. M. (1996) Mass tourism and international tour companies, in Barke, M., Towner, J. and Newton, M.T. (eds) *Tourism in Spain: Critical Issues.* Wallingford, CAB International, pp. 119–33.

Williams, A. M., King, R. and Warnes, A. M. (1997) A place in the sun: international retirement migration from Northern to Southern Europe, *European Urban and Regional Studies*, 4(2), pp. 115–34.

Williams, A. M. and Patterson, G. (1998) An empire lost but a province gained: a cohort analysis of British international retirement migration in the Algarve, *International Journal of Population Geography*, 4(2), pp. 135–55.

—————————— *11* ——————————

Mediterranean concentration and landscape: six cases

Elio Manzi

—————

Mediterranean landscapes and sustainability

The Mediterranean Sea is surrounded by densely populated countries endowed with an extraordinary range of natural-human landscapes and a cultural heritage that is unique in the world. The landscape itself – an ambiguous but irreplaceable concept – forms part of this heritage. This landscape appears at first sight to be the product of a particular physical environment or ecosystem (morphology, the sea, climate, and so on) but in fact it is heavily humanised. At once mythical setting and economic resource, it has been shaped by a complex and contradictory combination of processes. The concentration of people and settlements around the Mediterranean continues to accelerate, with effects that are threatening the very scenic and cultural resources that are the basis for tourism in the region. There can be no doubt that the changes that have taken place in the last 30 or more years exceed any parameter of sustainable development, with extreme environmental stresses inflicted upon the entire Mediterranean geosystem. Since landscape usually functions as a 'slow indicator' of social and environmental changes, it might seem difficult to interpret it as a sign of the human dimension of global environmental crisis. Yet surely there is no 'dimension' less human than landscape, the solid and visible setting for those meanings and values with which we invest what we see.

The research on which this chapter is based was made possible by a grant from MURST and CNR, Dipartimento Storico Geografico, University of Pavia, Italy, for research projects on 'Human Aspects of Global Environmental Change in the Mediterranean' and 'Global Change, Sustainable Development and Spatial Dynamics in the Coastal Regions of the Mediterranean'.

We need to go a little bit back in time if we wish to find out whether an authentic 'Mediterranean landscape' has ever existed. The landscape of tourism, especially mass tourism, tends to be made up of stereotypes, symbols and commonplaces; nevertheless, these clichés have origins more ancient than might commonly appear, as Allan Williams shows in **Chapter 9**. Does a 'Mediterranean landscape' exist, or are there, rather, different landscapes? When did the guide-book images of the Mediterranean come into focus? Four primary determinations constitute the basis of our inquiry. One is climatic and morphological, one is aesthetic, one a historical, cultural and archeological study of settlements (which also overlaps with the second category), and one the domain of mass-cultural stereotypes. This last domain includes, simplifies and trivialises the other three, transfiguring them into the fantastic components of an imaginary Mediterranean. At the end of the eighteenth century certain aesthetic canons came into force which regulated the perception of the Mediterranean according to a set of symbols and conventions. In the course of the nineteenth century these degenerated into stereotypes and commonplaces, which suffered further trivialisation throughout the twentieth century.

Between 1814 and 1824 William Henry Smyth, Rear Admiral of the Royal Navy, undertook an extensive survey of Sicily and the Mediterranean in the aftermath of the Napoleonic wars. Smyth later summarised his findings and impressions in a book (Smyth 1854) which is especially revealing of the author's expertise in matters geographical, hydrographical and cultural; besides his professions of sailor and cartographer, Smyth was an enthusiastic amateur archaeologist. More effectively than many more famous travellers and scholars, Smyth sums up the principal reasons for the renown of the Mediterranean and its coasts:

> The undertaking, though heavy, is nevertheless not wanting either in interest or importance: the Mediterranean Sea, so secondary in extent compared with others, being, *per se*, of vast surface, with many of its characteristics on the grandest scale. Besides, viewing it as the actual site where the intellectual culture to which we are most directly indebted was first developed, it cannot but be regarded for its portentous historical occurrences; nor will a sailor forget that it is the sea whereon the fleets of Carthage, Greece and Rome contended in former days, and those of Spain, France, Italy and England in later times. The grand object of travelling . . . is to see the shores of the Mediterranean. On those shores were the four great empires of the world: the Assyrian, the Persian, the Greek and the Roman. All our religion, almost all our arts, almost all that sets us above the savages, has come to us from the shores of the Mediterranean. It might appear strange that a coast of such paramount interest should still have required surveying in the present day. (Smyth 1854, pp.vi–vii).

The landscapes of the Mediterranean have been, as they are still, symbolic landscapes, interpreted in different ways according to different historical, social and cultural conditions. Nevertheless, certain ideas have remained constant, attaining the status of symbol or commonplace, according to their positive-aesthetic or negative-perceptual evaluation (Appleton 1996; Cosgrove 1984; Tuan 1989).

While many factors combine to determine the patterns of coastal settlement and population growth, especially in recent times, there can be no doubt that these symbolic landscapes, so universally renowned that their image has become more real than their territorial reality, have contributed much to the current congestion. In some cases, where the site has yielded particular benefits, the congestion is historically deep-rooted, stratified, chaotic and perhaps irremediable. In other cases, however, it is quite recent, deriving from a nostalgic (and delusive) need to recreate a lost world that once flourished on the Mediterranean. The quest for the 'lost sea' of Naples (lost because of pollution) on the coast of Calabria, to take just one example, has doomed the latter to being smothered in concrete. The Bay of Naples is one of the 'active symbolic systems that, over the last hundred years, has undergone an unlimited expansion – well beyond the borders of Italy – through the mass-cultural diffusion of visual models' (Fusco 1982, pp. 753–4). Of course, as Fusco goes on to point out, 'scenic or monumental *topoi* are generated from different conditions and motives'. The Bay of Naples belongs to the category of cases defined by ancient congestion and by the precocious, long-established fame of its landscape, now in decline. The Calabrian coast, in contrast, exemplifies the case of 'substitutive need' on a regional scale, attracting a national rather than an international tourism. Meanwhile, the case of the Côte d'Azur, the French Riviera of the first British *amateurs* in the nineteenth century, is linked to the recent development of the *fin-de-siècle* stereotype, and the wholesale destruction of the natural environment.

Three other cases present themselves and are chosen here for consideration: the central belt of the Spanish Mediterranean, between Valencia and Almería; the western shores of Sicily; and the Mediterranean coast of southern Turkey. In each of these regions, marvellous living remnants of the ancient landscape, authentically 'sustainable' in their relation to time and to the use of the land, alternate with zones devastated by speculative construction. Tens of thousands of holiday villas, built in a fake Mediterranean style, often encroaching right on the shore, bear witness to a human improvidence, driven by the desire for instant gain, which pays scant regard to any notion of sustainability –with consequences that are truly systemic, involving the future of the landscape itself. Figure 11.1 shows the location of the six coastal landscapes discussed in this chapter.

Coastal Mediterranean regionalisations

The north-west sector of the Mediterranean is enclosed by an arc of land, reaching from the Straits of Gibraltar to Sicily, which some French scholars have called the 'Latin Arc' (Cortesi 1995; Daviet 1994; Voiron Canicio 1994). This region is usually considered in terms of administrative territories, such as provinces (Italy) or *départements* (France), for reasons of comparative convenience and accessibility of statistical data, and despite some incongruencies. In fact we are dealing with a 3,500 km-long coastal stretch which forms 'a relatively heterogeneous unit. The size, economic importance and level of development of the regions which make up the coastline are very dissimilar' (Voiron Canicio 1994, p. 15). Studies of the region have almost always been geo-economic and geo-functional, and directly or indirectly determined by issues of territorial planning. Tourism, as might be expected, is one such issue. Despite the fact that the Mediterranean remains the most important

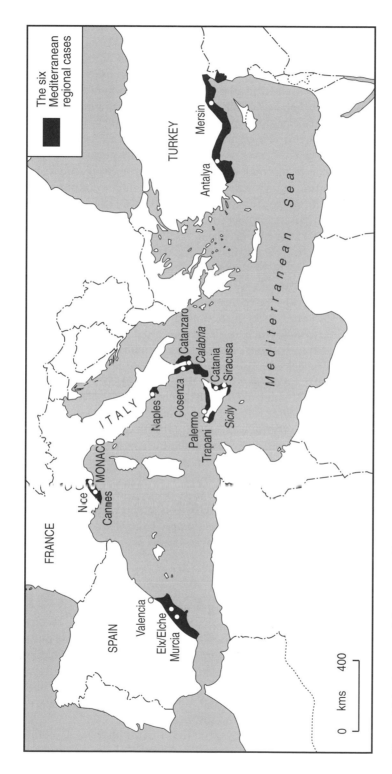

Figure 11.1 Location of the six coastal landscapes considered in the chapter

tourist region in the world, its landscape has been comparatively neglected. The prevailing tendency to view the area as a non-landscape has been reinforced by the questionable (if practical) decision to consider only those administrative regions directly bordering the sea, a decision in turn influenced by the Mediterranean Action Plan of 1976. In fact, the coastal provinces or departments are closely linked to the areas immediately inland, which are just as much gravitated towards the Mediterranean. It is the ancient and yet contemporary conflict of choices between, on the one hand, administrative regionalisations, within which statistical data are produced and local policy plans carried out and, on the other hand, regionalisations based on natural geographical, functional or urban-hierarchical criteria.

A different identification of regional space might take into account basic 'natural' elements: for example, drainage areas, which can be viewed as a type of ecosystem, in much the same way that geographers and others have long considered them as special regions (Vallega 1994). An excellent and little-known work on Southern Italy from the early nineteenth century by the engineer Carlo Afan de Rivera, director of the Administration of Bridges, Roads, Water, Forests, Hunting and Fishing in the Kingdom of the Two Sicilies, supports a regionalisation based on drainage areas (Manzi 1977). Nevertheless, the strong direct relation between the Mediterranean and the small rivers of the Italian peninsula is much diminished in the case of major water-courses such as the Rhône and the Nile. Would river basins appear more suited to the hypothesis of sustainable development? The regional scale for sustainable development can also comprise 'regional seas' (as opposed to the oceans, which can in turn be divided up for this purpose), such as the Caribbean, the Arabian Gulf and the Mediterranean. 'Sustainable development' is one of those notoriously ambiguous concepts we cannot do without, like a necessary utopia. Can we imagine, alongside 'sustainable development', the concept of a 'sustainable landscape', representing the perceptible index of such a development? 'Landscape' is likewise a concept that is ambiguous, polysemous, at times difficult to interpret and, for all that, charged with significance (Olson *et al.* 1988).

If landscape is the visible sign of a society, its projection onto the land, then the cases we have already mentioned (the Bay of Naples, Calabria and the Côte d'Azur) represent in large part an *unsustainable landscape*. Even if we consider landscape from the narrow perspective of the past, that is, simply as a panorama to be admired aesthetically, the conclusion is the same.

Landscape, planning and sustainability

The Mediterranean environment, although humanised over several millennia, is presently undergoing levels of stress, pressure and congestion which are unprecedented throughout its long territorial history. For all its conceptual ambiguity, a landscape reflects its present society; but the Mediterranean, besides that, bears a burden of memory, desire, a more or less unconscious myth transformed into popular reality. It is both stereotype and resource.

In 1993 Italy's official planning body, the Interministerial Committee for Economic Planning (CIPE), drew up, under Agenda 21, a 'National Plan for

Sustainable Development', which identified, among several priorities, the possible connection between its recommendations and tourism, since in Italy tourism has always been an important source of revenue. The plan calls for a 'sustainable tourism' which, rather than consuming the resources that constitute its own *raison d'être* as a social, cultural and economic phenomenon, will rediscover, enhance and conserve those resources for the enjoyment of future generations. Agenda 21 provided for a co-ordination among the regions responsible for administering tourism, and proposed several measures for achieving its objectives. These included the 'revision, in the narrow sense, of the present laws for the protection of the land-scape and the cultural heritage', and the 'promotion of a national campaign for the understanding and protection of the landscape, and incentives for tourist activities that will contribute to the conservation of natural, agrarian and historical-cultural landscapes' (CIPE 1994, pp. 93–4). In theory, landscape is awarded a primary role. Transforming theory into practice will be far more difficult, especially in densely populated coastal areas.

In France, territorial planning takes place via general directives which are applied through various administrative bodies, among them the Délégation à l'Aménagement du Territoire et à l'Action Régionale (DATAR). DATAR divides the nation into seven areas, among them the Mediterranean, which includes the regions of Languedoc-Roussillon and Provence-Alpes-Côte d'Azur. The latter is highly urbanised along the coast and is one of the regional hubs of the Mediterranean. Here the contradiction between the reality and the intentions expressed in the development and conservation plans of DATAR is starkly apparent: for example, the need 'to protect the richness and the fragility of the natural and historical environment, especially in the coastal and maritime zones' (Capineri *et al.* 1995, p. 84). The contrast is especially evident in the Alpes Maritimes, the *département* most famous for its tourist sites.

Naples, the French Riviera, the Calabrian Coast

Population density is a rather rough index of concentration for the larger adminis-trative regions. Nevertheless, at the provincial or departmental level (those of the Latin Arc) it permits us to make some comparisons and to highlight the significance of special areas of concentration.

In France, the density of the Alpes Maritimes department is 226 inhabitants per sq km, which is exceeded only by the Ile-de-France, by Lille in the old heavy-industrial North, by the Rhône (including Lyon), and by the Bouches-du-Rhône (including Marseille). The point of maximum concentration is the Principality of Monaco, which has a density of around 15,000 inhabitants per sq. km, one of the highest in the world, even taking into account the special circumstances of the prin-cipality (a small state with a surface area of less than 2 sq. km). The province of Naples, with more than 2,500 inhabitants per sq. km, has an even more striking density, since this figure refers to a territory with a surface area of 1,171 sq. km, 600 times that of Monaco. The comparison becomes clear if we consider only the municipality of Naples: 117 sq. km with around 9,000 inhabitants per sq. km, a truly unusual concentration for Europe, despite substantial recent population shifts toward outlying municipalities (in 1981 the density exceeded 10,000 inhabitants per sq. km). In several old districts in the centre the density, despite the considerable loss of population because of decentralisation, remains greater than that of Monaco (Cortesi *et al.* 1991).

In Calabria, the provincial population densities are not particularly high. The figures for the provinces lying along the Tyrrhenian Sea are: Cosenza, 113 inhabitants per sq. km; Catanzaro, 160; Vibo Valentia, 158; Reggio Calabria, 181. These are similar to those of the other Mediterranean regions, for example Spain. It must be noted, however, that the growth of the coastal population in Calabria is a relatively recent phenomenon, little more than a century old, like the Côte d'Azur, although occurring within a profoundly different context, as we shall see presently. The history of the Neapolitan concentration is much longer and practically uninterrupted.

The concentration of inhabitants along the coast of the French Riviera, from Menton to Saint-Tropez and, in particular, along the central arc from Nice to Cannes, received strong encouragement, beginning in the mid-nineteenth century, from the winter tourism of the élite, followed by summer tourism (including visiting the casinos) and, finally, attendance at congresses and cultural events. A considerable part of the former urban population did not live along the coast but inland, although not far from the coast, as we see in the case of Grasse and the small conurbation of Cannes (Cannes, Le Cannet, Mougins, Mandelieu-La Napoule).

Upper-class British tourism constituted a powerful propagandistic force, especially after the opening of the coastal railway, which arrived in Cannes in 1862, in Nice in 1864, and in Menton in 1869. With the railway the first hordes of tourists and day-trippers overwhelmed the upper-class British visitors, who had come there mainly by yacht. The term 'French Riviera' was replaced towards the end of the nineteenth century by 'Côte d'Azur', due partly to the pressures of French linguistic nationalism. Even today, a short distance from the Hôtel de Ville in Cannes, one can see the monumental fountain dedicated to Lord Brougham, who had the first holiday villa built on the pleasant hills overlooking the bay flanked by the Lerins Islands between 1835 and 1839. Already at the end of the nineteenth century the Côte d'Azur was characterised by a splendid (though in large part man-made) Mediterranean landscape; and the Principality of Monaco already stood out because of its greater concentration of buildings, even if this had little in common with the present skyscraper skyline.

In contrast, the coast of Calabria at the end of the nineteenth century was not very different from the way it was described by Giuseppe Maria Galanti, a scholar sent by the King of Naples to reconnoitre the area at the end of the eighteenth century, or by W.H. Smyth: for the most part wild, devoid of cities, accessible almost solely by sea. The railway arrived from Naples along the Tyrrhenian coast of Calabria only at the end of the nineteenth century, with the aim of linking the former capital city of the Mezzogiorno with the cities of Sicily, and it encouraged coastal settlement only after several decades (Kish 1953). In the meantime the natural environment of the Calabrian peninsula was rapidly deteriorating, especially due to the extensive deforestation which occurred after Italian unification, between 1880 and the turn of the century. Calabria was known to a small group of scholars for several natural-historical peculiarities, such as its geological structure and its climate. The Calabrian climate, although as yet little known to the general public, typified a Mediterranean which had already become associated with certain commonplaces (sun, sea, perennial summer). The impressive mountains rising near the coast, the winter snow cover, the abundant rains during autumn and winter along the Tyrrhenian slopes and (more heavily) on the Ionian side, contrasted with subtropical features in the valley bottoms and along certain coastal tracts; these elements aroused the curiosity of some scientists, but did not dispel the general ignorance about this isolated peninsula in the middle of the Mediterranean.

The particular location of Calabria, central within the Mediterranean Sea yet peripheral to the main arteries of European and Italian communication, which terminated at Naples, meant that coastal settlement remained sparse for a long period of history. This was also due in part to the historical relationship of the population with the sea: most Calabrians were mountain people, confined to the interior of the region. It might seem strange that a peninsula which extends so prominently into the Mediterranean should have such a limited relationship with the sea. Yet we observe a similar phenomenon in the mountainous islands of Corsica and Sardinia, whose sea-girt populations generally turn their backs on the sea. This mountain ecotype is, nevertheless, as typically 'Mediterranean' (see Chapter 12) as the more conspicuous maritime cultures of the Bay of Naples or the Côte d'Azur.

Until its recent human modification by linear settlements, the coastal landscape of Calabria was composed of mountainous spurs, overhanging the sea and covered at upper altitudes by luxuriant forests; of narrow plains, formed by brief and violent floods and by wide debris-choked river-beds (the so-called 'fiumare'); of long, lonely beaches with scattered settlements and poor fishing villages; of picturesque promontories and deserted sea coves. It called to mind until a few decades ago (the middle of the twentieth century) the landscape of a Mediterranean lost in time and untouched by tourism. This was the Calabria that attracted nobody but a few geographers and natural historians (such as Kanter 1930), and had the reputation of being a peninsula more isolated than Sicily which, although it was an island, possessed an important coastal urban network.

The Principality of Monaco, past and present

In the mid-nineteenth century the Principality of Monaco extended over a much larger territory than at present, including the territories of Menton and Roquebrune; the small area that remains today was the capital, almost entirely concentrated on the *Rocher*, the rocky promontory of Monaco-Ville. At that time the Principality was surrounded by the Kingdom of Sardinia, of which it was a protectorate. Its main resources, concentrated in Roquebrune and Menton, included the production of olive oil and the cultivation of citrus and other fruits. The palace of Carnolés, in Menton, was at that time the summer residence of the Grimaldi family, the rulers of Monaco. But in 1848 Menton and Roquebrune, which at that time had 4,900 inhabitants, rebelled against the authority of the Principality, proclaiming themselves free communes. By a unanimous vote the population opted for annexation to the bordering Kingdom of Sardinia; this was accepted at the end of the year, but was not ratified by the parliament in Turin. The main causes of the rebellion were the heavy tax on exports of olive oil, citrus fruits and essences, and the monopoly on flour and bread. After twelve years of provisional government Menton and Roquebrune accepted annexation to France, and were incorporated into the region of Nice, which the Kingdom of Sardinia had ceded to Napoleon III in exchange for help in the war of independence against Austria, which led to the unification of Italy (Rendu 1867). The French protectorate replaced the Sardinian one.

The loss of the two largest communes, depriving the Principality of almost all its resources, was a determining factor in the subsequent rapid urban expansion and transformation of the landscape. The loss of some of its traditional Mediterranean features foreshadowed the situation in recent years: a downtown consisting of a cluster of skyscrapers overlooking the sea, representing a massive concentration of capital as well as of people. The decisive transformation occurred during the reign of Charles III, with the creation of the *Societé des Bains de Mer* and the construction first of the famous casino in 1863 and then of the great hotels: the Hôtel de Paris in 1864 and, later on, the Hôtel Hermitage. All of this was favoured as well by the railway. The casino was the centre of activities meant to attract wealthy foreigners. The gardens of the Condamine, at the foot of the Rocher in front of the harbour, quickly became urbanised; while the farmland of the Spélugues was the site for the casino and, soon after, for another urban district which, in 1866, was called Monte-Carlo, a name which would become even more famous than Monaco.

> Expansion was so rapid that it exceeded all expectations and even spread to neighbouring regions; building specifications were not laid down early enough, nor planning regulations . . . It is also to be regretted that the *Société des Bains de Mer* did not create a wide area of lawns and flower gardens all around the Principality, beyond which very handsome buildings could have been sited, with wide avenues leading beyond them to working-class neighbourhoods. (Labande 1922)

The population increased rapidly from the 3,443 inhabitants of the 1868 census to 9,684 in 1888, subsequently reaching 23,418 inhabitants in 1921. The number of foreign visitors grew meanwhile from 158,831 in 1870 to no fewer than 1,767,983 in 1913. The natives of Monaco ceased to emigrate, and increasingly favourable laws regarding residency and citizenship helped bring back the population and encouraged immigration. In the meantime banking and tax laws tended more and more to favour the influx and transit of capital in the Principality.

However, the 'genetic mutation' of the landscape occurred only at the beginning of the 1960s: with almost all the territory built up, and with the elimination of the citrus and olive groves and subsequently of the gardens of the nineteenth-century villas, entire blocks of old buildings were demolished one after another. Skyscrapers rose up in their place, underground parking lots and service tunnels were excavated in the rock, land was reclaimed from the sea (such as at Fontvielle), at first for light industry and then to satisfy the demand for legal residency by individuals, financial corporations and other businesses. Land became an extremely rare commodity. It is easy to find analogies with the land reclamations in Hong Kong, or at Osaka and Kobe in the Japanese megalopolis, even if the underlying reasons were different. Thus was created a concentration of buildings symbolising the power of international capital, different from anything else along the Côte d'Azur. In this way Monaco, whatever its special circumstances, clearly anticipates the tendency towards dense construction on the Mediterranean littoral, especially the Franco-Ligurian coast, in vivid contrast to the sparser settlement in the hinterland. The French communes bordering on Monaco, such as Cap d'Ail, La Turbie, Roquebrune and, above all, Beausoleil, felt the effect, beginning in the 1920s, of the population trends of the Principality, and began to form a small and dense conurbation. Replacing the human and natural elements of the former landscape of the Principality – the Rocher, the rocks and the beaches, the olive and citrus groves, the villas – there appeared a highly dense urban network of high-rise buildings and connecting roads reaching to the underground parking lots, mingled with those few residual features of the Belle Epoque that were considered indispensable (at any rate for the time being). This North American-style downtown represents, whatever its landscape features, an actual and functional reality. The 'Americanisation' of the Côte d'Azur is also evident elsewhere, in street planning and decentralisation, which have led to a further concentration of productive and commercial activities along vast stretches of the coast which are easily accessible by automobile.[1] The first thing one notices, of course, is the intensive construction activity, some of it devoted to turning the large Mediterranean gardens of old and elegant villas, now demolished, into condominium residences.

The case of Calabria

The long Calabrian coast, extending for over 750 km, represents another situation, especially on the Tyrrhenian Sea. The small, scattered coastal settlements began to spread in the 1960s, with some of them coalescing into modest holiday resorts. This reflected an increase in individual mobility and the needs of the residents of the urbanised northern coast of Campania. The extension of the motorway from Salerno to Reggio Calabria and improvements in the ordinary roads along the Tyrrhenian have induced many inhabitants of Campania (as well as of Lazio) to move down to the Calabrian coast in search of that clean sea which cannot be found near their own place of residence, and of empty land to colonise with second homes. The Calabrian coast is now covered with tens of thousands of houses which are empty for 9–10 months of the year. According to the news media, the influence of organised crime is considerable and the enforcement of town-planning regulations scarce or non-existent. There is a similarity with the situation along the southern Campanian coast (Cilento), where, according to official census data, the number of coastal houses, most of them second homes, has greatly increased since 1951 (Viganoni 1988, 1995). This trend continued unabated into the 1990s, with an enormous destruction of coastal space. The Calabrian case has been summed up in the following words by Mura (1995, p. 26):

> In Calabria, on account of the form of the space, human pressure on the coastal zone
> is particularly perceptible and it has its most visible aspect in the urbanisation of the

coast. The urbanisation of the coastal belt has not been accompanied by adequate measures for the protection of the environment from pollution or for the protection of the coast from erosion and the effects of inland development. Every year, for the development of cities, ports and industries, 1–2 per cent of agricultural land is lost. In Calabria, the total loss in the last twenty years has been almost 14 per cent of the total regional surface!

Over much of its length the coastal landscape of Calabria is very far from being that 'sustainable landscape' which we imagined earlier. It is, rather, an 'incompatible landscape'. It represents a 'phantom concentration', containing macroscopic visual landscape features but lacking a real human presence, even though real, local, Calabrian coastal settlements have grown over the years.

The Bay of Naples: from myth to unsustainability

The panorama of the Bay of Naples is so famous, so standardised in the common perception, as to have become a worn-out stereotype, at least as regards some of its more celebrated features such as Vesuvius. In fact a volcanic morphology characterises much of the central coastal area of Campania, from the Campi Flegrei to Vesuvius itself. The enchanting landscape, which has fascinated so many foreigners and inspired countless paintings (as would, later, the French Mediterranean), is the result of an ancient, complex interaction between an exceptional natural endowment and a dense human presence which for the most part adapted itself to the natural setting. A combination of association of ideas and conceptual reductionism produced the stereotype, with its twin expressions of the picture-postcard and the search for local colour – meaning, usually, human congestion and poverty – well into the twentieth century. The classical view from the sea, still unsurpassed for its general view of the density of settlements and the overall landscape, has gradually been replaced by a view from the hills overlooking the bay, with Vesuvius in the background (Fusco 1982).

While there has always been a concentrated settlement around the Bay, this has spread from an historic core – the old city and some of its surrounding hills[2] – to cover the entire area with an uninterrupted density of buildings. Thus the classical Neapolitan landscape, which existed between reality and a stereotype made up of aesthetically pleasing elements, perceived above all as *natural*, has been greatly transformed in the last 100 years, and drastically so in the last 50 years. What the old tourist guide-books called the 'extraordinary beauty of the sights' was always in large part 'constructed'.

The deluge of cement that so disfigured the appearance of the Bay and worsened the quality of life (in any case not very high) during the 1950s and 1960s subsequently slowed down, although it did not cease altogether. The abatement was due to the increasing difficulty of finding land that could be built on apart from the coastal and inland peripheries to the north-east and south-east. The most recent concentration of buildings is the new service-sector complex of the Centro Direzionale, raised in the 1980s and accompanied by a thorough urban renewal of the old industrial district bordering it in the east of the city. This project, however, has created a downtown profile of skyscrapers which are conspicuous from the hills surrounding the city, with a strong impact on the landscape (Citarella 1988).

The hyper-concentration of people in Naples represents only the starkest symptom of a greater concentration over the central coastal area of the region of Campania, which has increased the disequilibrium between the inland portion of the region and the narrow central area near the sea. However, between 1971 and 1991, the increase in construction exceeded the increase in population threefold in the provinces of Naples and Caserta (bordering the province of Naples to the north-west); while in the province of Salerno (to the south-east) it has been five times as great. Construction has also increased in the central coastal towns of the Naples urban region, where the population has only slightly increased or, in some parts, diminished through processes of counterurbanisation (Emanuel *et al.* 1995). Automobile traffic is at a maximum: the province of Naples contains more than half of the vehicles in the whole of Campania. There are high levels of air and noise pollution, especially in the lower city and the coastal conurbation.

This 'deconcentration' is only apparent, however, since a more widespread concentration has taken place over a larger area at the expense of the remaining coastal agricultural land, which has by now been greatly reduced. The famous 'historical-natural' landscape can only be perceived on a small scale (as a general view, seen from above) or on a greatly enlarged one, in the sense of considering several 'islands' of the landscape: celebrated, conspicuous monuments, such as the Castel dell'Ovo by the sea at the centre of the coastal arc (Mautone 1994), or the Castello S. Elmo and Certosa di San Martino, a small preserve of 'historic green space' dominating the city of Naples. These are symbolic sites, prominent in all the paintings of views from the sea in the last few centuries.

Time passes, but deeply-rooted symbolic perceptions endure. The Bay of Naples is among the most popular sites in the false-colour satellite images from the 1980s. From the Landsat can clearly be seen the urban congestion that creeps upwards even across the slopes of Vesuvius, an area of grave environmental risk. An excellent example is provided by the plate of the Bay of Naples in the volume edited by Charles Sheffield entitled *Man on Earth* (Sheffield 1983). Despite the visual evidence of excessive urban concentration, the accompanying description emphasises the area's traditional, natural and historical-archaeological features: the volcanoes, Pompei and Herculaneum. In any event, it is significant that the plate appears in the 'Commerce and Trade' section, together with Chicago, San Diego, New York, Sydney, the English Channel and South Wales, Hamburg, Cape Town, Montreal, Rio de Janeiro and Vancouver. None of those cities or regions possesses the ancient geographical history of Naples: in 1995 UNESCO recognised the Greco-Roman area of the city's historical centre as one of the world's heritage sites.

Sicily, Spain, Turkey: a comparison

The analogies between Spain and Sicily are sufficiently clear to anyone familiar with the historical-geographical forces that have shaped the lands around the Mediterranean for centuries. Such analogies range from empty cliché through statements of the obvious to a somewhat more sophisticated reflection upon the formative dynamics of place.

One must always be on one's guard against stereotypes. Stereotype and the empirical facts of landscape are obsessively combined in our present culture of consumerism and the instant image, and consequently in the representations of organised tourism, journalism, and (in part) the school curriculum. Certain images define Sicily and Spain: a hot, sunny climate, with perpetually blue skies; the strong influence of Moorish or Arabic culture; a social and economic system bound to an obsolete folklore; and so on. White houses with pseudo-Mediterranean décor, all the effects of bad taste and building speculation, fit the picture all too well. It is a picture that often ignores the closeness of mountain and tableland to the coastal plain, and the importance of the historical interrelations among these features, at once the cause and effect of settlement. Without their specific morphology, Sicily and central-southern Spain would be far more arid than they are in fact. The image of vast deserts may entertain the shallow imagination of those unaware that the Mediterranean is a sea surrounded by mountains and highlands, where snow, and especially rain, can fall for six months of the year; the precipitation flows down to

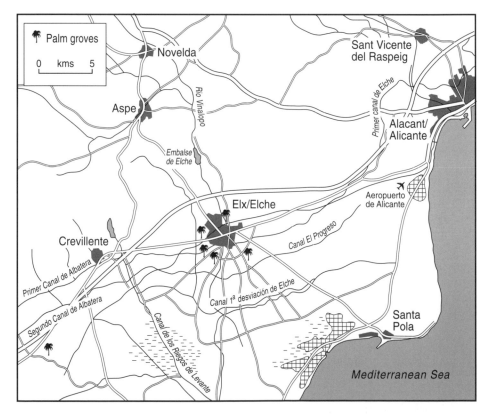

Figure 11.2   Alicante, Elche and surroundings in south-east Spain

the valleys and the sea, allowing life to flourish for thousands of years in a land where rain is scarce.

Such considerations might seem to refer to the past, to a remote Mediterranean of faded photographs from the turn of the last century. In fact they are entirely contemporary, because the sites of settlement and human activity are largely the same; except that the settlements are not adapted to their situation, but to the needs of visitors from distant metropolitan centres. All that has changed is the scale, which has dwindled in relation to these greater distances.

The hilly coast between Alicante and Almería is very dry. The aridity index here is the highest in the whole of the Iberian peninsula and one of the highest in Europe (Manzi 1991, p. 315), appreciably exceeding the driest parts of the Sicilian coast, such as Punta Granitola, the Gela plain, the headland of Capo Passero and all of the Plain of Catania (Manzi and Ruggiero 1973, pp. 21–6). This has profoundly influenced the natural landscape: the coast between Almería and Alicante exhibits strongly steppe-like features, such as alfalfa bushes dried by the hot winds that blow periodically out of Africa (cf. the Sicilian *scirocco*), and the difficulty of cultivating any crop without artificial irrigation. This was the region where certain film companies, in the 1960s and 1970s, chose to direct low-budget Westerns, including some of the Italian 'Spaghetti-Westerns', which went on to enjoy considerable box-office success and became cult films; the arid hills and gullies of Spain passed plausibly enough for the high plains of northern Mexico and the deserts of the American South-West (Collin-Delavaud 1987).

Tourists who come here, whether in transit or as residents in a holiday village or cottage, do not perhaps realise how exceptional this landscape is, since it corresponds to the most common stereotype of an arid Mediterranean coast. The region (figure 11.2) contains a jewel of the ancient Mediterranean landscape, representing an original example of environmental sustainability: the marvellous palm groves of Elche (Elx in Catalan). This former Roman colony of Iulia Ilici Augusta

> is famous for its palms, the only kind in Europe to bear dates. The trees, up to 30 metres high, provide shelter for lower trees (pomegranates) and the cultivation of cereals. These are not confined to a single grove, but are found in many walled gardens (*huertas*). Planted by the sides of small canals, the palms are irrigated with water drawn from a nearby artificial lake (*pantano*). The whitewashed houses, with their roof terraces and narrow shuttered windows, also exhibit an African appearance. (Migliorini 1964, p. 85)

This dated but evocative description, translated from Italian, adapted in turn by the geographer Elio Migliorini from earlier accounts of the 1920s and 1930s, has the air of a vanished epoch. The description of a 'Promenade à Alicante et à Elche' by Mme Marthe Mallié, published in the travel magazine *Le Tour du Monde* in 1892, takes us back to a still more remote time and space:

> There are nearly one hundred thousand palm trees, of which thirty-five thousand produce fruit . . . at their feet may be grown alfalfa, cotton, hemp and especially pomegranates. The pomegranates grown in Elche are the most highly prized in Spain . . . Their sale, and that of dates, make up the principal wealth of the area.

Today the magical landscape of Elche still survives, although, like other cele-brated Mediterranean landscapes, much of it has suffered dismemberment and burial.

The *Palmeral* of Elche consisted of more than 100,000 palms (even more, according to other authorities, although we may be dealing with traditional infor-mation transmitted through inertia), with many female trees bearing the famous dates that are still sold to tourists. The connoisseur of Mediterranean landscapes who visits Elche today must be prepared for some bitter reflections, along with a few pleasant surprises. The bitterness comes from the evidence of the destruction of many of the palm groves, turned into wastelands of speculative construction, or else, with the break-up of the old irrigation network by roads and buildings, into the 'provisional' but enduring landscape typical of the unsustainable landscapes of our time. The pleasant surprise comes from what remains, which is more than enough to give one an idea of the ancient splendour: the comprehensive extent of the palm grove, a huge oasis in the surrounding steppe; the large areas that are still intact, the *huertas* of palms and other kinds of vegetation, the appearance of the communal gardens reproducing the grove in a form suited to recreational needs, and lastly, 'El Huerto del Cura'. A thousand palm trees grow in this prestigious botanical garden, distinguished as a 'Jardín Artístico Nacional', among them the famous Palmera Imperial, at least 150 years old. Many other exotic and beautiful plants, imported by the Spaniards from the Americas, also flourish in the Huerto del Cura.

The palm grove of Elche comprises a precious 'landscape-park', unique in Europe, of immense environmental and historical interest. Nevertheless it is not one of the 'ecosystems' dear to naturalists, in so far as it has been modified by humans – indeed, it was planted by humans and modified across the centuries, with a city in the middle of it. Today the grove represents not so much the ancient Arab-Islamic culture, as might easily be supposed, as the still more ancient Mediterranean of the Greeks and the Phoenicians, when the landscape formed part of a recognisable cultural unity between the two shores, taking into account the obvious local differ-ences of climate. Limited areas of this sustainable landscape are currently protected, and it is to be hoped that protection will be extended over most of the region (Boira i Maiques 1995). A comparison between the 1975 edition of the *Mapa Militar de España* of the Servicio Geográfico del Ejército (1:100,000) and earlier maps in the series confirms the necessity for protective measures to be taken on behalf of the palm grove, as it shows a diminution in the area of rural dwellings within the network of irrigation, forming a semi-circle from west to south-east of the urban centre. This area marks the presence of the live palm trees (not registered as such on the map), which remain agriculturally useful, bearing their winter date harvest. Another map, the *Mapa Provincial de España* 1:200,000, published by the Instituto Geográfico y Catastral, 1976, shows more synthetically and clearly the principal irrigation channels, from which branch off the smaller canals: the *Primer Canal de Elche*, the *Desviaciones 1° y 2° de Elche* from the two canals of Albatera. These form half-moon arcs south of the inhabited zone, although the smaller network of irri-gation extends for roughly 1 km north of the inhabited zone, and the palms with it. While the military map brings the contour lines, on that of the Geographical Institute the relief is represented with only the main contour lines and artistic shading, allowing the reader to appreciate the situation of the landscape of Elche, on a plain lightly elevated towards the north, almost at the base of the *sierras* which herald the morphological gradations of the central tableland, the Meseta. This characteristic position is not exclusive to the Spanish Mediterranean: it recurs in Anatolia and Sicily.

The impact of tourism on much of the Turkish Mediterranean coast (figure 11.3) has been recent and violent, resembling in some respects the 'descent to the sea' of settlement in Calabria. Like this, it is in part 'fake', with ugly houses standing empty for most of the year and many provisional, unfinished and partly-constructed buildings. The German geographer Udo Sprengel, who has studied coastal settlement around Antalya, Avsallar-Incekum and Mersin, describes the hasty and often illegal construction and the ambiguous intensification of a service economy devoted to the new residents and visitors:

> Everywhere completely new urban centres are developing, which not only reproduce the well-known mixture of hotels, seasonal hotel-residences, and tourist markets, but are increasingly taking over the functions of supplying the rapidly increasing, economically and socially heterogeneous population of Turkish immigrants. In its geographical, socioeconomic and infrastructural aspects the situation more and more resembles the mining camps and boom-towns of the nineteenth-century American West. In the early 1990s the two southern provinces most involved in the tourist boom, Antalya and Ilce (Mersin), ranked fifth and eighth out of seventy-three Turkish provinces in the volume of new construction, up to 95 per cent of which was devoted to detached houses, apartment blocks and commercial buildings. (Sprengel 1999, p. 166)

Like the Iberian Peninsula, Sicily possesses a central plateau and an urban network developed around its periphery. The situation is more emphatic in that the island does not have a dominant central core like Madrid; all the major towns are

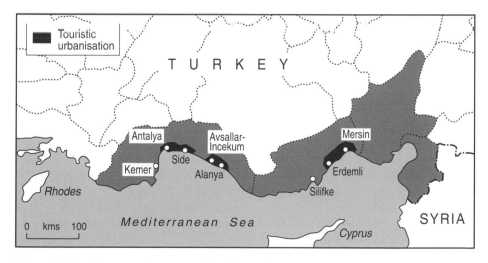

Figure 11.3 Touristic urbanisation along the south coast of Turkey

found along the coast, a phenomenon linked not only to Sicily's island status but to its central position in the Mediterranean and its historical-geographical relations, variable but intense, with the sea's western, northern and eastern quarters. Arab-Islamic expansion absorbed the island for some 250 years, but subsequent geopolitical changes made the relationship with Africa a difficult one, leading to the relative stagnation of the urban settlements along the southern coast. Speculative building began to invest the shores of Sicily in the 1960s, reaching a devastating scale in the 1970s and early 1980s, affecting previously remote areas as well as districts near the major cities. Three cases can be considered here: the coast between Mondello and Scopello, along the Gulf of Castellammare west of Palermo; the stretch of coast between Selinunte and Sciacca in the south-west; and the stretch between Gela and Capo Passero, in the south-east.

In the first of these cases, the motorway between Palermo and Punta Raisi has provided an axis for new construction, speculative and otherwise, filling up the space between the city's western suburbs and the airport almost without a gap. Viewed from the hills behind it, the plain of Carini resembles a city where no city exists; the plain is covered with hundreds of condominiums and single-family houses, many of them second homes owned by residents of Palermo. From an administrative point of view those buildings occupy various municipalities' territories, reaching right down to the sea. Illegal construction has been extremely common, not least of all due to powerful Mafia interests (Robb 1998). Recent legislation by the Italian state has brought much of this construction within the law in exchange for payment of taxes; many landlords, while waiting for legal normalisation to take effect, have not hesitated to extend construction here and there.

An analogous phenomenon presents itself along the Gulf of Castellammare, at one time a splendid Mediterranean riviera with fishing, agriculture, and scattered settlements owned by proprietors from Palermo, Alcamo and other nearby centres. Problems with the water supply, with the removal of refuse in summer and with marine pollution have for a long time signalled the environmental unsustainability of a settlement system subject to a rapid increase of density. Recent years have yielded some evidence of improvement in the situation, perhaps due to the legalisation of much of the new construction and more decisive action on the part of the local communal administrations (Rao 1992).

Individual cottages are more numerous on the south-west coast, where they sometimes form entire residential centres, for example around Porto Palo. These centres tend to be insufficiently served by commercial businesses, partly because of the proximity of older, well-supplied urban centres such as Sciacca. In this region too unregulated construction has spoilt some once-enchanting sites: speculative building has not spared the mouths of the Belice and Platani rivers, close to the ancient Greek cities of Selinunte and Eraclea Minoa.

On the south-east coast, however, the *marine* specially developed over the last 20–30 years represent a more complex phenomenon. These are villages designed for organised vacations, or other settlements made up of rows of modest individual cottages, or bathing establishments converted from the old fishing hamlets that were once the coastal filiations of the historic urban network of the southern section of this part of Sicily: Vittoria, Comiso, Ragusa and Modica. This stretch of the coast retains much of its charm, despite a certain amount of tampering, and provides some foundation for hope that the trend towards unsustainability might be arrested and reversed.

Conclusion

None of the landscapes we have examined appears really 'sustainable', and for different reasons. Monaco, with its unique concentration of capital manifested in the concentration of buildings; the Calabrian coast, with its concentration of second houses and its late and unbalanced tendency toward coastal settlement, resembling certain stretches of the Spanish coast, or southern Portugal or Turkey; and Naples, a symbol of the drastic acceleration of the 'human dimension of global environmental change' in the Mediterranean. Yet this ancient city, with the degradation of its famous landscape, has recently exhibited some positive signs, pointing in the difficult direction of sustainable development and a sustainable landscape. The large iron and steel foundries at Bagnoli, a small coastal plain at the base of the promontory of Posillipo, closed for good in 1994. Built at the beginning of the twentieth century, the industrial complex had gradually transformed the appearance of an area famous as one of the most pleasant sites along the coast of the Campi Flegrei. Current reclamation and restoration projects, aimed at turning the area over to service industries and tourism, may be sending a faint but nonetheless important message that, if we want to recreate a 'sustainable landscape', it is not enough to dream of a Mediterranean of the past, such as that of Captain W.H. Smyth of the Royal Navy.

Notes
1 For Monaco, the term Hong Kong-isation is perhaps more appropriate.
2 See the wonderful large-scale (1:5,800) map of Naples drawn up in 1775 by Giovanni Carafa, Duke of Noja.

References
Appleton, J. (1996) *The Experience of Landscape.* Chichester: Wiley.
Boira i Maiques, J. V. (1995) Gli strumenti della pianificazione ambientale e territoriale in Spagna con particolare riferimento alla protezione ambientale del paesaggio mediterraneo: aspetti fisici e umani, in Brandis, P. and Scanu, G. (eds) *La Sardegna nel Mondo Mediterraneo. Vol. 8: I parchi e le aree protette.* Bologna: Pàtron, pp. 61–75.
Capineri, C., Lazzeroni, M. and Spinelli, G. (1995) Mediterranean France: in search of a balance, in Cortesi, G. (ed.) *Urban Change and the Environment: the Case of the North-Western Mediterranean.* Milan: Guerini, Geo & Clio Series, Vol. 3, pp. 65–92.
Citarella, F. (1988) Centro direzionale e struttura urbana a Napoli, in Celant, A. and Federici, P.R. (eds) *Atti del XXIX Congresso Geografico Italiano, Geothema 3: Nuova città, nuova campagna.* Bologna: Pàtron, pp. 95–105.
Collin-Delavaud, C. (1987) Paysages, photographie et cinéma, *Hérodote,* 44, pp. 94–105.
CIPE – Comitato Interministeriale per la Programmazione Economica (1994) Piano nazionale per lo sviluppo sostenibile in attuazione dell'Agenda XXI, *Gazzetta Ufficiale della Repubblica Italiana,* 26 February.
Cortesi, G. (1995) Introduction: identity and pattern of urban change in the North-Western

Mediterranean, in Cortesi, G. (ed.) *Urban Change and the Environment: the Case of the North-Western Mediterranean.* Milan: Guerini, Geo & Clio Series Vol. 3, pp. 15–35.

Cortesi, G., Marengo, M. and Spinelli, G. (1991) Il nuovo assetto demografico delle aree metropolitane di Milano e Napoli, in Petsimeris, P. (ed.) *Le Trasformazioni Sociali dello Spazio Urbano. Verso una nuova geografia della città europea.* Bologna: Pàtron, pp. 129–47.

Cosgrove, D. (1984) *Social Formation and Symbolic Landscape.* Beckenham: Croon Helm.

Daviet, S. (1994) L'arc latin, histoire et problématiques d'un concept, *Méditerranée,* 79(1), pp. 3–6.

Emanuel, C., Frallicciardi, A. M. and Sbordone, L. (1995) Campania: a region around Naples, in Cortesi, G. (ed.) *Urban Change and the Environment: the Case of the North-Western Mediterranean.* Milan: Guerini, Geo & Clio Series, Vol. 3, pp. 199–229.

Fusco, A. M. (1982) Il 'luogo comune' paesaggistico nelle immagini di massa, in De Seta, C. (ed.) *Storia d'Italia. Vol. 5, Il Paesaggio.* Turin: Einaudi, pp. 751–801.

Kanter, H. (1930) *Kalabrien.* Hamburg: De Gruyter.

Kish, G. (1953) The 'marine' of Calabria, *Geographical Review,* 43(3), pp. 495–505.

Labande, L. H. (1922) *Histoire de la Principauté de Monaco.* Monaco: Imprimerie Nationale.

Mallié, M. (1892) Promenades à Alicante et à Elche, *Le Tour du Monde,* 64, pp. 209–24.

Manzi, E. (1977) I problemi del Mezzogiorno nel pensiero di Carlo Afan de Rivera, *Rivista Geografica Italiana,* 84(1), pp. 23–72.

Manzi, E. (1991) L'Europa del Sud, in Manzi, E., Melelli, A. and Persi, P. *L'Europa Occidentale.* Turin: UTET, pp. 302–403.

Manzi, E. and Ruggiero, V. (1973) *I Laghi Artificiali della Sicilia.* Naples, Memorie di Geografia Economica e Antropica, Vol. 8.

Mautone, M. (1994) Spazio vissuto e been culturale: Castel dell'Ovo, una emergenza ritrovata, in Caldo, C. and Guarrasi, V. (eds) *Beni Culturali e Geografia.* Bologna: Pàtron, pp. 113–33.

Migliorini, E. (1964) *Profilo Geografico dell'Europa Occidentale.* Naples: Libreria Scientifica Editrice.

Mura, P. M. (1995) At the bottom of the arc: the Calabrian region, in Cortesi, G. (ed.) *Urban Change and the Environment: the Case of the North-Western Mediterranean.* Milan: Guerini, Geo & Clio Series, Vol. 3, pp. 231–67.

Olson, J., Nakaba, K. and Sutton, B., eds (1988) *Sustainable Landscapes.* Pomona, CA: California State Polytechnic University.

Rao, S. (1992) Dimore secondarie nell'Europa mediterranea: il caso della Sicilia, in Manzi, E. (ed.) *Regioni e Regionalizzazioni d'Europa: oltre il 1993.* Pavia: Università di Pavia, Dipartimento Storico Geografico, pp. 237–43.

Rendu, A. (1867) *Menton & Monaco (Alpes Maritimes). Histoire et description de ce pays suivies de la climatologie de Menton.* Menton: Amarante; Paris: Lacroix, Verboekhoven.

Robb, P. (1998) *Midnight in Sicily.* London: Harvill Press.

Sheffield, C. (1983) *Man on Earth: How Civilization and Technology Changed the Face of the World.* New York: Macmillan.

Smyth, W. H. (1854) *The Mediterranean. A Memoir Physical, Historical and Nautical.* London: J. W. Parker.

Sprengel, U. (1999) Luck and bane of tourism: the 'Turkish Riviera' between economic valorization and destruction of landscape, in Manzi, E. and Schmidt di Friedberg, M. (eds) *Landscape and Sustainability. Global Change and Mediterranean Historic Centres: from Rediscovery to Exploitation.* Milan: Guerini, Geo & Clio Series, Vol. 4, pp. 165–75.

Tuan, Y. F. (1989) Surface phenomena and aesthetic experience, *Annals of the Association of American Geographers,* 79(2), pp. 233–41.

Vallega, A. (1994) La scala geopolitica regionale dello sviluppo sostenibile, in Citarella, F. (ed.) *Studi Geografici in Onore di Domenico Ruocco.* Naples: Loffredo, Vol. 2, pp. 583–98.

Viganoni, L. (1988) Il rapporto turismo-ambiente nella sezione settentrionale della costa calabra, in Celant, A. and Federici, P. R. (eds) *Atti del XXIX Congresso Geografico Italiano, Geothema 3: Nuova città, nuova campagna.* Bologna: Pàtron,. pp. 583–90.

Viganoni, L. (1995) Insediamento costiero nel Mezzogiorno: un prolungato spreco del territorio, in Federici, P. R. and Zunica, M. (eds) *Lo Spazio Costiero Italiano. Problemi di crescita, sensibilità ambientale.* Florence: Società di Studi Geografici, pp. 27–45.

Voiron Canicio, C. (1994) A la recherche d'un arc méditerranéen, *Méditerranée,* 79(1), pp. 15–23.

12

Mountain communities and environments of the Mediterranean Basin

Don Funnell and Romola Parish

Mountains form an essential ingredient of the Mediterranean landscape. According to McNeill (1992a, p. 12), 'Except for the desert between Tunisia and Sinai one is almost never out of sight of the mountains if within sight of the Mediterranean coast.' Similarly Braudel (1972, p. 25) remarked, 'It is above all a sea ringed by mountains. This outstanding fact and its many consequences have received too little attention in the past from historians.'

The extensive distribution of mountains surrounding the Mediterranean is shown in figure 12.1. The main ranges include the Moroccan Rif and the Algerian Tell to the south; in the east, the mountains of the Lebanon, the Taurus in Turkey, the Pindus in Greece and the Dinaric Alps; in the north, the Italian Apennines and the Maritime Alps of both France and Italy; and in the west, the Pyrenees and the highlands of the Spanish Sierra, including the Sierra Nevada. Within these mountain ranges, the most notable peaks are Toubkal (High Atlas) 4,165 metres, Mount Corno (the Italian Apennines) 2,912 metres, peaks in the Lebanon reaching 3,088 metres and, in the Sierra Nevada, Mulhacen at 3,481 metres. These mountain ranges have certain broad similarities, particularly in the economic traditions they supported in the past, although more recent history reveals significant disparities. The mountains are the wettest parts of this essentially semi-arid region, and thus hold certain advantages for settlement and agriculture over the lowlands. However, apart from high relative relief and usually steep slopes, all these ranges have distinctive physical characteristics which include their climate, soils and relief. Other factors such as seismicity are also important in the triggering of large disturbances at infrequent intervals.

Mediterranean climates are characterised by hot dry summers and warm wet winters. Rainfall is therefore concentrated in a few months of the year, and in the summer occurs as intense downpours onto sun-dried, bare soil surfaces. Precipitation falls as snow on the mountain peaks, and this forms an important source of summer water for irrigation agriculture. In more recent times, snow has

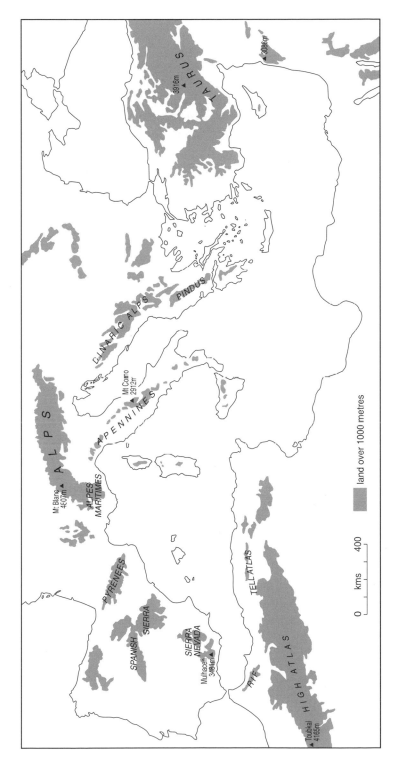

Figure 12.1 Main mountain ranges of the Mediterranean Basin

also provided the basis for lucrative tourist development. The disparity in rainfall totals between highland and lowland increases inland and to the south of the Mediterranean region. As a result of the climate, vegetation growth is highly seasonal, both on agricultural land and on pastures. This seasonality is intricately integrated into the traditional agro-pastoral activities on mountain zones, which benefit from a wide range of ecological niches which can be exploited by diverse activities. Altitudinal zonation of land use and vegetation is largely determined by climate, and local variations are significant and important in determining potential activities.

Much has been written about individual ranges within the Mediterranean Basin but there are few data that give a clear overall picture of the socio-economic status of the communities within them. The seminal work of McNeill (1992a) provides an important starting point, but his account focuses only on the historical evolution of five mountain zones[1] and pays very limited attention to contemporary policy.

Why are mountains important?

Within the Mediterranean Basin, mountains form an important element of the natural resource base of many countries; they are home to communities which traditionally managed these natural resources and form a distinctive component of the cultural identity of many people. In addition, the forests and water resources of the mountain slopes are of considerable significance for the populations which live in adjacent lowlands. Consequently, the appropriate management of the mountains plays a vital role in the economic development of the lowlands as well as directly contributing to the control of hazards associated with flooding and landslides. At various times mountains have provided the most favoured landscape for settlement, for example in seventeenth-century Italy, when the lowlands were beset with drought, disease and civil strife.

Although it is a mistake to assume that these mountain communities were totally isolated, it has been the development and spread of capitalism that has gradually altered the relationship between mountains and lowlands. Improved economic opportunities in the lowlands have meant that almost all mountain areas surrounding the Mediterranean have now experienced some outflows of population, but the scale and nature of this process are highly variable. As economic development proceeds, so the mountain areas became increasingly incorporated into a commercial/capitalist economy. In the hotter, drier parts of the Mediterranean to the south and east, mountains have provided the best environment for extensive cereal cultivation, livestock production, both for food and raw materials like hides and skins, and arboriculture. Alongside land clearance for settlement and cultivation, mining and forestry activities developed, which together have been held guilty of causing excessive denudation of many watersheds.

Today, there is a strong revival of interest in mountain areas of the Mediterranean because of their importance in the biosphere and also as a location of tourist activity. This is particularly marked in those countries on the north shore which are now members of the EU. In the last decade there has been a notable increase in 'mountain-based' initiatives which will be examined below. To the south and east, in the Taurus, the Dinaric Alps, Lebanon and North Africa, similar developments have occurred but with much less significance in relation to other over-riding problems of their respective economies.

Much of the work on Mediterranean mountains uses the concept of 'carrying capacity' well-known by geographers and ecological anthropologists. McNeill (1992a) uses this concept as an underlying thread through his discussion of five mountain areas. He argues that mountain populations in the Mediterranean Basin have risen to levels which 'overshoot' the environmental resources of these areas and that this has been followed by a collapse and depopulation as a combination of 'push' and 'pull' factors drove people away. By World War II this process was largely complete in the Alpes Maritimes; during the early postwar decades it spread through Italy and, more recently, it has strongly affected the mountain regions of Spain, Greece and the former Yugoslavia. It is yet to emerge as a determinate feature of the Moroccan Rif. It is possible to argue that 'carrying capacity' can be viewed dynamically so that changes in crops, techniques or institutional management may set 'new limits' to population growth. However McNeill's main thesis is that, in the Mediterranean, it is the environmental decline through soil erosion and deforestation, itself a function of population growth, which has effectively led carrying capacity to spiral downwards.

Whilst this idea is intuitively attractive, it carries with it the conceptual baggage of 'determinism' and, as such, is open to serious question. It assumes that there exists, and that we can usefully detect, limits to the exploitation of a given resource, beyond which the resource will deteriorate and disappear. However, as many writers have suggested (Dhondt 1988; Street 1969), the concept has many weaknesses, being based upon an argument of comparative statics, so that with a given resource, exploited in a given way for a specific period of time, we will be able to determine an equilibrium rate of exploitation, namely the carrying capacity. In addition there is a danger that, in the present day, policy-makers seek to determine the 'carrying capacity' of an area in order to use it as a template for 'sustainable' development policies. Such an approach, although it has merits of simplicity, does not work. Conceptually, we now recognise that society–environment interactions are complex and often indeterminate (Thompson 1997). Constant reflexive adjustments are made as human experience and perceptions change and with them environmental interactions. In order to explore this issue it is necessary to dissect the four main processes most writers claim explain the present state of mountain areas, namely deforestation, fire damage, degradation and depopulation.

Deforestation

Much of the Mediterranean has been stripped of its forest cover in historical times, and the reconstruction of the original species composition has concentrated on the analysis of pollen cores and of remnants of original forest remaining around sacred sites. The importance of timber in the classical and historical world for construction, charcoal and ship-building has meant that many documentary sources give detailed accounts of the pattern of deforestation, forest management and species. Timber stands were a much sought-after, fought-over strategic resource, as attested by Braudel (1972), Meiggs (1982), Thirlgood (1981) and others. Deforestation is considered a vital link between human activity and the physical environment.

At the end of the Holocene, much of the Mediterranean was wooded. Mountains represented refugia for tree species whilst the lowlands were dry steppe. As the climate became warmer and wetter, trees spread to the lowlands. Clark (1996) notes two periods of deforestation of the mountain areas. The first can be dated to the period between 200AD and 600AD, the second, more recently, in the twentieth century, especially since World War II. Regional differences in species composition occurred; in the European Mediterranean pine and birch woodlands were succeeded by oak woodlands, including evergreen oak, cork oak, ash and hazel. The effects of human activity such as clearance and grazing resulted in the replacement of oak forests with a *maquis* vegetation of low shrubs including heath, buckthorn, myrtle and juniper. In areas such as Turkey and Corsica the slopes have had a *maquis* vegetation for at least 500 years. In other areas this has been further degraded into *garrigue*, a mixture of aromatic shrubs which remains on many uplands.

After initial clearance, grazing, particularly by goats, has prevented successful regeneration of forest species, and subsequent removal of soil in exposed areas prevents this for ever. However, not all deforestation gives rise to higher rates of soil erosion; where clearance has been followed by terracing soil erosion is arrested. In addition, some areas are better protected by dense low-growing shrubs than by open forest. Clark (1996) and Naveh (1982), for instance, point out that fire may increase nutrient levels in the soil by at least 50 per cent and the interaction between fire, grazing and natural regeneration is an important element in watershed protection. Throughout history there has been a constant fluctuation in forest cover as human populations have increased and abandoned the mountains, as farming practices have changed and as plantations developed or were abandoned. Thornes (1996) notes that in some areas of abandoned grazing the regeneration of the land with scrub and woodland can be observed.

Fire

Fire is an increasingly serious hazard in Mediterranean regions. In 1998 for instance, parts of upland Greece experienced particularly serious forest fires resulting in loss of life. According to Vélez (1990) economic losses from forest fires in the period 1982–85 totalled just over $1 million in Italy, and $2 million in Spain. Between 1988 and 1997, the states of southern Europe experienced a total of 570,000 forest fires covering an area exceeding 4.4 million hectares.[2] Many of the incidents occurred in highland areas, where fire-fighting appliances have difficulties of access. The number of fires has been increasing gradually, although the area of each incident has decreased. These forest fires are not only a threat to the ecological balance, they are highly dangerous to people and property in the affected zones.

Fire has been used as an essential tool of agricultural management at least since the Carthaginians dominated the Mediterranean in the sixth century BC, and the Romans had strict legislation providing penalties for unauthorised burning. It was the principal mechanism for clearing land for settlement and agriculture. In addition, for centuries, shepherds in almost all mountains around the Mediterranean have continued to use fire to stimulate the growth of new vegetation to feed their flocks.

However, it has become a recognised hazard for three main reasons. In the first instance, as noted above, it is a direct danger to settlement and other human activities. Secondly, it damages important economic resources such as commercial plantations and, finally, it is a part of the deforestation and denudation process whereby the removal of vegetation is an important element in aggravating soil loss.

The particular character of the region's climate exacerbates the fire hazard. The long, dry, hot summers can reduce the moisture content of leaf litter to less than 5 per cent, so that a tiny spark, match or cigarette can initiate a conflagration. In addition strong, dry winds such as the Mistral in the Rhône valley, the Tramontana in Catalonia and the Sirocco in the Maghreb may result in widespread devastation. While it is acknowledged that fire can result from natural causes, for instance lightning, the main cause of fires today is the result of human activity. In the past, most of the blame was placed on shepherds and their burning practices, and although this threat remains, more and more emphasis is directed to careless behaviour of tourists or urban residents sometimes wilfully starting a blaze. Vélez (1990) even suggests that some fires have been started deliberately by members of the fire services seeking the added payment they receive when on duty!

Most of the Mediterranean states have increased investment in fire prevention and containment practices. These range from educational and publicity campaigns aimed at the general public, to the use of highly sophisticated GIS systems that handle data provided from air- and space-borne sensors. The production of fire danger maps and fire simulation software has become an increasingly useful tool in planning the deployment of fire-fighting resources. However, climate change, bringing even longer, drier periods, and increasing human settlement in the region are expected to raise the profile of this hazard in the future and demand further investment to prevent or contain the outbreaks.

Degradation

The Mediterranean region has always suffered relatively high rates of degradation, and this has been recognised and documented at least from classical times. The physical nature of the Mediterranean climate and landscape gives rise to these naturally high rates and the imposition of human activity upon the landscape has a tendency to increase the vulnerability of these regions to accelerated degradation. Degradation of landscapes includes not only the erosion of soils and exposure of bare rock surfaces, but also the impoverishment of soils, in terms of their potential productivity, by the reduction in nutrient cycling and the loss of organic matter and structure.

Soil erosion and degradation are not new issues in the Mediterranean. Classical literature makes numerous references to the impoverishment of the hills, and the construction of terraces and effective soil management techniques for slopes were long established by the time of the Roman Empire (Foxhall 1996). Plato in the third century BC was bemoaning the despoliation of the hills around Athens. The younger fill identified by Vita-Finzi (1969) was attributed to the collapse of terrace systems following the dissolution of the Roman Empire and the widespread resurgence of pastoralism throughout the Mediterranean. Recent interest in the degradation of the mountains, therefore, needs to be considered in the context of longer time-scales. Concern has intensified particularly in the last decade (Poesen and Hooke 1997), partly as a result of past catastrophes (notably flooding in the lowlands), and partly reflecting the potential effects of future climate changes. In particular, investigations are focusing on the effects of downstream flooding and sedimentation and on the impact of urban centres on development and of tourist and industrial facilities on floodplains. In the uplands, concern is directed towards 'desertification' and the impact of changing climates on soil productivity. However, economic conditions are more likely to determine the success or otherwise of farming and upland soil conservation measures than climate.

Work that has been done has concentrated primarily on the modelling of processes on small plots or on limited hillslopes. Projects such as the MEDALUS initiative have increased the understanding of process on hillslopes in selected test areas in the northern Mediterranean countries, and promote the need for longer-term monitoring.[3] However, these initiatives fail to address the problem of extreme events, in part because by their very nature such events are almost impossible to predict. Also they all fall short of effectively accounting for the variation in rates of erosion so typical of all mountain regions. The models derived from test sites run the risk of being applied across the Mediterranean Basin in all mountain areas, by policy-makers who seek quantifiable answers to policy questions. One cannot blame them for doing so, and the complexity of the issue is beyond effective explanation except in either a very general way across the Basin, or in great detail for very limited areas. With the development of more sophisticated and complex modelling approaches, a more rapid assessment of watershed-scale processes will be possible, but until then, problem areas need to be tackled on an individual basis with particular policy questions in mind.

Depopulation

The mountains of the Mediterranean have undergone marked depopulation with significant consequences for upland agriculture, forest management and settlements. In fact, whilst there are numerous generalisations about demographic change in the mountain areas, the data on which these are based are hard to find in more than a few of the mountain ranges under examination. There is as yet no comprehensive survey of the demography of Mediterranean mountains, in part because of the fact that existing census material is not acquired in units that clearly differentiate mountain areas. Whilst this is not a problem for census units in the geographical heart of mountain ranges, the perimeter zones pose a much bigger problem. The result is that much of our knowledge is drawn from detailed case studies, and, in the case of carefully researched studies of demographic dynamics, is confined to a few countries – notably Italy and France – where exercises in historical demography are possible. However, as a generalisation it is possible to draw a distinction between those mountain ranges to the north and west of the Basin where populations have experienced a sharp decline over the last century and those to the south and east where relatively high population densities remain. For example crude densities of 5 persons/km² in the Pyrenees in Spain contrast with those of between 50 and 200 persons/km² in the Moroccan Rif and Algerian Tell.

A landscape of deserted or declining settlements remains very strongly imprinted on public perceptions of Mediterranean mountains. However, this is not the case everywhere and the process has varied dramatically according to locality and historical period. Some examples serve to illustrate this process. Figure 12.2, taken from Romano (1995), shows the historical pattern of population change in three different Italian mountain communities. First, Courmayeur, in the Valle d'Aosta in the western Alps, shows population decline from 1861 to 1921, followed by an increase in population, which became very rapid after 1951, due to the growth of ski tourism. Roccaraso, in the Abruzzo Apennines east of Rome, shows a later onset of decline, from 1891, and a later recovery, after 1951, again due to winter tourism. Finally S. Stefano di Sessanio, also in the Abruzzo but with no significant involvement in tourism, exhibits a profile of continuous decline since the end of the nineteenth century. Figure 12.3 shows the spatial pattern of population change in the South Italian region of Basilicata; municipalities in the eastern mountain core of the

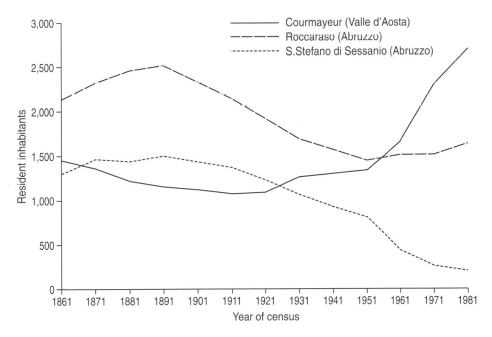

Figure 12.2 Population trends between 1861 and 1981 in three Italian mountain communities

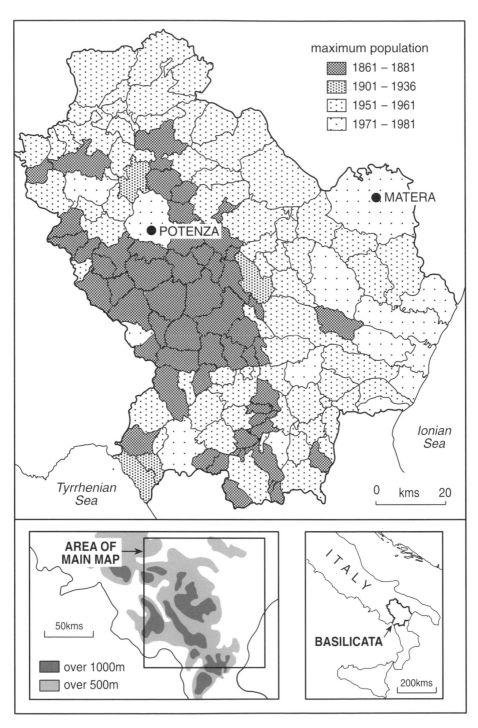

Figure 12.3 Basilicata: spatial pattern of maximum population, by municipality, 1861–1981
Source: Touring Club Italiano (1999, fasc. 32).

region, where much of the land is above 1,000 metres, reached their maximum population as early as the nineteenth century, whereas at lower altitudes decline has been more recent, mainly since 1951. Population growth is confined to the coast and to the two main towns.

Starting at a somewhat later date, evidence from the Spanish Pyrenees suggests a similar story, with most of the Pyrenean valleys experiencing depopulation since World War II, in some cases becoming a mass exodus by the 1970s. According to Daumas (1986), villages located on the slopes of valleys, such as the Val de Benasque, have lost over half their population since the 1940s, whilst those in the valley bottoms have been less severely hit but have seen at least a 20 per cent decline. Moreover, most of the villages now have ageing populations as the young seek opportunities elsewhere. Casabianca (1987) cites the case of Corsica where, after heavy losses of males during the First World War, the mountain communities of this island have experienced long-term decline. In Greece, depopulation in some areas of the Pindus can be traced to the troubled period at the end of the nineteenth century and recovery has been hampered by continued economic and political problems (McNeill 1992a). However, not all mountain regions on the north coast of the Mediterranean have suffered recent population decline. In the Taurus mountains of southern Turkey, population remains very sparse except in areas adjacent to good water supplies. These mountains have been occupied by pastoralists seeking good grazing in the summer. In the valleys there is small-scale cultivation, often on terraced strips, and whilst there are places where these have been abandoned there is little evidence of deserted villages. Thus whilst this century has witnessed considerable mountain depopulation in some areas, so too other mountain zones illustrate how changing economic fortunes such as the growth of tourism or the opening of mines can reverse population decline.

By way of contrast we can examine the pattern of population change in the belt of mountains on the southern rim of the Mediterranean, including the Rif and pre-Rif in Morocco, the Tessala, Beni-Chougrane and western Tell in Algeria, and the mountains of Tunisia, the Dorsale and Haute Tell (Maurer 1993). Within this extensive mountain chain, there are considerable differences in population density, but most sections have experienced rapid expansion of population. In the Tell range of north-west Tunisia rates of growth have been between 1.6 and 2.0 per cent per annum since 1975. Even in Algeria, after the dislocation of the Wars of Independence, many mountain areas began to show an increase in population from 1977. The Moroccan Rif has experienced fluctuations in population levels over a long historical period. Refass (1992) argues that high population densities in the Rif date back to the fifteenth century, as the population of the narrow coastal belt retreated into the mountains in the face of Spanish attacks. Later, the flux reversed and there was emigration to other areas in the seventeenth century, but by the end of nineteenth century the mountain areas once again experienced expansion. The last 30 years have witnessed nearly a doubling of population, especially in the eastern Rif (Fadloullah 1990), which appears to have been sustained according to preliminary evidence from the 1994 census.[4]

Economic development and agrarian change

Depopulation is largely a response to the changing fortunes of the mountain economies, which are in turn a function of the increasing integration of these areas into wider economic relationships. Traditional cereal cultivation in upland areas provided food for both animal and human populations. However, with a combination of improved arable technology and new crops, farmers in mountain areas have found it more sensible to import grain. This effect has been felt throughout the Mediterranean and is even present in the Atlas mountains in Morocco. At the same time, urbanisation and industrialisation in lowland regions have generated job opportunities for both seasonal and permanent migration. This in turn alters the demographic composition of mountain settlements.

In Italy, approximately 35 per cent of the land is classified as mountainous (official statistics define as mountains land over 600m in the north and over 700m in the centre and south). Viazzo (1989), working in the headwaters of the Valle d'Aosta, in northern Italy, indicates how developments in the lowlands, combined with the evolution of agricultural practices in the mountains, have led to periods of population stability followed by emigration from the mountain villages. Various causes can be pinpointed: the problems associated with inheritance of land; low productivity and hence incomes; the attractions of the lowland town in an industrialising nation; and for some, overseas emigration.

A considerable literature has also accumulated about agricultural change in the Spanish Pyrenees (Puigdefabregas 1990). The complex systems of land use depended upon abundant labour and transhumance of large flocks of sheep from the Ebro valley into the mountains during the summer months. Following patterns previously experienced in France and Italy, most of the Pyrenees valleys have experienced depopulation since the 1950s. Whilst the traditional systems of land use attempted to integrate livestock management, crop cultivation and other economic activities that, at least for certain periods, provided some ecological stability, the decline in population and associated social and economic changes has meant that new patterns of activity are currently being established. In terms of land use, there is clear evidence of a marked decline in the cultivated area which now often only exists in close proximity to the villages, whereas in earlier times this had expanded almost to the upper limit of cultivation. In some cases, these abandoned areas have become zones of wasteland covered with scrub and sometimes subject to severe erosion (García-Ruíz and Puigdefabregas 1982). However, the decline in cultivation is not *per se* a result of the fall in population but of the fact that the economics of cereal cultivation evolved in such a way that it became cheaper to purchase grain for both food and feed (García Ruíz and Lasanta-Martínez 1990).

The livestock economy in the Spanish Pyrenees has also seen considerable changes in the same period. Post-Second World War developments in the Ebro valley, particularly associated with irrigation and accentuated by EU financial assistance, have led to the conversion of former winter pastures to more valuable cropping systems. Thus not only do herders have to vacate pasture areas as early as February, but the rents have increased considerably. At the same time, fewer persons are attracted to herding as a lifestyle. Where sheep remain important, they are now often kept in stalls whilst waiting for the higher summer pastures to open. The decline in sheep has been partially replaced by the development of beef and milk production. However, whilst the valley bottoms are used intensively and also benefit from heavy manuring, there are problems with summer pasturage for cattle. Cattle need high volumes of forage (compared with sheep) and can graze intensively in the lower slopes, but other areas once suited to sheep have become derelict. Thus the lower pastures are in danger of over-exploitation. There has been a large increase in the production of milk supplying the expanding urban areas in the lowlands. This provides an excellent example of the impact of specialisation in a mountain region which previously operated an integrated agricultural economy utilising all the ecological niches provided by the landscape. Today, the economics of this are such that mountain residents cannot generate an adequate living without following the market trends and so, in addition to changes to agriculture, the development of tourism has played an important part in bringing new sources of employment.

A very clear case of mountain depopulation due to linkages with the wider market is that of Corsica (Casabianca 1987). As we noted earlier, Corsica experienced depopulation in the first decades of the twentieth century. In 1900, one area of this mountainous island, Castagniccia, had one of the highest densities of rural population in Europe, based upon the integrated production of animal husbandry, intensive tree cultivation and a variety of artisanal trades. As the name of this region implies, the dominant tree crop was chestnuts. With the loss of population after 1914, most of the local trades and the viability of local agricultural production collapsed as improved links were established with the mainland. At the same time, the import of cheap wheat undermined the viability of cereal cultivation, and the price of chestnuts, an important export, fell by 300 per cent. This example vividly illustrates the impact of the exposure of a mountain economy to a wider market, the more so because of the fact that local production was stimulated briefly during the isolation of the 1939–45 war. By 1975, the population had fallen to one third of the level of 1900.

In those mountain areas where population continues to increase, particularly in North Africa, it is necessary to examine whether they are simply at another point in the trend towards 'overpopulation' as suggested by McNeill (1992a), or whether their economic structure will see a very different scenario emerge. Some areas, especially the Grande Kabylie of Algeria, were marked by an increase in population after the War of Independence, but there has been a decline in the proportion of households in which agriculture plays a dominant role. This mountain region represents an important resource in the overall pattern of Algeria's semi-arid landscape. Maurer (1993, 1996) points out that this area is adjacent to the industrial development and urbanisation of Algiers, and the late 1980s have seen a revitalisation of agricultural activity as families return to their deserted fields, refurbish terraces, introduce new systems of irrigation and invest in tractors.

To the west, in the Moroccan Rif, there has been a very rapid increase in the total area cultivated, but there are marked variations from one part of this mountain range to another. As might be expected given the figures for population growth noted above, it is in the eastern Rif and the pre-Rif that we find figures as high as 70 per cent or more of the land in cultivation. In the west, 15–25 per cent is more common. McNeill uses the Rif as the best example of present-day 'overshoot' and gives a grim warning about the extensive environmental damage resulting from clearance of vegetation cover. However, it may also provide us with an example where we can observe another adjustment mechanism emerging. As McNeill himself documents (McNeill 1992b), the same period of rapid population growth has, in some parts of the central Rif, been matched by an extraordinary increase in the cultivation of Kif (cannabis), which probably involves at least 200,000 persons and covers some 70,000 ha. (Maurer 1993). Kif produces revenues at least 20–30 times higher than a crop of cereals; together with other downstream processes this is equivalent in revenue terms to a change in the 'carrying capacity' of the area. Perhaps this can be equated with the introduction of the potato into the Alps in the seventeenth century, which also represented a new level of 'carrying capacity'. However, the success of Kif production depends upon heavy manuring, and the cheapest form of providing this is through grazing goats. These accelerate the removal of vegetation cover which, in turn, exacerbates the erosion process.

Development policies

Mountain areas in the Mediterranean have always caused problems for national governments in the form of political dissension, economic decline and environmental damage. Broggio (1992) points out that many of the northern Mediterranean countries had developed a *'politique de la montagne'* well before the advent of the European Community and interventionalist policies have a long tradition elsewhere such as the afforestation policies to protect watersheds instigated by the French colonial government in Morocco. In this respect, the Italian experience is interesting because of the long public recognition (if not successful action) of the need to provide support for mountain areas. In the Fascist period (1920s and 1930s), legislation set up an institutional framework to provide special help to the mountain valleys. After 1952, further legal provision was made, introducing *'consorzi'* and *'comprensori di bonifica montana'*, special areas and organisations for mountain improvement. Overall, these measures had only a patchy impact, mainly due to their political purpose, and it was only in 1971 that the *comunità montane* were established. These are administrative agencies for a mountain region which are created by the aggregation of mountain communes into a federal organisation with responsibility for the development of the locality. According to Romano (1995), total government financing for this initiative has reached US $1.25 billion, apparently just for administration. However, through these agencies, EU and other government funding has been obtained for improvements to roads, land reclamation, irrigation and afforestation. There are some 352 such units in Italy, containing 10 million people (Casabianca 1987).

Barberis (1992) has reviewed some of the recent research on Italian mountains and concludes that whilst it is now reasonably certain that at least two-thirds of the *comunità montane* have stabilised their populations, this has been at the expense of agriculture which has generally shown a marked decline. Between 1981 and 1990, of 3,524 communes surveyed, about 80 per cent showed either some growth in population or a stable condition. The mountain road building programme, dating from the Fascist period, has meant that by the 1980s, many mountain communes are accessible and approximately two-thirds of the *comunità* show signs of positive developments. Barberis notes, furthermore, that the disposable income of mountain areas is only about 10 per cent below that of the Italian average; that there is almost the same degree of entrepreneurship (measured by registered business initiatives) as the national figure; and that various other aggregate indices suggest that the mountain areas are today much less 'backward' than popular imagination perceives them to be. Whilst some of this advance may be illusory, resulting from distortions of the statistics in mountain regions, there can be no doubt that the policies discussed above have played their part. Nevertheless, the decline in agriculture has continued unabated despite the portfolio of initiatives put in place, and it is interesting to note that recently much of the emphasis has switched to encouraging 'pluri-activity', of which tourism is a critical element. Although the evidence would suggest that a combination of Italian and EU initiatives have assisted mountain communities, success has not been uniform. There remains about one third of all mountain *comunità* that fall significantly below national levels of development and welfare and show signs of a falling population. Most of the success has been in the north whereas in the south the evidence of success is patchy. This difference is attributed to the persistence of clientelism which serves to undermine any new initiatives which conflict with old alliances and institutions.

However, Romano feels that the relatively recent Italian legislation (1991) which has established new national parks may offer some hope. The Abruzzo mountain area contains both attractive landscapes but also small settlements of historic interest. Whilst many residences have now been abandoned or used seasonally, as in parts of the French and Italian Alps, he suggests that the framework of the national parks provides a new format for rethinking the way in which a local economy is stimulated. He emphasises that 'typical' tourist investment, as seen elsewhere in the Italian Apennines, is not desirable. Most of the benefits leave the area and a massed tourist influx in mountain regions often produces either a damaged landscape or one so highly protected as to be more like a 'living museum'. In his view, the funds available should enable the location of the relevant agency offices in existing mountain settlements and encourage much greater use of existing residential capacity for tourists. However, Romano argues that much available investment (from European funds) is misguided because it aims to create 'elite' tourist centres, entirely out of keeping with the social structure of the existing community. Thus any new initiatives must spring from local involvement, in line with recent International Union for the Conservation of Nature suggestions (IUCN 1993). Developing family businesses and ensuring that 'soft tourism' predominates are just two such strategies which have had success elsewhere. Consequently, the onus lies with the enabling institutions to provide the right stimulus to local activities.

In many cases the most significant interventions are those that are not specifically targeted on mountains but deal with wider issues such as investment and agricultural development. This has been particularly evident for those countries of the Mediterranean which are members of the European Union, for a range of EU policies impinges on the livelihoods of mountain areas (Lowe 1992). For example, since 1975, the Community has been paying compensatory sums to farmers of 'less favoured areas', which naturally includes mountains. This policy developed from earlier initiatives in France where the principle of compensation for the physical and locational difficulties facing farmers (relief, climate, accessibility etc.) had been established (Broggio 1992). However, since the 1980s, and the admission of new members into the EU, the approach has changed somewhat. Whilst mountain areas have been increasingly recognised as important zones for community-based action, particularly because they often represent boundary areas (e.g. the Pyrenees), the financial payments now emerge as structural funds paid as a result of regional policy to counteract relative poverty. Mountain areas as such are no longer recognised, although payments are now made under Objectives 1, 2 and 5b, particularly to mountainous areas in countries bordering the Mediterranean.[5] It is now recognised that environmental factors need to be considered within overall resource management and there is now greater emphasis on the development of enterprises which will stimulate the local economy, including tourism, local artisanal activity and assistance to co-operative activities.

Elsewhere within those mountain areas subject to EU policy, agricultural development has been associated not just with the supply of funds but with institutional factors. One of the most significant has been the use of *Appellation d'origine contrôlée* to provide protection for high-quality mountain products. The idea was initiated most successfully in the 1950s in the Beaufort Valley in France where subsequent expansion of traditional cheese production has resurrected the local dairying industry. At the same time, new technologies such as portable milking sets and cooling containers have been introduced, enabling milking to take place on high pastures, thereby facilitating supplies. The producers have developed 'micro-markets' which provide some measure of success in economies which are increasingly dominated by large food manufacturers (Price 1996; Vivier 1992). Although originally developed to help these areas avoid depopulation, such initiatives also have an important environmental impact through their support to continuous management of mountain landscapes. This has been of particular value during a period in which the Common Agricultural Policy has been dominated by productivist policies, but there are now some doubts as to the resilience of even the micro-markets in the face of global competition which may follow recent changes in EU agricultural policy.

Although the Blue Plan of 1988 did not specifically address mountain issues it does contain a strong statement to the effect that authorities in the mountain regions of the Mediterranean should give a higher priority to soil maintenance measures to combat overgrazing and deforestation (Grenon and Batisse 1989). Subsequently, many Mediterranean mountain areas qualified for provisions under Integrated Mediterranean programmes for the analysis of desertification and afforestation of marginal land. Studies under the MEDMONT and MEDALUS programmes noted earlier have been the focus of research initiatives (see Chapter 14). However, there is yet to be any agreed action plan which embraces all Mediterranean mountains despite the recognition of similar problems.

More recently, following the Rio Conference in 1992 which developed Agenda

21, the EU countries surrounding the Mediterranean have been developing new initiatives focused upon the needs of mountain areas. The Green Paper on the Alps (Pils *et al.* 1996) specifically deals with the question of tourist development and the associated landscape issues set within the framework of 'sustainable development', whilst IUCN and other agencies are developing plans for projects to protect mountain landscapes. Also indicative of a shift in attention was a conference of European NGOs associated with mountain areas across Europe, held in Toulouse in 1996.[6] Again the main interests were tourism, conservation, agriculture and forestry. The result of these initiatives is due to be discussed shortly at a UN-sponsored conference which aims to re-establish mountain areas as a critical focus for attention within the needs of sustainable development policies.

All these issues are well recognised by those countries on the southern shores of the Mediterranean but, as yet, they do not assume a similar level of significance due to far more important questions associated with economic development. In Algeria, the recent period of civil strife has seriously undermined efforts to develop a moun-tain-based policy, whilst in Morocco broader questions associated with maintaining economic growth and coping with Structural Adjustment programmes have taken priority. Nevertheless, it is possible to observe a number of important developments which mirror those to be found on the northern shores (see, for example, Bencherifa 1988; Messerli and Winiger 1993; Swearingen and Bencherifa 1996). For instance, mountain areas in Morocco have long been subject to legislation concerned with afforestation and the protection of landscapes from erosion (Dressler 1982). In this case the prime purpose has been to limit the rate of siltation of the dams which are a crucial element in the country's development programme. For example, the Lalla Takerkoust dam, built in 1935 on the N'fis river which drains from the High Atlas, has lost 28 per cent of its capacity due to siltation (World Bank 1995). The importance of water development to the national economic strategy in Morocco has been evident since 1914 when rights to water were enshrined in the state. The Atlas mountain range is the principal source of water for Morocco's economic develop-ment plans which are heavily dependent upon irrigation. Although traditional access has been allowed, the past 80 years have seen an increasing number of schemes which extract water for power, irrigation and potable supplies. In the next 30 years, 51 dams are scheduled for construction including that at Al Wahda which will be the second largest dam in North Africa after the Aswan. Apart from the investment involved, such schemes have important implications for mountain people, as the planned sites for many of the constructions are in the densely-settled and intensively-utilised foothill zones. Whilst the benefits may be clear to those downstream of the new dams, questions concerning loss of valley-bottom grazing and arable land, and the even more delicate issue of population resettlement will need careful resolution.

The Moroccan mountains have experienced outmigration for most of this century, although in the Rif the spatial and historical patterns of migration and population change have been more variable, as noted earlier. Consequently, the state has encouraged various initiatives aimed at improving the incomes of mountain households in the hope that this will arrest the outflow. Given the prob-lems of rapid urbanisation faced by Casablanca and other cities, this remains an

231

important issue. Unlike the EU, there is no real support for mountain agriculture, whose fortune depends crucially upon the climate and, to a lesser extent, the state of the market. However, there have been an increasing number of initiatives aimed at developing tourist potential as the mountains are seen to offer a new source of attraction away from the increasingly 'overdeveloped' coast. At present, the mountains immediately bordering the Mediterranean (the Rif) are not heavily patronised by tourists in comparison with similar locations in Southern Europe. In part, this reflects the fact that the coastline itself is not very hospitable, but is also related to the fact that historically the Rif has never been 'welcoming' to tourists and today, with the problem of Kif, remains relatively isolated. This is not the case for the Middle or High Atlas where tourism, both by Moroccans and foreigners, has been encouraged. In the 1920s French mountaineers developed a strong interest in the climbing and ski-ing opportunities; ski resorts were developed at Mischlifien, near Ifrane in the Middle Atlas, and at Oukaimeden, south of Marrakech in the High Atlas. The Moroccan authorities want to expand the tourist potential of these mountain areas to match the current popularity of the coastal regions. This, of course, raises many issues akin to those under discussion in other Mediterranean mountain areas. At present, only those areas which have good road access feature on the itinerary of the large number of tourist buses that operate from the big cities. Most of the visits are of very short duration, often just a stop at a *souk*. Where hotels have been built, they often remain empty for large parts of the year. For the enthusiastic mountaineer/walker a programme was developed in the Azilal valley (south of Marrakech) which envisaged the establishment of a network of *gîtes* to foster greater 'participatory mountain tourism'. According to recent work (Balloui 1996), this scheme has been partially successful but the benefits from income (guides, mule-drivers etc.) remain confined to a few households. Certainly the experience of some of the European initiatives would be of value, but the cultural values of these mountain communities must be more fully appreciated before extensive intervention takes place.

The Moroccan experience highlights an important factor often overlooked in many discussions of mountain development, namely that many mountain communities have developed their own distinctive cultural values which, whilst not always apparent superficially, often determine the particular way in which individuals and communities react to new pressures. The accounts of Southern European mountain areas noted above often include references to traditional community values but in Morocco at least, if not elsewhere also, these values are very significant. For example, Moroccan mountains remain dominated by Berber-speaking peoples whose language is not that of the state in which they live. Until very recently Berber culture was largely marginalised by the predominately Arabic-speaking state, even though Morocco's history remains a complex amalgam of both peoples. As this view has changed so mountain communities and their ways of life will perhaps be less marginalised; but equally, the incursion of modern capitalism is likely to be just as destructive unless special provision is made to preserve the Berber institutions. However, as in the case of the Italian Apennines reported earlier, it is important to avoid creating 'museums' and it is this aspect that provides the greatest challenge to the Moroccan authorities.

Conclusions

Mountains provide an important and highly visible feature of the Mediterranean landscape. With few exceptions, mountain ranges are rarely far from the coastline. In the long history of Mediterranean civilisations mountains have been present, sometimes as barriers, sometimes as refuges for groups seeking protection from invading armies. The physical environment is often described as fragile, exacerbated by the summer drought and likelihood of intensive rainfall. In the drier zones, especially on the south shore, the additional rainfall and coolness of the mountains provide important areas which are suitable for agriculture.

Throughout the long history of human settlement in mountain regions, commentators have frequently written about the processes of deforestation and degradation. Ancient civilisations removed much of the tree cover in Tunisia to supply timber for shipbuilding. Today the concerns about deforestation and degradation link the various causes with the increasing population densities and overgrazing. Whilst individual countries have undertaken programmes of afforestation, there are no Basin-wide initiatives that have gone beyond the discussion table. Despite a plethora of studies, the extent of degradation remains uncertain; much research is confined to localised case studies and the links between human activities and degradation remain contested. The main thrust of environmental research must be the scaling-up process whereby our understanding of localised processes can be linked to wider environmental variables as well as different forms of social organisation. The attempt by McNeill (1992a) to portray the environmental history of Mediterranean mountains as a linear process associated with overshooting the carrying capacity remains questionable. In particular, it presupposes that those areas in which high population densities still remain will, of necessity, have to pass through a phase of environmental collapse.

The population question has been central to many discussions of Mediterranean mountain ranges. For many, but not all, of the mountains to the north, depopulation has been the main experience of the last 150 years. This has been attributed to a process of 'peripheralisation' in which both agricultural activity and the opportunities for off-farm employment have concentrated in the lowlands. Consequently, out-migration has been a key element of the demography of mountain settlements. Abandoned settlements can be found in the Apennines, the Dinaric Alps, the Pyrenees and elsewhere. By contrast, on the south shore, there are areas of high and growing population concentrations in the Moroccan Rif and the Algerian Tell. These mountain ranges, situated in countries where problems of economic growth remain problematic, present different challenges to those elsewhere.

In many of the countries surrounding the Mediterranean, governments have attempted to develop policies to tackle both the environmental and the economic and social problems in mountainous regions. In France and Italy, for example, there has been a long history of government initiatives which have attempted to counterbalance the depopulating pressures experienced by mountain communities. The success of such policies has been varied but, since the formation of the EU, these initiatives have had much greater impact. Indeed, there is evidence to suggest that,

233

in some mountain regions, populations have been stabilised through a combination of state incentives and changes in the economy which offer advantages to a more decentralised society. In some areas, the mountain landscape itself has been the most important resource, with tourism offering new opportunities for generating incomes in upland communities. However, this also raises important questions about the nature of the activities and the extent to which mountain people actually benefit from tourism. The scenery itself can be damaged both by excessive use and through the construction of facilities which do not enhance the character of the landscape. Without careful planning and the involvement of the local community, there is a danger that the mountains become 'zoos' or 'museums' offering no more than a glimpse of a faded lifestyle and fossilised landscape. Interestingly, the problem is not confined to the 'more developed' economies; mountains in countries of the southern shore face similar questions despite the urgent need to develop new sources of income in the communities.

The last decade has been marked by an upsurge in the political debate about the management of mountain areas which has resulted from pressures for 'sustainable development', and in particular the impact of Agenda 21 policies. EU mountain states, including those of the Mediterranean, are developing initiatives which focus on both the problems of the physical environment and the social and economic role of mountains. The Alpine Convention, the NGO consultative meeting in Toulouse and the European Inter-Governmental Consultation on Sustainable Mountain Development are examples of this process in the EU. Elsewhere, under UN auspices, policies influencing mountain areas are being drafted. A notable shift has been the increasing attempt to ensure that mountain communities themselves play a much stronger role in policy formation. Nevertheless, there remains a large gap between policy and praxis. At the Mediterranean level such initiatives, whilst important, have yet to embrace all the countries of the Basin and provide valid policies which may be implemented in regions which still face serious problems of development. Highly-populated mountain areas, especially in the relatively poor countries of North Africa and the eastern Mediterranean, do not benefit from these EU policies; indeed many suffer as attempts to generate new markets for mountain products (fruit for instance) meet market barriers. In this respect the mountains of the Mediterranean therefore remain highly diversified and call for similarly varied policy initiatives.

Notes

1   NcNeill's five sample regions are the Taurus mountains of southern Turkey, the Pindus in western Greece, the Lucanian Apennines in southern Italy, the Sierra Nevada in southern Spain and the Moroccan Rif.
2   Figures from FAO/ECE, Forest Fire Statistics 1995–97, *The Timber Bulletin*, 51(4), 1998.
3   MEDALUS stands for Mediterranean Desertification and Land Use. See also Brandt and Thornes (1996) and Chapter 14 of this book.
4   While this is true at a general level, at a local level there are variations, with population decline in the mountains and settlement growth in the valleys.
5   Under regulations for the application of EU Structural Funds, Objective 1 refers to lagging regions where development indicators score less than 75 per cent of the EU average, Objective 2 covers regions affected by industrial decline and having high rates

of unemployment, and Objective 5b refers to rural areas with low incomes and high levels of agricultural unemployment, and often with low population densities too.

6 See European NGO Consultation on Sustainable Mountain Development, Recommendations of NGOs and Mountain Populations to Governments and the European Union, Toulouse, 1996.

References

Balloui, A. (1996) Tourisme et développement local dans le Haut-Atlas marocain: questionnement et réponse, *Revue de Géographie Alpine*, 84(6), pp. 15–23.

Barberis, C. (1992) La montagne ou les montagnes italiennes, identités et civilisation, *Revue de Géographie Alpine*, 80(4), pp. 65–76.

Bencherifa, A. (1988) Demography and cultural ecology of the Atlas Mountains of Morocco: some new hypotheses, *Mountain Research and Development*, 8(4), pp. 309–13.

Brandt, C. J. and Thornes, J. B., eds (1996) *Mediterranean Desertification and Land Use*. Chichester: Wiley.

Braudel, F. (1972) *The Mediterranean and the Mediterranean World in the Age of Philip II*. London: Fontana.

Broggio, C. (1992) Les enjeux d'une politique montagne pour l'Europe, *Revue de Géographie Alpine*, 80(4), pp. 26–42.

Casabianca, F. (1987) Y-à-t-il place pour un développement des zones montagneuses en Méditerranée? *Peuples Méditerranéens*, 38–39, pp. 209–18.

Clark, S. C. (1996) Mediterranean ecology and an ecological synthesis of the field sites, in Brandt, C. J. and Thornes, J. B. (eds) *Mediterranean Desertification and Land Use*. Chichester: Wiley, pp. 271–301.

Daumas, M. (1986) La redistribution géographique de la population dans les hautes vallées montagnardes: l'exemple du Val de Benasque, *Revue de Géographie Alpine*, 74(1–2), pp. 189–96.

Dhondt, A. A. (1988) Carrying capacity: a confusing concept, *Oecologia Generalis*, 9(4), pp. 337–46.

Dressler, J. (1982) The organisation of erosion control in Morocco, *Quarterly Journal of International Agriculture*, 21(1), pp. 62–79.

Fadloullah, A. (1990) Evolution récente de la population et peuplement au Maroc, in Bencherifa, A. and Popp, H. (eds) *Le Maroc: Espace et Société*. Passau: Passavia Universitatverlag, pp. 75–84.

Foxall, L. (1996) Feeling the earth move: cultivation techniques on steep slopes in classical antiquity, in Shipley, G. and Salmon, J. (eds) *Human Landscapes in Classical Antiquity*. London: Routledge, pp. 44–67.

García-Ruíz, J. M. and Lasanta-Martínez, T. (1990) Land use changes in the Spanish Pyrenees, *Mountain Research and Development*, 10(3), pp. 267–79.

García-Ruíz, J. M. and Puigdefabregas, J. (1982) Formas de erosión en el flysch eoceno surpirenaico, *Cuadernos de Investigación Geográfica*, 8, pp. 85–130.

Grenon, M. and Batisse, M. (1989) *Futures for the Mediterranean Basin: the Blue Plan*. Oxford: Oxford University Press.

Lowe, P. (1992) Préface, *Revue de Géographie Alpine*, 80(4), pp. 8–19.

Maurer, G. (1993) L'agriculture de montagne dans les pays rifains et telliens au Maghreb, in Bencherifa, A. (ed.) *African Mountains and Highlands*. Rabat: Faculté des Lettres et des Sciences Humaines, Université Mohammed V, Série Colloques et Séminaires, 29, pp. 35–54.

Maurer, G. (1996) L'homme et les montagnes atlasiques au Maghreb, *Annales de Géographie*, 587, pp. 47–72.

McNeill, J. (1992a) *Mountains of the Mediterranean World*. Cambridge: Cambridge University Press.

McNeill, J. (1992b) Kif in the Rif: an historical and ecological perspective on maruyana, markets and manure in Northern Morocco, *Mountain Research and Development*, 12(4), pp. 389–92.

Meiggs, R. (1982) *Trees and Timber in the Ancient Mediterranean World*. Oxford: Clarendon.

Messerli, B. and Winiger, M. (1993) Climate, environmental change, and resources of the African mountains from the Mediterranean to the Equator, in Bencherifa, A. (ed.) *African Mountains and Highlands*. Rabat: Faculté des Lettres et des Sciences Humaines, Université Mohammed V, Série Colloques et Séminaires, 29, pp. 3–33.

Naveh, Z. (1982) Mediterranean landscape evolution and degradation as multivariate biofunctions: theoretical and practical implications. *Landscape Planning*, 9, pp. 125–46.

Pils, M., Glauser, P. and Siegrist, D., eds (1996) *Green Paper on the Alps*. Vienna: Nature Friends International.

Price, M. (1996) Appellations contrôlées in the Beaufort region of France, in Preston, L. (ed.) *Investing in Mountains*. Franklin, WV: The Mountain Institute, pp. 24–5.

Poesen, J. W. T. and Hooke, J. M. (1997) Erosion, flooding and channel management in Mediterranean environments of Southern Europe, *Progress in Physical Geography*, 21(2), pp. 157–99.

Puigdefabregas, J. (1990) Preface, *Mountain Research and Development*, 10(3), pp. 197–9.

Refass, M. (1992) Historical migration patterns in the Eastern Rif mountains, *Mountain Research and Development*, 12(4), pp. 383–8.

Romano, B. (1995) National Parks policy and mountain depopulation: a case study in the Abruzzo region of the central Apennines, Italy, *Mountain Research and Development*, 15(2), pp. 121–32.

Street, J. M. (1969) An evaluation of the concept of carrying capacity. *Professional Geographer*, 21(2), pp. 104–7.

Swearingen, W. D. and Bencherifa, A., eds (1996) *The North African Environment at Risk*. Boulder CO: Westview Press.

Thirlgood, J. V. (1981) *Man and the Mediterranean Forest*. London: Academic Press.

Thompson, M. (1997) Security and solidarity: an anti-reductionist framework for thinking about the relationship between us and the rest of nature, *Geographical Journal*, 163(2), pp. 141–9.

Thornes, J. (1976) *Semi-Arid Erosional Systems*. London: LSE Department of Geography, Special Publication 7.

Touring Club Italiano (1990) *Atlante Tematico d'Italia*. Milan and Rome: Consiglio Nazionale delle Ricerche.

Vélez, R. (1990) Mediterranean forest fires: a regional perspective, *Unasylva*, 41(162), pp. 3–9.

Viazzo, P. P. (1989) *Upland Communities: Environment, Population and Social Structure in the Alps since the Sixteenth Century*. Cambridge: Cambridge University Press.

Vita-Finzi, C. (1969) *The Mediterranean Valleys: Geological Changes in Historical Times*. Cambridge: Cambridge University Press.

Vivier, D. (1992) Les micro-marchés des produits de qualité: un atout pour le développement des montagnes d'Europe, *Revue de Géographie Alpine*, 80(4), pp. 168–84.

World Bank (1995) *Kingdom of Morocco: Water Sector Review*. New York: World Bank Report 14750-MOR.

13

Marginal or nodal? Towards a geography of Mediterranean islands

Russell King and Emile Kolodny

It is logical that this chapter on Mediterranean islands should follow one on Mediterranean mountains, since most of the major islands in the Mediterranean are mountainous in character and even the small ones have steeply sloping relief. Some islands contain prodigiously high peaks: Sicily's Mount Etna, for example, rising to 3,323 metres, rivals the highest points in the Sierra Nevada or the Italian Apennines.[1] Mediterranean insularity has a quasi-continental form, with a multitude of mountainous islands spread over an almost enclosed basin surrounded in turn by high mountain ranges. And just as McNeil's (1992) concepts of 'carrying capacity' and 'overshoot' were shown to be relevant (with qualifications which were outlined in the previous chapter) to Mediterranean mountain environments, so too does the obviously fixed resource of an island's limited geographical area impose limits on the numbers of people that can be supported, depending on the economic system prevailing. Thus islands, like mountains, have been deforested, degraded and depopulated.

Scholars and travellers have long been fascinated by islands, and by Mediterranean islands in particular (Wilstach 1926). Early work by famous French geographers such as Brunhes (1920) and Vidal de la Blache (1926) emphasised the heuristic role of islands as spatial laboratories where population–environment relationships could be studied in microcosm. For instance Brunhes suggested that students of human geography should first try their hand at the study of islands and, after mentioning several Mediterranean examples, exclaimed 'how many islands have been made the subject of monographs!' (Brunhes 1920, pp. 415, 499). Later, Perpillou, another prominent French geographer, wrote that 'islands may be regarded as little regions held as in a matrix', remarkable both for the uniformity of their natural conditions and for their tendency to conserve endemic, archaic forms of life; following Le Lannou (1941), he pointed to the survival of the Roman wooden plough and the near-Latin dialect on Sardinia (Perpillou 1966, p. 18). Meanwhile, for Braudel, studying the Mediterranean islands was 'one of the most rewarding ways of approaching an explanation of Mediterranean life' (1972, p. 161).

From an ecological perspective, islands offer an opportunity to study at close quarters the 'classic' Mediterranean environment and its symbiotic interrelation-ships of climate, relief, soil, vegetation, agricultural systems and settlement forms. As with mountains, today there is a revival of interest in Mediterranean islands because of their special environmental and political importance, their existence as semi-closed systems, their desirability as havens of tranquillity, and the problems they face in controlling pressures from tourism and other economic activities.

This chapter consists of four parts. First, we present some factual data about the size and importance of the Mediterranean island realm, a domain which is polarised between the overwhelming areal and demographic importance of a handful of large islands and a myriad of small islands which are clustered in groups or strung out along coasts. Second, we examine the various historical roles that Mediterranean islands have played: as centres of civilisation and commerce; as stepping-stones for navigators and traders; as centres of colonisation and plantation agriculture; and as isolated spaces peripheral to the mainstream of Mediterranean life. The third part of the chapter is a study of the demography of Mediterranean islands. Contrasts are drawn between the larger islands whose populations have generally increased, and the much more demographically fragile smaller islands; case studies are made of Corsica and the Dalmatian Islands. Tourism appears to be the crucial factor in the recent economic and demographic survival and revival of most Mediterranean islands, and this is the subject of the fourth section of the chapter. The example of the Balearic Islands is used to illustrate both the economic benefits of the mass tourism model and some of the potential pitfalls, which lie more in the realms of culture and the environment.

Island facts and figures

In round terms, the Mediterranean Basin contains about 200 permanently inhab-ited islands with a combined area of 103,000 km^2 and a total population of some 10.6 million – in area and population terms roughly equivalent to the size of Portugal or Greece.[2] Figure 13.1 maps the main islands and island groups, together with some population data. The map shows that most islands lie close to the northern, European shore: with the exception of a small cluster of islands off the coast of Tunisia, virtually the entire North African and Levantine shores are islandless. The national 'ownership' of Mediterranean islands is highly concen-trated. Italy and Greece account for three-quarters of the island area and 80 per cent of the total insular population. Of these two, Italy is the dominant country, having double the area and five times the population of the Greek islands. On the other hand, Greece heads the 'insularity index' table (see table 13.1), with 19 per cent of its national area in islands (cf. Italy 17 per cent) and 14 per cent of its population (Italy 12 per cent). In general, the island share of national populations has been falling, due to out-migration and stifled urban growth: as recently as 1920, one quarter of Greece's population lived on its islands. But elsewhere trends are variable, as we shall see later. We should also note that only two islands, Malta and

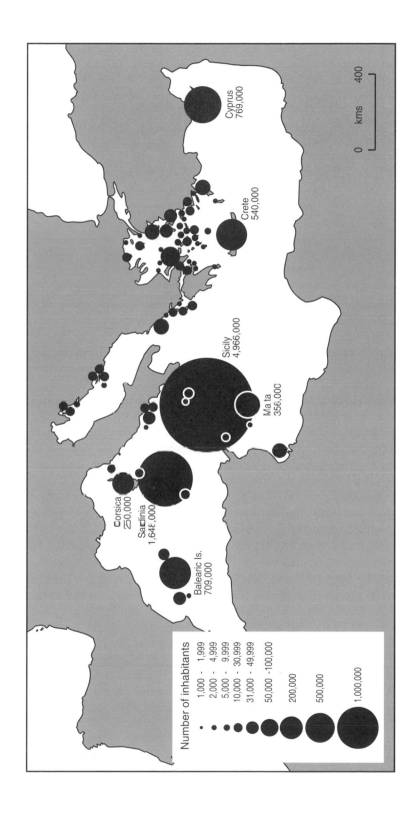

Figure 13.1 The Mediterranean's island population, c. 1990

Table 13.1 Mediterranean islands and their shares of national area and population, 1991

Country	Island area km²	%	Island population no.	1991 %	1960 %
Greece	25,150	19.1	1,475,980	14.4	16.9
Italy	50,159	16.6	6,966,386	11.8	12.3
Croatia	3,180	5.6	124,727	2.6	3.4
France (1990)	8,681	1.6	250,371	0.4	0.4
Spain:					
Balearics only	5,014	1.0	709,138	1.8	1.4
Balearics and Canaries	12,287	2.4	2,202,922	5.7	4.6

Source: Kolodny (1997, p. 771), with modifications.

Cyprus, ex-colonies of Britain, are independent countries.[3] Today, these two island states comprise one tenth of the total island population of the Basin.

Islands occupy 4 per cent of the area of the Mediterranean Sea, rising to 12 per cent in the island-rich Aegean, and to 8 per cent in the western basin, where most of the larger islands are located. The nine largest islands, each with an area in excess of 1,000 km², make up 83 per cent of total island area and 85 per cent of the insular population. Almost two-thirds of the population are concentrated in the two largest islands, Sicily (5 million) and Sardinia (1.6 million), both major semi-autonomous regions of the Italian state. Economically, islands are usually 'poor relations' of their respective national territories. Sicily and Sardinia, part of the Italian Mezzogiorno, have economic and welfare indicators which are well below the Italian average, and Corsica is one of the least developed parts of France. An exception to this general rule is presented by the Balearic Islands, now one of the richest parts of Spain due to the economic effects of mass tourism, and with a rising share of national income and population (table 13.1).

Perhaps the most obvious contrast within the island world of the Mediterranean is between the relatively small number of large islands, each with its hierarchy of urban and rural settlements and developed transport systems, and the smaller islands where habitation is generally limited to scattered agricultural settlements and small fishing ports and landing-stages (Kolodny 1974 and 1992). Braudel (1972, p. 148) describes the larger islands as 'miniature continents' with their own regional structures and major towns. Four islands have their own railway networks.[4] Sicily now has its own extensive motorway system connecting the largest cities – Palermo (695,000), Catania (327,000), Messina (234,000) and Syracuse/Siracusa (126,000) – as well as some of the smaller towns. Other large island cities (over 100,000 inhabitants) are Cagliari and Sassari (Sardinia), Palma (Majorca), Nicosia and Limassol (Cyprus), and Iraklion (Crete). Generally the cities are growing at the expense of the smaller settlements and scattered dwellings. Universities exist at Palma (Majorca), Cagliari and Sassari (Sardinia), Corte (Corsica), Palermo, Messina and Catania (Sicily), Chania, Iraklion and Rethymnon (Crete), and on

Malta and Cyprus. The new University of the Aegean has its various departments scattered on four different islands, and there is a separate university on Corfu.

Mediterranean islands in history

Islands have played a variety of roles in the economic and political history of the Mediterranean (Braudel 1972, pp. 148–67). Some have remained remote and isolated; others discharged a more central, nodal function; and yet others were islands of passage, stepping-stones for colonial ventures or for commercial traffic between different parts of the Basin. Many islands performed several roles at different periods of time. These differentiated histories have also combined with diverse geological origins to produce marked contrasts in island landscapes and population densities. Ninety years ago Semple observed that, in the Mediterranean, 'the small islands . . . of volcanic origin show the greatest production and hence marked density of population'. On volcanic islands like Lipari, Salina, Pantelleria, Ponza and Santorini (and, outside the Basin proper, in the 'Mediterranean Atlantic' islands of Madeira and the Canaries) elaborate terrace systems, often supplemented by irrigation channels, create fertile slopes which 'lift vineyards, orchards of figs and plantations of currants to the sunny air' (Semple 1911, pp. 450–1). In most of these islands, subsequent emigration and depopulation led to the abandonment of much of this carefully manicured and terraced farmland, as we shall see later in this chapter. In the non-volcanic islands, geology has generally been less kind. In the Aegean Sea and along the Dalmatian coast there are many islands with thin, stony soils based on limestone and other porous rocks. Irrigation is difficult and farming has traditionally been based on the grazing of sheep and goats, some specialised plots of olives and vines, and the small-scale cultivation of cereals and vegetables. The populations of these islands have generally remained low, or have turned to fishing (as on Lampedusa) and other maritime activities (e.g. sponge diving on Kalymnos and Symi, ship-building and recruitment of sailors on Hydra, Syra, Chios and Cephalonia) to compensate for the poverty of the land.[5]

Isolated, poor and precarious: islands as marginal places
Braudel (1972, p. 154) wrote that the fate of many Mediterranean islands was 'a precarious, restricted, and threatened life'. Isolation, poor soils and a harsh version of the Mediterranean climate explain the historical poverty at the heart of Mediterranean islands, even the larger and richer ones. Famine is a recurring theme in the history of these islands. Few, if any, were assuredly self-sufficient: 'the great problem for them all, never or only partly solved, was how to live off their own resources, off the soil, the orchards, the flocks, and if that was not possible, to look outwards' (Braudel 1972, p. 152). In the latter half of the sixteenth century, Corfu, Crete and Cyprus, Venetian islands of the Levant, were frequently threatened by famine. On Malta, too, food was often short, despite regular imports of wheat from Sicily. Curiously, the situation was often less precarious in the poorer and more backward islands – those which had a lower population density and which had not been subjected to a colonisation of specialised crops grown for export. Sardinia

exemplified this: for centuries it remained a virtually closed world of hardy pastoralists and farmers who lived inland and turned their backs on the sea (Le Lannou 1941). Still today, despite the land reclamation, industrialisation and tourism of the last few decades, it remains a 'world apart' and preserves the strong cultural character of its language, customs and folklore (Houston 1964, pp. 623–34).

Famines have also afflicted Mediterranean islands within living memory. The Greek famine of 1941–42, produced largely by the German, Bulgarian and Italian occupying armies and exacerbated by the temporary blockade exercised by Allied forces, had its most devastating effects on islands. The results were that 8 per cent of the population of Hydra and Hermoupolis (Syra) died of starvation; mortality on Lesbos almost quintupled (1,400 deaths in 1942, compared to 390 in 1939), and the death rate on Mykonos increased by more than ten times, from 13 per thousand in 1939 to 144 during the famine (Hionidou 1996; Kolodny 1974).

Another concomitant of the isolation of many Mediterranean islands is their frequent use as prisons, lazarets and places of exile. According to Semple (1911, p. 440), there is scarcely an island in the Mediterranean without this sinister vein in its history. The islands of the Aegean were constantly receiving consignments of political exiles from continental Ancient Greece. Napoleon's exile to Elba, and thence to St. Helena, is well-known. Throughout its history Italy has used its many minor islands as convenient places to banish people classed as convicts, *mafiosi* and political exiles. During the Greek Civil War (1946–49), concentration camps were established at Makronisos and Yaros, and Yaros and Leros prisons were re-utilised during the Colonels' dictatorship (1967–74).

Finally, the marginal or peripheral location of certain islands or island groups can change as a result of external economic or geopolitical circumstances. War may give islands a special, temporary strategic status. In the case of the islands of the eastern Aegean, the reverse happened. The endless unification of Greece (1821–1947) provoked mass population exchanges and the exodus of more than 1 million Greeks from Asia Minor in 1922. This transformed the archipelago from a bridge between Greeks on two 'mainlands' to a peripheral appendage of island cul-de-sacs up against the Turkish coast (Kolodny 1974).

Islands as places of civilisation and development

Counterposed against those islands whose main historical role has been as marginal spaces are others whose function has been nodal, at least for certain defined periods. Since prehistory, the Mediterranean islands attracted population thanks to their resources and their high accessibility on 'the common highway of the ocean'. Semple termed such islands 'thalassic', serving as mid-sea markets and contact-points, so that 'sailors and traders, colonists and conquerors flock in from every side', demonstrating the power of islands 'to attract, preserve, multiply and concentrate population' (Semple 1911, pp. 424, 451). Hence, in contrast to the cultural and linguistic stagnation of Sardinia's highland shepherds, positioned off the beaten track as regards the main currents of Mediterranean history, 'polyglot Malta tells the story of successive conquests, a shuttlecock history reflecting its role as a meeting ground for north and south in the Mediterranean – even in its language' (Semple 1911, p. 428).

Yet these roles too have changed through time, as the following brief examples show. Many large islands which in recent centuries were populated by farmers and shepherds, with little urban dynamism, were in ancient times the centres of advanced civilisation, as witnessed by the survival of numerous stone monuments (especially in Corsica, Sardinia, Malta and the Balearics) dating from the Neolithic, Bronze and Iron Ages. According to the archaeologist Glyn Daniel (1963, p. 82), these islands of the west Mediterranean were the main centres of megalithic and cyclopean tombs in prehistoric times. Sicily, inhabited since the Palaeolithic, became a power under Ancient Greece to rival Athens itself. The Minoan civilisation on Crete remains one of the richest, yet least understood, of the ancient island cultures. Lipari's plentiful resource of obsidian – a flint-like black volcanic glass – gave this island a special role in the prehistory of the Mediterranean.

Located along the main routes of trade and colonisation in the Mediterranean, islands also functioned as stepping-stones for sailors and conquerors. Mountainous and volcanic islands have special importance in this regard: Homer wrote of the 'clear-seen islands' which enabled ancient navigators, fearful of the open sea, to find their way around the Mediterranean (Semple 1932, p. 589). Corsica and Sardinia, and Sicily and Pantelleria, provided navigational bridges between Europe and Africa; whilst the Balearic Islands were the bridge to the west, to Spain. The frequent eruptions of Stromboli, every few minutes, made the island a natural lighthouse, whilst south of Sicily the little islands of Lampedusa and Lampione were so named because of the custom of lighting fires to warn sailors against shipwreck.

For the Adriatic and the eastern Mediterranean, Braudel (1972, p. 149) evocatively describes the Dalmatian Islands, Corfu, Crete and Cyprus as Venice's 'stationary fleet', mapping the axis of Venetian power to the east. Italy tried to reconstitute this insular axis through the annexation of the Dodecanese (1912) and, during the Second World War, the occupation of the Dalmatian and Ionian Islands. British naval hegemony over the Mediterranean was exercised via three strategically-sited colonies: the near-island of Gibraltar at the entrance to the Basin, Malta at the centre, and Cyprus to the east, not forgetting the Suez Canal to the south-east. During the seventeenth and eighteenth centuries, Britain was also present on Menorca and Corfu.

Not surprisingly, islands were frequently fought over, resulting in fortified citadels and chains of castles and coastal watch-towers. Many Mediterranean islands functioned as the starting-points for imperial projects extending far afield. Hence the Phoenicians started by colonising Cyprus, then progressed to Rhodes, Crete, Thera (Santorini), Malta, Pantelleria, Sicily, Sardinia and the Balearic Islands (Perpillou 1966, p. 483). Rome first expanded to Sicily, Sardinia and Corsica before incorporating the entire Basin and much of Europe in its vast empire. Islands express intensively-layered histories featuring all the maritime peoples who cruised the Mediterranean. For Corsica and Sardinia, the list starts with the Etruscans and Carthaginians and proceeds via the Greeks, Romans, Vandals, Saracens and Catalans to the medieval Italian powers of Pisa and Genoa and, eventually, to the modern French and Italian states.

Islands as colonial plantations

The role of islands as beads strung out along the paths of commerce and colonial relations within (and beyond) the Mediterranean often led insular economies to be taken over by specialised functions. Braudel (1972, p. 155) asks, 'How many islands were invaded by foreign crops, whose justification lay solely in their position on Mediterranean or even world markets?' Grown mainly for export, these crops – wheat, wine, currants, sugar cane, olive oil, cotton – regularly threatened the equilibrium of the island economy, leading sometimes to famine and more often to emigration.

One small Mediterranean island whose economic and demographic history has been deeply affected by monocrop plantation cropping is Salina, the second-largest island of the Aeolian archipelago. For centuries Salina remained bereft of a stable population because of the depredations of pirates. Once piratical raids had been brought under control by the British and French fleets, the Church, owner of most of the island, leased plots of land to incoming farmers who settled, terraced the fertile volcanic slopes, and specialised in the production of malmsey wine, which was much in demand on North European markets during the late eighteenth and the nineteenth centuries. By the latter decades of the nineteenth century practically the whole island was covered in vines and virtually all the inhabitants were engaged in viticulture and the production and the shipping of the wine. Disaster struck at the end of the century when a phylloxera epidemic wiped out the entire vinestock. Lacking capital to replace the vines, many islanders emigrated. In twenty years the population declined 40 per cent from 7,200 (1891) to 4,300 (1911); continued emigration until the end of the 1960s reduced the population to less than 2,000 at the 1971 census. Now, tourism beckons a new future, but one which brings renewed problems (Buffoni 1987; King and Young 1979, pp. 197–200).

Braudel (1972, pp. 155–8) gives plenty of other examples of plantation islands both within and beyond the Mediterranean. One of the classic examples was the chain of 'sugar islands' which extended from Madeira and the Canaries to Cape Verde and São Tomé. All were ultimately ravaged by this monoculture which edged out more sustainable cropping regimes, as north-east Brazil and the Caribbean islands were also to discover. Within the Mediterranean Braudel cites the wheat-growing invasion of Sicily, the monopoly of olives on Djerba, and vines and olives on Corfu, Zante and Crete. Again, all were single (occasionally dual) crop economies produced as a result of foreign intervention, and often harmful to the islanders' long-term survival. On the other hand, Mediterranean crops such as oil, wheat, vine fruits and vegetables make up the genuine 'Cretan diet' and permitted the population of this island to survive under great duress during the last war.

According to Vernicos (1987, p. 103) the extensive vine, olive and other cash-crop monocultures which have been introduced on Mediterranean islands have had three important environmental effects: they have altered the landscape, reduced ecological variety and stability, and interfered with the water cycle. But island specialisms are not limited to agricultural products. They also include the operation of harbour and military facilities (as with Rhodes and Malta under the Knights of St John and British rule), mineral exploitation (such as salt on Ibiza and Milos) and, in recent decades, the 'plantation industry' of tourism. Like monocrop agriculture, such specialisms are implemented to cater for the demands of external markets, on which they are highly dependent and therefore vulnerable, and they usually involve limited consideration given to environmental protection and to long-term strategies of economic and resource diversification. As the Caribbean islands demonstrate most clearly, once economic diversity is destroyed by plantation cropping, and once the 'natural' resources have been exhausted, the export of labour remains the only option. For Sicily, Schneider and Schneider (1976) have persuasively linked the export of wheat, the plantation crop of this island since Roman times, to the export of labour, first in the early decades of the twentieth century towards the United States, and then in the 1950s and 1960s towards Northern Europe.

The complex demography of Mediterranean islands

A discussion of the evolution of population in the Mediterranean islands is a potentially complex debate because of the variety of trends observable and the fact that population change itself is far from straightforward, being based on interactions between fertility, piracy and wars, mortality and migration. Migration alone is a complex phenomenon for there is a great variety of flows to be observed, both within the island realm (interior to coast, small islands to larger islands) and external migratory linkages to the mainland and abroad, which may or may not be balanced by return migration and the immigration of 'outsiders'.

Two basic observations underpin the following discussion. The first is that demographic phenomena on islands are inherently unstable and subject to wide fluctuations (Cruz *et al.* 1987). Vulnerability to famine and epidemics accounts for some of this demographic variability, but there is a more constant and all-embracing factor which is islands' traditional openness to external influences and events – be they natural occurrences such as storms, volcanic eruptions and earthquakes, or political processes such as warfare and colonialism, or regional and global market forces (Connell and King 1999). During the sixteenth century, dozens of islands were completely deserted after their plundering by the warlord Barbarossa. In the early 1820s, the populations of Samothrace, Chios and Psara were slaughtered or sold as slaves by the Turks.

This leads to the second observation, which is that the reconstruction of demographic trends in these islands can be seen to be closely related to the progress (or collapse) of their economies (Carli 1994). Persistent economic problems include smallness of scale, lack of technical modernisation, higher transport costs, and dependency on external markets. To take agriculture as an example, most Mediterranean islands do not have the size or capacity to develop large-scale farming operations, mechanisation, land reclamation or irrigation schemes; hence farming, and agricultural employment, have been in steady decline over several decades.[6]

Setting aside for a moment the migratory option, this decline in agricultural employment has been largely offset by a rise in employment in tertiary activities, above all in tourism and the public sector (civilian and military administration, education, health etc.). Industrial employment has remained stagnant, although there have been some successes in the development of light manufacturing in Malta and Cyprus. Government-sponsored attempts to foster large-scale industrial development in Sicily and Sardinia in the 1960s and 1970s had only partial success. For the most part these complexes were based on oil-refining and petrochemicals, and they have been in crisis for the past twenty years, as was noted in Chapter 3.

Recent population increase in the major islands

Let us turn to an examination of some population data for the main islands and island groups. Figure 13.2 graphs the evolution of island populations at approximately ten-year intervals between 1820 and 1990. A number of trends are apparent. Most of the larger islands – Sicily, Sardinia, Cyprus, Crete, Malta and the Balearics – show a more or less uninterrupted pattern of growth in their populations. Where interruptions to this growth have occurred, they have been short-lived. In the case of Crete, the bloody insurrection of 1820–29 cost half the island's population, which fell from over 200,000 in 1820 to 107,000 in 1834. Further uprisings against the Ottomans, who held Crete from 1669 to 1898, led to the progressive expulsion of the so-called 'Turkocretans', the last remnants of whom were exchanged for Greeks from Asia Minor in 1923 (Kolodny 1996). Thereafter the Cretan population grew steadily, apart from during German occupation and a small decline in the 1960s, the latter due to the excess of net out-migration (to Athens and Germany) over natural increase. The same decade also witnessed temporary population loss due to mass emigration from the densely populated islands of Malta and Sicily. Sicily's emigrants went mostly to mainland Italy and to Switzerland and Germany; Malta's moved within the anglophone world, to Britain, Australia, Canada and the United States.

The Balearics constitute a further variant within this cluster of major islands. Here population stagnation occurred around the turn of the century (1890–1910), due to a combination of out-migration (to Catalonia, Algeria and the Americas) and low birth-rate; later the Balearic archipelago showed accelerated population increase after 1960, in contrast to most other Mediterranean islands where the 1960s were a decade of population stagnation or loss due to net out-migration. As we shall see later, tourism is the key to the post-1960 Balearic boom, which is both demographic and economic, but local demographers also insist on the existence of a specific Balearic demographic model, distinct from the rest of Spain (but shared by Barcelona and its hinterland), characterised by a very early demographic transition to low fertility and low mortality (Vidal Bendito 1994).

The steadiest evolution is shown by Cyprus during British rule (1878–1960). Cyprus avoided the large-scale population exchanges between Greece and Turkey, damage and famine during the Second World War, and the Greek Civil War. During the mid-1970s, however, the internal demography of the island underwent a major upheaval as a result of the partition of Cyprus into Turkish and Greek sectors.

Corsica: an island of mountain depopulation

Even more unique is the case of Corsica which, alone of the larger Mediterranean islands, has a demographic history dominated by long-term decline, arrested only

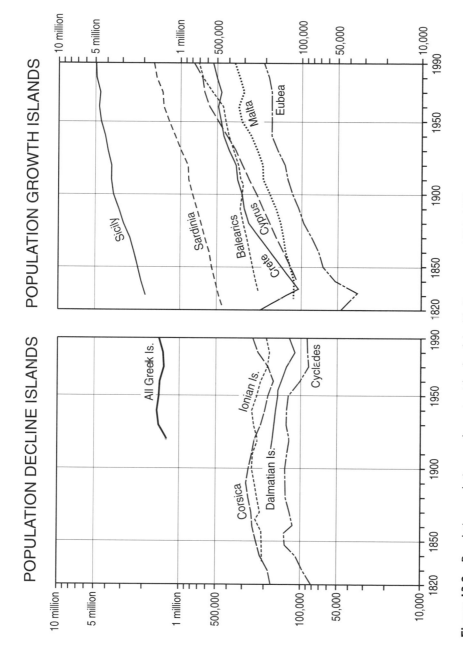

Figure 13.2 Population evolution on the major islands of the Mediterranean, 1820–1990

Source: Kolodny (1976), updated.

in recent decades. For Braudel (1972, pp. 158–9), Corsica was the island of emigrants *per excellence*: 'the population, too great for the island's resources, swarmed in all directions' – to Genoa, Venice, Rome, Leghorn, Algiers, Constantinople and, above all, Marseilles, which became the effective 'metropolis' of Corsica. Braudel was writing of the sixteenth century; more recent data confirm the tradition of Corsican emigration. As figure 13.2 shows, Corsica's population reached a maximum of about 285,000 in 1890, after which a long period of decline ensued lasting until 1960, when the population stood at around 160,000, a decrease of 44 per cent. There were many causes of this population loss (including military casualties during the First World War), but the key one was the failure of an over-populated agro-pastoral economy to cope with the market penetration of the modern age, above all cheap imports of wheat and other staples (Thompson 1977). Given the mountainous nature of the island and its pattern of settlement, concentrated inland on hill-tops and steep valley-sides, this was both a case of mountain depopulation and a specifically insular problem in that, up to the advent of the modern steamship, Corsica's island status had preserved the viability of a self-sufficient economy. Corsica illustrates the confrontation of island autarky with the economy of the most developed country of the Mediterranean – France. The counterpart was the availability of many employment opportunities in the administration and the army on the so-called 'continent' and in the former French colonial empire (Kolodny 1962).

Since the late 1950s some regional development measures, especially for land reclamation and tourism, have produced a modest economic recovery and a demographic turnaround. Although some young people still tend to migrate to mainland France, there have been several incoming migration flows (*pieds noirs*[7] from North Africa, Maghrebi labour migrants, Corsican returnees and long-term residential tourists) and the population has risen to 227,000 in 1975 and to 250,000 in 1990 (Kolodny 1997, p.776). Nevertheless the broad historical contrasts with other Mediterranean islands must be stressed. In 1840 Corsica had around twice the population of Cyprus, Crete and Malta; by 1990 these islands had respectively three times, twice and 40 per cent more than the Corsican total.

Population trends on smaller islands: the case of the Dalmatian Islands

In many respects Corsica's historical population profile is more akin to that of the smaller islands of the Mediterranean such as the Dalmatian Islands or the Cyclades. These islands, over the long term, have hosted stagnant or decreasing populations, with some recent population increase, albeit spread unevenly between islands.

Starc's (1987) study of population change in the Dalmatian Islands provides a good example of intra-archipelago variation according to factors such as population size and distance from the mainland shore. This elongated archipelago, stretching between the Istrian peninsula and Dubrovnik, saw its maximum population attained in the early decades of the twentieth century (1910: 173,263), having grown from the 117,481 enumerated at the first complete census of the islands in 1857. After 1921, the population fell to 112,208 in 1981, before recovering to 126,447 in 1991.[8]

The main period of population growth, the second half of the nineteenth century, corresponded to a period of prosperity when many islands specialised in the production of wine. Easy access to the main market, Austro-Hungary, was gained via Trieste, Rijeka (Fiume) and Zadar (Zara). Towards the end of the century, Dalmatian wine enjoyed privileged access to this market, for the competitor vineyards in France and Italy were already diseased. In 1894, however, phylloxera arrived in Dalmatia and, as it spread through the islands, the trade fell away. The dry limestone islands were unable to replant with new stock and were incapable of developing alternative forms of intensive production.[9] Heavy population loss ensued due to out-migration to the mainland and emigration overseas: a familiar story, recounted earlier in our summary of the situation on the Italian island of Salina. Much early emigration was to North America; after World War Two the islanders joined in the main Croatian emigration which was to Germany. This emigration continued until the 1970s, by which time deserted cultivation terraces and picturesque but half-empty villages were the only witnesses to the prosperity of the past (Roglić 1964, p. 560). In fact the sparsity of population in the Dalmatian Islands can be readily seen by some comparative summary data: in 1991 the islands accounted for 5.6 per cent of the area but only 2.6 per cent of the total population of Croatia; and for 3.1 per cent of the total area of the Mediterranean islands, but only for 1.2 per cent of their population.

Tables 13.2 and 13.3 explore the geography of population change on these islands, using selected censuses.[10] Table 13.2 divides the islands into three groups according to the size of the population in 1981. The table shows that the larger islands lost 30 per cent of their population over the period 1910–81, the middle-range islands (500–2,000 inhabitants in 1981) lost 52 per cent, and the smallest islands lost 70 per cent. The data indicate that population loss occurred first on the larger islands (from 1910 onwards), whereas the smaller islands held on to their population rather better during the period 1910–53, then suffering emigration at a far higher rate than the larger islands in the postwar decades. Over the postwar

Table 13.2 Dalmatian Islands: evolution of population by size of island, 1857–1981

Census year	Island groups by population size 1981			Total population
	>2,000	500–2,000	<500	
1857	97,177	8,981	11,323	117,481
1910	144,837	13,895	14,531	173,263
1931	137,334	13,932	14,358	165,624
1948	125,626	12,957	13,252	151,835
1961	117,585	11,244	10,662	139,491
1981	101,222	6,629	4,357	112,208
% change 1910–81	−30.1	−52.3	−70.0	−35.2

Source: Starc (1981, p. 150).

Table 13.3 Dalmatian Islands: evolution of population by location of island, 1948–81

Census year	Islands linked to mainland by bridge	Other inner islands	Outer islands	Total population
1948	34,602	91,241	25,992	151,835
1953	33,919	90,772	25,382	150,073
1961	30,732	86,915	21,844	139,491
1971	28,207	82,156	16,700	127,063
1981	27,277	73,321	11,610	112,208
% change 1948–81	−21.2	−19.6	−55.3	−26.1

Source: Starc (1987, p. 150).

period (1948–81), the respective annual rates of population decline for the large, medium and small islands were 0.9 per cent, 2.0 per cent and 3.4 per cent.

Table 13.3 makes a different threefold division of the islands: those connected to the mainland by a bridge (Vir, Pag, Murter, Čiovo, Krk), other 'inner' islands located close to the mainland shore and with frequent ferry connections, and 'outer' islands (more distant from the coast and with less than one boat per day). The data show a clear contrast between inner and outer islands (the latter have been losing population at four times the annual rate of the former – 2.4 per cent as against 0.6 per cent), but no difference between the bridged islands and those without a fixed connection.[11]

A third dimension of the geography of population change in the Dalmatian Islands is the shifting distribution between coastal and inland settlements. Census micro-data for 1948 and 1981 are again very revealing (Starc 1987, p. 151). Of 279 settlements in the Dalmatian Islands listed in the 1981 census, 155 are coastal and 124 are in the interior. During the intercensal period in question, the coastal settlements lost one fifth of their populations, whereas the interior locations lost one half. Only 32 settlements increased their populations: all were coastal, including the largest settlement, Mali Lošinj (5,244 in 1981, 6,566 in 1991), a garrison of the Yugoslav army and fleet. In 1981 nearly 80 per cent of the Dalmatian island population lived in coastal settlements; the remaining one fifth in the interior consisted of elderly people.

Case studies of individual islands reveal more of the variations in the dimension, timing and nature of the population processes on different islands. For instance, during the earlier period of decline, from 1910 to 1953, when there was a 13 per cent decline in the total population of the archipelago, some of the biggest losses occurred on the larger islands which were remote from the new economic and administrative centres on the mainland coast. Hence, to the north Krk, Cres and Lošinj lost 21, 41 and 45 per cent respectively, and Brač to the south lost 36 per cent (Roglić 1964, pp. 572–4). On several smaller islands the long-term population decline threatens to prove fatal: Unije slumped from 783 inhabitants in 1921 to 81 in 1991, Susak from 1,629 in 1948 to 188 in 1991, and Zlarin from 1,980 in 1921 to 359 in 1991. More and more islands are seeing their populations fall below the 500 mark – roughly the minimum size upon which a sustainable isolated community can be built.[12]

Before the outbreak of the Yugoslav wars in 1991, tourism was considered as the economic salvation of the Dalmatian Islands, not only because of the direct financial and employment effects, but also because of beneficial multiplier effects on the local agricultural and fishing sectors. Starting in the 1970s, the northern islands (Krk, Cres, Rab) enjoyed an economic and demographic revival because of their accessibility for tourists coming from Northern Italy, Germany and Austria. Much of the capital for investment in tourist enterprises came from the savings of returning migrants (Bennett 1979). By the mid-1980s there were 165,000 tourist beds on the islands, catering to 1.5 million annual visitors and 14.5 million bednights. Whereas in the 1950s there was no island where tourism accounted for more than 10 per cent of the income, by the 1980s the share was 25 per cent on average, and much higher on some of the northern islands like Rab and Cres (Starc 1987, p. 151). On Krk, where the population grew by 23 per cent during 1981–91, key factors were the building of the bridge and the opening of the airport for Rijeka at Omišalj, on the northern tip of the island.

The islands' incipient tourist boom was brought to an abrupt end by the break-up of Yugoslavia. During late 1991 and early 1992 islands became temporary homes for refugees, many of them housed in requisitioned tourist accommodation. Hvar hosted 1,150 refugees from Vukovar, Korcula took more than 2,000 residents of Dubrovnik during the bombardment of that city, and thousands more were taken to Cres, Lošinj, Krk, Pag, Rab, Murter, Ugljan, Zlarin and Dugi Otok (Starc 1992). At the same time most of the non-Croatian population resident on the islands departed.[13] After 1995 and the Dayton Accords, the islands, having absorbed further contingents of displaced Croats from Bosnia-Herzegovina, started to recover some of their touristic vocation.

This detailed examination of population trends on the Dalmatian Islands precludes more than a brief mention of other archipelagos made up of smaller islands, where many similar trends of long-term decline and selective recent recovery can be observed: the cases of the Aeolian Islands and the Cyclades, analysed by the authors, have been cited already (King and Young 1979; Kolodny 1991 and 1992); mention should also be made of the detailed study of postwar population trends in the Aegean Islands by Siampos (1994). Above all, the case study presented here points to the crucial role of tourism in regenerating many of these smaller Mediterranean islands which were in economic and demographic decline, and in contributing to the economic diversification of larger islands with already quite developed economies.

Tourism: a major factor in recent economic development

Some insights into the role of tourism in the economic geography of Mediterranean islands, and of the position of islands in the geography of tourism flows in the

Mediterranean region, have already been given in two earlier chapters. In **Chapter 9** it was pointed out that the first inclusive air package holiday was to Corsica in 1950 – but that now Corsica is the most expensive Mediterranean region for holiday-makers. It was also shown that Cyprus and Malta, with 2.1 million and 1.2 million tourists arriving annually (1994), are the Mediterranean countries where tourism has the greatest relative impact, whether this is measured by the ratio of tourist arrivals to the resident population (about 3:1 in both cases), or by tourist receipts per capita ($2,341 and $1,770 respectively in 1994). On the other hand, both Cyprus and Malta are far more dependent than any other Mediterranean country on the supply of tourists from a single country – the UK. This is yet another example of the fragile and dependent nature of island development. In **Chapter 10** a special section dealt with the Balearic Islands as a destination for North European residents who have settled for climatic and lifestyle reasons, often upon retirement: in 1995 these islands hosted more than 60,000 European residents, more than three-quarters of them German or British. The Canary Islands too have been heavily affected by North European mass tourism and residential settlement for business and retirement.

Outside of Cyprus, Malta and the Balearics, the other main complex of tourist islands in the Mediterranean are the Greek islands, especially Corfu, Crete, Rhodes and the Aegean islands. On some of the Greek islands, such as Mykonos, Skiathos and Santorini, tourism overwhelms and saturates virtually all aspects of island life. Also important are Corsica, Sardinia and Sicily, where the size of the islands makes tourism a complementary economic activity rather than the dominant one, and the Isle of Capri which has a history as a holiday island dating back to the Romans.

Islands and mass tourism: the case of the Balearic Islands
The link between economic and demographic changes on islands, which has been a recurrent theme in this chapter, is nowhere seen more clearly than in the Balearic Islands, where population increase has occurred very much in parallel to the islands' tourist development (Salvà Tomàs 1991). According to Bull (1997, p. 137), the Balearic Islands provide the quintessential example of development based on the 'constantly renewable resource of tourism', achieved with 'stunningly successful results'. The economic impact has been 'truly remarkable', making the archipelago – a Spanish autonomous region – one of the most prosperous regions of Southern Europe and well above the EU average on most socio-economic indicators. However, the 'Balearic model' of Mediterranean development is not entirely un-problematic, as we shall see later.

The statistical story of postwar population change, migration and tourism arrivals is set out in table 13.4. Unlike many other Mediterranean islands, where a high rate of natural population increase continued until quite recently, in the Balearics the demographic situation had already reached a mature stage by the early decades of the twentieth century (Vidal Bendito 1994). Hence overall population change during the postwar period is less a reflection of natural change, and more attributable to net in-migration. Migrants have come from three main origins: first from Andalusia and other less-developed parts of mainland Spain; then from North

Table 13.4 Balearic Islands: trends in population and tourism, 1950–98

	Majorca	Menorca	Ibiza	Total
Population				
1950	339,716	41,412	38,154	419,282
1960	363,202	42,305	37,225	441,732
1970	438,656	48,817	45,473	532,946
1981	534,511	57,243	64,155	655,909
1991	568,187	64,412	76,546	709,145
1996	609,150	67,009	84,220	760,379
Migration balance				
per thousand per year				
1950–55	−0.34	2.04	−12.28	−1.16
1955–60	3.02	−8.56	−1.78	2.85
1960–65	9.01	−0.74	6.77	7.91
1965–70	12.76	8.09	14.71	12.50
1971–75	13.38	7.20	11.88	12.69
1976–81	9.52	5.48	30.47	11.10
1981–86	2.31	4.39	7.28	2.99
1986–91	3.75	9.58	11.12	5.05
1991–96	12.66	5.65	14.36	12.22
Tourist arrivals ('000)				
1960	361	8	31	400
1965	960	14	106	1,080
1970	1,853	56	363	2,272
1975	2,765	121	549	3,435
1981	3,065	211	613	3,889
1985	3,747	330	873	4,950
1990	4,881	546	1,010	6,437
1995	6,131	813	1,449	8,393
1998	7,460	970	1,666	10,096

Source: Salvà Tomàs (1991), updated.
Note: Ibiza includes Formentera.

Africa, especially Morocco, whose labour migrants find work in low-grade occupations in tourism and construction; and thirdly from Northern Europe, the aforementioned residents and seasonal occupants of second homes. This in-migration has two effects: a direct and obvious contribution to increasing the populations registered at each census, and an indirect contribution as a result of their young age structures and high fertility potentials.[14]

The first part of table 13.4 shows that the total population increased by 81 per cent from 419,000 in 1950 to 760,000 in 1996. The rate of increase was highest for Ibiza (121 per cent), despite an initial decrease in population between 1950 and 1960, and lowest for Menorca (62 per cent) – respectively the islands most and least affected by the 'tourist boom'. For all islands, the fastest rates of increase were registered during the two decades 1960–80, when the pace of tourist development was at its most frenetic: during these two decades the annual population growth rate was 2.2 per cent (Majorca 2.2, Ibiza 2.8, Menorca 1.6). The pace of population growth slackened off during 1981–91, concomitant with the deceleration in the rate of tourist expansion; but then demographic growth accelerated, due to a new wave of in-migration, during 1991–6.

The middle section of table 13.4 sets out the migration component of this population change. After an initial phase of slight out-migration during the 1950s, net in-migration dominates the islands throughout the rest of the period under study. During the years 1965–81 and again during 1991–6, net in-migration added well over 1 per cent per year to the population – and over 3 per cent in Ibiza during 1976–81. Nearly all these migrants have settled in the main urban and tourist areas, joining the local currents of rural-urban migration within the islands. By 1990 the main towns (Palma, Ibiza Town, Mahón and Cuitadella on Menorca) accounted for well over half the total population of each island (Vidal Bendito 1994, p. 130).

The final section of the table shows the dramatic development of tourist arrivals over the postwar decades: a tenfold increase during 1965–98, from 1 million to 10 million. This expansion faltered somewhat in the 1980s, reflecting price- and product-competition from other tourist countries, both in the Mediterranean and beyond, within a context of global recession which put a squeeze on holiday expenditure. Growth then resumed strongly during the 1990s, especially in Menorca. It is interesting to note inter-island variations in tourist growth profiles, particularly between Ibiza, where there is evidence of a 'boom-bust' cycle of migration and tourist development, and Menorca where both the tourist and demographic expansion have been more modest and evenly paced.[15]

As well as contributing strongly to economic and demographic growth, tourism has also had the effect of drastically reducing the insularity of the islands, above all by the density and proliferation of air connections to Spain and Northern Europe. Palma has 90 flights each week to Barcelona, 57 to Madrid, 27 to Valencia, 23 to Alicante and smaller numbers of direct flights to Bilbao, Seville, Tenerife and Gran Canaria. Scheduled services also fly to London, Paris, Marseilles, Frankfurt, Geneva and Zurich. But it is the swarm of charter flights that makes Palma airport, in the summer months, one of the busiest in the world, with flights landing and taking off every couple of minutes. In 1992, 16 million passengers entered the islands by air, 4.2 million on scheduled services, mainly from within Spain (3.7 million), and the rest on charter flights from a wide variety of North European airports, all within two hours flying time.

Questioning the tourism model of development

However the tourist-led transformation of the Balearics has not been without its disadvantages. To some extent, these are the problems attendant on virtually all areas of mass tourism, and they have been thoroughly reviewed in the literature on the geography of tourism (see, for example, Pearce 1995). But the island setting, both in the Balearics and in other Mediterranean islands where coastal landscapes have been smothered by tourist development, throws these issues into unusually sharp relief.

First, there is the dramatic change in population distribution, settlement patterns and built forms. Traditionally, the population of the Balearics, like the inhabitants of many other Mediterranean islands, lived in clustered village settlements in the interior. They located thus for reasons of security against invaders and pirates, and because the interior plains contained much of the best agricultural land. The coast was deserted save for a few ports. Now, as a result of unbridled tourist development, 'a monotonous built form encircles much of the coastline' consisting of 'slab and high-rise hotel and apartment blocks . . . joined together by . . . strips of bars, cafés [and] souvenir shops' (Bull 1997, p. 145). This is a landscape which reaches its most highly-developed form around the Bay of Palma on Majorca, but it can also be seen on Ibiza, along the south coast of Cyprus (especially at Limassol and east of Paphos and Larnaca), and on Crete (between Iraklion and Ayios Nikolaos) and Malta (north of Valletta). Pearce (1995, p. 119) assembles data from various holiday islands worldwide which show that the tourist accommodation density (rooms per square kilometre) is higher in Majorca than any other major tourist island, whilst from table 13.4 we can see that the number of annual tourist arrivals, 10.1 million in 1998, is 13 times the resident population.[16]

Second, there is the overwhelming reliance of the economy of tourist islands on one employment sector. Bull (1997, p. 138) quotes figures which suggest that up to 70 per cent of all employment in the Balearics derives either directly or indirectly from tourism. This over-specialisation on tourist employment has further prob-lematic features: much is low-status, low-paid work; and there is a high degree of seasonality involved – three-quarters of foreign tourists holidaying in the Balearics arrive in the peak summer months. A further element of the structural dependency of the Balearic tourist trade is the fact that more than 60 per cent of the foreign tourists originate from two countries, the UK and Germany.

Market specialisation can also be observed on islands, which further narrows their interaction with the supply of tourists. Ibiza has placed itself firmly within the youth tourism market: for some years now it has been at the pinnacle of the European clubbing scene, but at some cost to other sectors of the tourist market. Menorca offers a type of holiday experience which appeals more to families with small children and to older people. Similar specialisms can be observed amongst various Greek islands – appealing to sailing enthusiasts, backpackers, gay people etc. Capri, Hydra, Symi and Gozo are much affected by day-trip tourism, with obvious effects on these islands' touristic economies and rhythms of life. The inci-dence of national internal tourism is important too, including islanders coming back for holidays.

Aware of the dangers of over-specialisation, many economic strategists and policy-makers are seeking ways to escape the specialist tourist monoculture by diversifying to other types of tourism (for instance, walking tours in Majorca's beautiful limestone mountains) and to other economic activities, including those which attract wealthy and high-status residents. Environmental improvements, including tree-planting, new esplanades, pedestrianisation and even the dismantling of obsolete hotels, are also being implemented by the Balearic regional government in an effort to upgrade facilities and re-position the islands away from the lower end of the mass tourism market (Bull 1997; Morgan 1991). Similar efforts are being made to remodel island tourism elsewhere in the Mediterranean, notably in Malta and Cyprus where history, culture and (in the case of Cyprus) mountain scenery and skiing can be included as part of the 'tourism product' (Lockhart 1997). However, in the smaller Mediterranean islands, the combination of smallness and insularity limits the diversification of tourism beyond the standardised 'sun, sand and sea' product (Pearce 1995, pp. 118–35).

A final problem with the 'tourism and development' discourse is the implication that tourism offers the only path to development for Mediterranean islands and the further assumption that, before tourism 'happened', islands were in an underdeveloped state. Vidal Bendito (1994) criticises the common division of Balearic development into 'pre-tourism' and 'tourist' periods, because it implies that in the pre-tourist phase the islands were in a primitive state from which tourism performed a rescuing, modernising role. Economic and demographic data show that the modernisation of Balearic society was well under way before the advent of tourism – and this fact probably accounts at least in part for the overall success of the tourist expansion as well as for the generally liberal attitudes of the local population towards the behaviour of the tourists. We can assume the same for Cyprus, where irrigated crops for export, banking and offshore activities, and light industry utilising high technology offer a prosperous standard of living to the Greek sector.

Conclusion: 'insular fatality' does not exist

Synthesising the material presented in this chapter on the economic history, demography and current state of tourism on Mediterranean islands, we can divide the island realm into two categories. These island types are not mutually exlusive. One island or island group can move from one type to another over time, and it is also possible for both types to co-exist on the same island, perhaps reflecting the contrast between the coast and the interior, or between summer and winter.

Firstly, there are islands which are economically and demographically in decline. These islands tend to be small, remote and without significant economic resources. Cruz *et al.* (1987) describe them as *dominantly emigrant islands*, characterised by an ageing labour force from which emigration has removed most of the younger, skilled and educated people. These islands need human and economic resources for innovation, investment and diversification, but such critical inputs are lacking. Economic activity is limited to residual cultivation of smallholdings and some fishing. The declining population leads to a fading away of social and community activity as the thresholds necessary to sustain various functions (schools, shops, associations etc.) are no longer met. Typically, such islands are also distinguished by particular landscape characteristics – near-derelict villages, crumbling terraces, abandoned fields and olive groves, overgrazing and vulnerability to fire. Migrant remittances and pensions are the main sources of income.

Second, there are islands with rising populations and growing economies, although the long-term health of the economy may be open to vulnerability through over-reliance on one or very few sectors, such as tourism. These are *dominantly immigrant islands* (Cruz *et al.* 1987) such as the Balearics or Rhodes, or their population buoyancy may be the result of a continued pattern of natural increase, as with Sardinia, Crete, Euboea and Cyprus. Other problems faced by at least some of this group of islands are high population densities and overcrowding, resulting in environmental stresses – shortage of land, water, energy etc., and difficulty of waste disposal, particularly if these resource issues are strongly exacerbated by seasonal mass tourist flows. Further problems arise in the social domain: the erosion of island identity, cultural conflict between 'insiders' (long-term island residents) and 'outsiders' (immigrants, tourists), and the breakdown of the decision-making consensus (e.g. as to the future direction of the islands' development).

Two remarks by way of final conclusion – one regarding the nature of economic and demographic development, and the other stressing the importance of the ecological dimension in these islands' future.

This chapter has repeatedly demonstrated the close interrelationship between economic development (and decline) and demographic trends. In addition, virtually all islands have heavy economic dependence on the outside world. In practice this means that, for the forseeable future, a large proportion of Mediterranean islands will continue to form a part of Europe's 'leisure periphery' along with other regions of the Mediterranean Basin (Montanari 1995). Within this broad regional area, islands may have distinctive roles, catering to different tastes and markets. While some travellers and writers will want to believe that they can still enjoy the remaining 'undiscovered islands' of the Mediterranean (Lancione Moyer and Willes 1990), it can be expected that most islands will conserve and even intensify their engagement with tourism, although the forms that tourism takes may change, and some economic diversification may occur in the future. Last but not least, we would point out that, unlike the 1950s and 1960s, Mediterranean islands are no longer to be regarded as poverty-stricken regions.

In most islands the opportunities for development are closely interlinked with the preservation or restoration of ecological balances which have often been badly disturbed in the past. Especially with the expansion of tourism, this need becomes paramount, since tourism depends on environmental quality more than any other economic activity. As Coccossis (1987) points out, the ecological approach – the effort to combine development with environmental protection – becomes central to the *problématique* of island planning. In both immigrant and emigrant islands, the overall problems of demographic instability and environmental fragility remain, greatly complicating the process of planning for sustainable development. However, we reject the hypothesis of 'insular fatality' – islands of the Mediterranean are diverse and changing, and not subject to deterministic fates just because of their island status.

Notes

1 Other high elevations on Mediterranean islands include Monte Cinto 2,710m (Corsica), Mount Ida 2,456m (Crete), Olympus 1,951m (Cyprus) and Gennargentu 1,834m (Sardinia). Moving outside the Mediterranean to other islands 'owned' by Southern European countries, we can note Pico de Teide 3,718m (Tenerife), the highest mountain in Spain; Pico Alto 2,351m (Azores), the highest mountain in Portugal; and Pico Ruivo 1,861m (Madeira).

2 Population figures in this chapter are for the early 1990s, unless otherwise stated.

3 Cyprus achieved independence in 1960, Malta in 1964. However, the former split into two *de facto* states as a result of the Turkish landings of 1974 and the resultant partition

of the island. The Republic of Cyprus (Greek Cypriot) occupies the southern 60 per cent of the island; to the north lies the self-proclaimed Turkish Republic of North Cyprus.

4   These are Sicily, Sardinia, Corsica and Majorca. Railways used to exist on Malta and Cyprus, but they have been closed. For details see Kalla-Bishop (1970).

5   A study of the Greek islands divides them into two groups: islands of 'land-people' – agriculturalists engaged in typical Mediterranean polyculture and stock farming – and islands of seafarers. Both sets of islands, however, have been deeply affected by tourism since the 1960s (Kolodny 1974, 1991, 1992).

6   There are some exceptions to this argument about the limitations on agricultural development. Large-scale irrigation works, often based on the construction of reservoirs in the hills, have enabled agricultural intensification in many of the larger islands, such as Corsica (the eastern lowlands), Sardinia (the Campidano), Sicily (the Plain of Catania), Crete (Messara Plain) and Cyprus (the southern coastal plains), whilst tourism has given a market boost to agricultural production on some islands, such as the Balearics. In spite of these developments, the numbers of people engaged in farming are everywhere in decline. For a detailed examination of the Sardinian case see King (1971).

7   French colonial people leaving North Africa, especially after the Algerian independence war; quite a lot of *pieds noirs* were of Corsican origin or ancestry. Many of them settled in the newly-reclaimed farms of the eastern plains of Corsica, south of Bastia (Thompson 1966). Corsican development was given fresh impetus by the granting of a measure of regional status in 1982.

8   Of the many hundreds of islands in the archipelago, only 66 are inhabited, and only 43 inhabited permanently (23 are occupied seasonally or are the sites of lighthouses, churches, monasteries etc.).

9   Replacement of infected vines with resistant American stock was very difficult, for several reasons: the small farmers lacked capital, the karstic soils were unsuited to the new varieties which required deeper soils, and the reconstituted French and Italian vineyards had already reclaimed the markets which had been supplied by the Dalmatian product (Roglić 1964, pp. 558–60).

10   We have taken 1910 as the census representing maximum population because the 1921 census, which recorded a fractionally higher total, was less reliable, being based on a partial updating of the 1910 figures at a time when some of the islands were under Italian occupation.

11   Of the five islands connected to the mainland, four are of medium size and have had their bridges for some 30–35 years without any appreciable impact on their migration trends. On Krk a bridge was built in 1982 and an immediate increase in traffic, tourism and population was noted. According to Starc (1987, p. 150) the population increase and economic prosperity experienced by Krk in the 1980s are generally explained by reference to the island now being 'part of the continent'.

12   The median population of the permanently inhabited islands in the Dalmatian group was only 657 in 1991, dramatically down on the figure for 1910, 1,163. By contrast, the median Greek island in 1991 had 1,960 inhabitants.

13   These were mainly Serbs (2,356), 'Moslems' or Bosnians (944), ethnic Albanians (548) and Montenegrins (199), many of whom had been linked to military establishments of the former Yugoslavia. It is also worth mentioning the more gradual diminution of the Italian population on the islands, from 18,000 in 1910 to 336 in 1991.

14   This is especially the case for Spanish migrants, who tend to be young adults and young families, originating from the poorer parts of southern Spain; it is less the case for Europeans, a high proportion of whom are elderly, and North Africans who are mainly single male workers.

15   The Menorcan pattern of more moderate tourist expansion produces, and is a result of, a more balanced economic structure on this island, with high-value agriculture (especially dairying), craft industries (leather, clothing, shoes, jewellery) and commerce. This more broadly-based and sustainable economic development reflects Menorca's

strength in established economic sectors rather than any inherent weakness of its tourism sector. Mari (1994, p. 125) claims that Menorca's early domination by the British favoured commercial liberalisation and capital accumulation for investment in industry, while Vidal Bendito (1994, p. 129) points out that the demographic transition and 'European' models of demographic behaviour started earlier on Menorca than the other Balearic isles.

16 Tourist densities are also very high in the Canaries, Spain's other holiday archipelago, and it is no accident that the growth of the Canarian population between 1960 and 1991 – by 58 per cent from 944,448 to 1,493,784 – is almost as high as the 60 per cent growth of the Balearic population over the same period.

References

Bennett, B. C. (1979) Migration and rural community viability in central Dalmatia, *Papers in Anthropology*, 20(1), pp. 75–83.

Braudel, F. (1972) *The Mediterranean and the Mediterranean World in the Age of Philip II*. London: Methuen.

Brunhes, J. (1920) *Human Geography*. London: Harrap.

Buffoni, F. (1987) Salina: economic development of an island in the Aeolian archipelago, *Ekistics*, 323–324, pp. 158–64.

Bull, P. (1997) Mass tourism in the Balearic Islands: an example of concentrated dependence, in Lockhart, D. G. and Drakakis-Smith, D. (eds) *Island Tourism: Trends and Prospects*. London: Pinter, pp. 137–51.

Carli, M. R., ed. (1994) *Economic and Population Trends in the Mediterranean Islands*. Naples: Edizioni Scientifiche Italiane (Collana Atti Seminari 5).

Coccossis, H. (1987) Planning for islands, *Ekistics*, 323–324, pp. 84–7.

Connell, J. and King, R. (1999) Island migration in a changing world, in King, R. and Connell, J. (eds) *Small Worlds, Global Lives: Islands and Migration*. London: Pinter, pp. 1–26.

Cruz, M., D'Ayala, P. G., Marcus, E., McElroy, J. L. and Rossi, O. (1987) The demographic dynamics of small island societies, *Ekistics*, 323–324, pp. 110–15.

Daniel, G. (1963) *The Megalith Builders of Western Europe*. Harmondsworth: Penguin.

Hionidou, V. (1996) The Greek famine of 1941–42: an overview, in Gentileschi, M. L. and King, R. (eds) *Questioni di Popolazioni in Europa: una Prospettiva Geografica*. Bologna: Pàtron, pp. 245–54.

Houston, J. M. (1964) *The Western Mediterranean World*. London: Longman.

Kalla-Bishop, P. M. (1970) *Mediterranean Island Railways*. Newton Abbot: David and Charles.

King, R. (1971) Development problems in a Mediterranean environment: history and evaluation of agricultural development schemes in Sardinia, *Tijdschrift voor Economishe en Sociale Geografie*, 62(3), pp. 171–9.

King, R. and Young, S. E. (1979) The Aeolian Islands: birth and death of a human landscape, *Erdkunde*, 33(3), pp. 193–204.

Kolodny, E. (1962) *La Géographie Urbaine de la Corse*. Paris: CDU-SEDES.

Kolodny, E. (1974) *La Population des îles de la Grèce*. Aix-en-Provence: Edisud.

Kolodny, E. (1976) Aspects d'ensemble de l'insularité méditerranéenne, *Bulletin de l'Association des Géographes Français*, 435–436, pp. 191–5.

Kolodny, E. (1991) La population d'un espace micronésique égéen: les Cyclades, in *Territoires et Sociétés Insulaires*. Paris: Ministère de l'Environment et de la Prévention des Risques Technologiques et Naturels Majeurs, Collection Recherches Environment 36, pp. 217–23.

Kolodny, E. (1992) *Un Village Cycladien: Chora d'Amorgos*. Aix-en-Provence: Publications de l'Université de Provence.

Kolodny, E. (1996) Chypre et la Crète: similitudes et contrastes d'évolution des deux îles principales de la Méditerranée orientale, in Métral, F., Yon, M. and Ioannou, Y. (eds)

Chypre Hier et Aujourd'hui entre Orient et Occident. Paris and Lyon: Travaux de la Maison de l'Orient Méditerranéen 25, pp. 29–52.

Kolodny, E. (1997) Insularité méditerranéenne et spécificité de la Corse, in Villain-Gandossi, C., Durteste, L. and Busuttil, S. (eds) *Méditerranée, Mer Ouverte. Tome II: XIXe et XXe Siècles.* Malta: International Foundation, pp. 767–87.

Lancione Moyer, L. and Willes, B. (1990) *Undiscovered Islands of the Mediterranean.* Santa Fe, New Mexico: John Muir Publications.

Le Lannou, M. (1941) *Pâtres et Paysans de la Sardaigne.* Tours: Arrault.

Lockhart, D. G. (1997) Tourism to Malta and Cyprus, in Lockhart, D. G. and Drakakis-Smith, D. (eds) *Island Tourism: Trends and Prospects.* London: Pinter, pp. 152–78.

Mari, S. (1994) The economic specificity of Menorca in relation to the Balearic Islands, in Carli, M. R. (ed.) *Economic and Population Trends in the Mediterranean Islands.* Naples: Edizioni Scientifiche Italiane (Collana Atti Seminari 5), pp. 111–28.

McNeill, J. (1992) *Mountains of the Mediterranean World.* Cambridge: Cambridge University Press.

Montanari, A. (1995) The Mediterranean region: Europe's summer leisure space, in Montanari, A. and Williams, A. M. (eds) *European Tourism: Regions, Spaces and Restructuring.* Chichester: Wiley, pp. 41–66.

Morgan, M. (1991) Dressing up to survive: marketing Majorca anew, *Tourism Management*, 12(1), pp. 15–20.

Pearce, D. (1995) *Tourism Today: A Geographical Analysis.* London: Longman.

Perpillou, A. V. (1966) *Human Geography.* London: Longman.

Roglić, J. (1964) The Yugoslav littoral, in Houston, J. M., *The Western Mediterranean World.* London: Longman, pp. 546–79.

Salvà Tomàs, P. (1991) La population des îles Baléares pendant 40 ans de tourisme de masse (1950–90), *Méditerranée*, 77(1), pp. 7–14.

Schneider, J. and Schneider, P. (1976) *Culture and Political Economy in Western Sicily.* London: Academic Press.

Semple, E. C. (1911) *Influences of the Geographic Environment.* London: Constable.

Semple, E. C. (1932) *The Geography of the Mediterranean Region: its Relation to Ancient History.* London: Christophers.

Siampos, G. S. (1994) Economics and population in the Mediterranean islands: the Aegean islands, in Carli, M. R. (ed.) *Economic and Population Trends in the Mediterranean Islands.* Naples: Edizioni Scientifiche Italiane (Collana Atti Seminari 5), pp. 59–93.

Starc, N. (1987) The islands of the Yugoslav Adriatic coast: development problems and prospects, *Ekistics*, 323–324, pp. 147–52.

Starc, N. (1992) Croatian islands at war – an agony in the Adriatic, *Insula*, 1, pp. 4–9.

Thompson, I. B. (1966) Some problems of regional planning in predominantly rural environments: the French experience in Corsica, *Scottish Geographical Magazine*, 82(2), pp. 119–29.

Thompson, I. B. (1977) Settlement and conflict in Corsica, *Transactions of the Institute of British Geographers*, 3(3), pp. 259–73.

Vernicos, N. (1987) The study of Mediterranean small islands: emerging theoretical issues, *Ekistics*, 323–324, pp. 101–9.

Vidal Bendito, T. (1994) The Balearic population in the twentieth century, in Carli, M. R. (ed.) *Economic and Population Trends in the Mediterranean Islands.* Naples: Edizioni Scientifiche Italiane (Collana Atti Seminari 5), pp. 129–54.

Vidal de la Blache, P. (1926) *Principles of Human Geography.* London: Constable.

Wilstach, P. (1926) *Islands of the Mediterranean.* London: Geoffrey Bles.

14

Environmental crises in the Mediterranean

John Thornes

Environmental crises are important problems whose resolution usually requires a significant effort of organisation and enterprise as well as an interdisciplinary approach. Such crises usually involve both humankind and the physical processes that shape the earth's surface. As such, they are not problems for which it is possible to legislate readily.

This chapter will critically analyse three Mediterranean environmental crises: water, land degradation and pollution. Although the situation is extremely serious on all three fronts, especially with regard to water, the Mediterranean Basin's most critical resource for the future, I seek to moderate the Armageddon view and to point to constructive measures for the future. My focus is mainly on the European part of the Mediterranean, and especially on Spain, where a considerable amount of research has been conducted in recent years.

Although it is widely held that the Mediterranean was deforested and that environmental problems started in earnest in classical times, Gilman and Thornes (1972) have shown that in Spain the land was already severely eroded 6,000 years ago. Nevertheless, there has been a more or less continuous onslaught against the natural environment ever since 4000 BP, as witnessed both by documentary evidence and from sedimentological evidence in the form of deltas built up from the products of soil erosion, such as the Guadalfeo delta in south-east Spain on which the town of Motril now stands. In many West Mediterranean areas, the long Moorish presence led to the development of elaborate irrigation systems and the introduction of many of the elements of contemporary agriculture. The two hundred years up to about 1950 were years of agricultural stagnation in Spain, if not in the rest of the Mediterranean, but the environmental onslaught began again with renewed vigour in the postwar period. This reflected the growth in tourism, the mechanisation of agriculture and the steady but inexorable rise in industrial and urban growth that have continued unabated ever since. The economic backwardness of the area led to a no-holds-barred gallop to modernisation of production and the start of the contemporary environmental crises.

The framework of rural change and environmental pressures

When the Mediterranean countries (Greece, Spain and Portugal) joined the European Community in the 1980s, they comprised the major 'Objective 1' regions, that is areas with a per capita GNP significantly below the average for Europe and therefore eligible to receive special subsidies under the Cohesion Fund, aimed at bringing new members up to the standard of living of the rest of the Community.[1] In the same period, Mediterranean farmers received generous support based on the Common Agricultural Policy. Both these factors contributed to the significant increase in pressure for agricultural production and consequent heavy pressures on land and water. Matters were exacerbated by a series of deep droughts which were set against a background of a steady reduction in rainfall since the 1960s in the Western Mediterranean. But, paradoxically, less rain means more intensive rain. In 1973 a very heavy storm affected the south-east of Spain, especially the provinces of Granada, Almería and Murcia. Similar devastating floods have affected other Spanish as well as Moroccan and Italian rivers in recent years. Such events have highlighted the sensitivity of the Mediterranean countries to environmental crises.

In the mid-1970s a new spectre loomed over the horizon in the shape of the threat of global warming. Although the procedure of modelling the reality and the impacts of global warming at the regional scale is fraught with difficulties, especially in an area as complex as the Mediterranean, the best available estimates indicate a reduction in annual precipitation in the next century of about 15 per cent, mainly affecting winter rainfall (Hadley Centre 1992; Palutikoff et al. 1994). When this is coupled with changes in carbon dioxide and temperature in the atmosphere, biological computer simulation models indicate a reduction of shrub and grass vegetative biomass in the order of 50 per cent in those areas having seven months or more without rainfall in the Mediterranean area (Woodward 1994). This is a quite dramatic change which is expected to lead to a marked increase in soil erosion, as we shall see later on in this chapter.

The culmination of all these environmental effects is the acceptance that the Mediterranean area is undergoing desertification (Mairota et al. 1998). In the words of UNEP, such desertification is:

> Land degradation in arid, semi-arid and dry sub-humid areas resulting from various factors, including climatic variations and human activities where 'land' means the terrestrial bio-productive system that comprises soil, vegetation and other biota and the ecological and hydrological processes that operate within the system. 'Land degradation' means the reduction and loss, in arid, semi-arid and dry sub-humid areas, of the biological or economic productivity and complexity of rain-fed cropland, irrigated cropland or range, pasture, forest and woodlands resulting from land uses or from a process or combination of processes arising from human activities and habitation patterns, such as soil erosion, deterioration of the physical, chemical and biological and economic properties of the soil and the long-term loss of vegetation.[2]

Following the Desertification Conference held in Nairobi in 1972, the world has become slowly aware that desertification is a global problem and to this end more than 50 countries ratified the International Convention on Combating

Desertification (ICCD) in 1996. The ICCD has a special Annex which recognises the particular desertification problems in the Mediterranean countries and argues that, unlike African countries, whose problems are regarded as an international responsibility, those of the Mediterranean are essentially the responsibility of the European Community. This is an important political statement that has been slowly accepted by the European Parliament, just at a time when support from the Common Agricultural Policy and the Cohesion Fund is beginning to be withdrawn and a number of new states are about to join the European Union.

Desertification has to be identified as a separate problem from desertion, though the latter is an important contributor to the former. By desertion is meant the loss of rural population to towns and cities and to overseas migration. This process has gone on all over Europe in recent centuries; in the Mediterranean Basin such rural depopulation has been particularly characteristic of mountain regions, as we saw in Chapter 12. In fact the cultural heritage of urbanism is still alive and strong throughout Southern Europe, in contrast to the idealisation of the countryside in the countries of Northern Europe (see Chapter 5; also Leontidou 1990). Massive rural depopulation has been a recurrent theme throughout Mediterranean Europe and a series of crises have led to this, such as the pan-European black death epidemic. There was also local economic change with the creation of institutionalised grazing and the abandonment of villages (Delano-Smith 1979). However, the total population of Europe has doubled since 1850, in spite of massive emigration. These changes in demography, coupled with changes in global markets for Mediterranean goods, have both increased and eased pressures on the environment over the years.

A good preliminary example of the fluctuating global influences on environmental degradation comes from the island of Lesvos in Greece (Bakker *et al.* 1998). The first written documents referring to Lesvos call it *Lasia,* which means 'full of dense forests'. During the Ottoman occupation (fifteenth to nineteenth centuries), economic activities were based on typical forms of Mediterranean agriculture and stock-breeding; olive groves expanded and pastures increased at the expense of woodland and vineyards. In recent years the olive has continued to expand in area. Since olive trees are mainly planted on terraces, often with a terrace for a single tree, this expansion has had important consequences for runoff of water and the erosion of hillslopes. However, as rural depopulation is now taking place, the terraces are breaking down through lack of maintenance. Apparently the Common Agricultural Policy has had little impact on the total olive oil production as indicated by the poor correlation between European Union subsidies and oil production on the island.

The oak, *Quercus macrolepis,* was extensively planted on the island in the past in order to exploit oil from the acorn. Today the area is much restricted in Lesvos and has been largely replaced by intensive grazing; the number of sheep increased steadily and then by 47 per cent between 1976 and 1981 (Kosmas *et al.* 1997). Not only did grazing pressure increase, but burning was also practised to control the spread of thorny shrub. These combined processes have led to intensive erosion of already thin soils, with more rock fragments at the surface and the development of erosion pavements (Bakker *et al.* 1998).

263

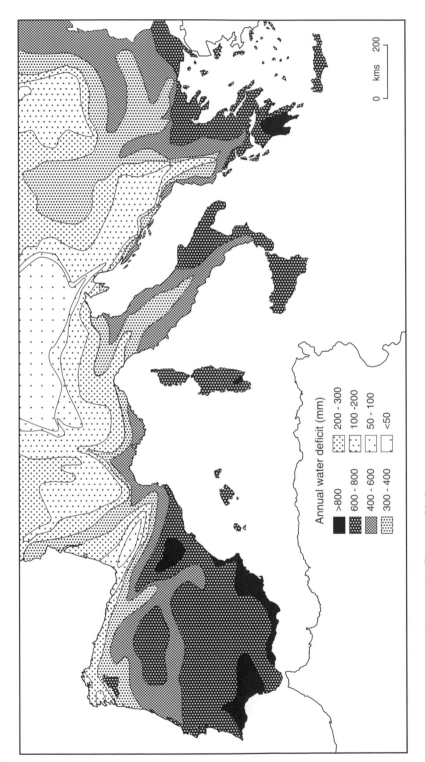

Figure 14.1 Water deficit over the European Mediterranean

These examples from Lesvos illustrate well the sensitivity of Mediterranean agricultural systems to global socio-economic conditions as well as to global climates, and highlight the importance of underlying socio-economic controls on the environmental crises. In the remainder of this chapter, three such contemporary crises are addressed: the water crisis, the land degradation crisis, and the pollution crisis.

The water crisis

Climatic controls on water resources

Figure 14.1 shows the water deficit over the European Mediterranean and is drawn from Thran and Broekhuizen's (1965) *Agro-Climatic Atlas of Europe*.[3] Water deficit is the difference between rainfall and evapotranspiration. This deficit is greater than 800mm in southern Portugal, south-east Spain and the coastal areas of Italy and Greece. Absolute amounts of rainfall vary from 200mm in the south-east of Spain to almost 1000mm in the Apennines of central Italy, reflecting the strong variations over space due to topography.

The second most important water characteristic is its seasonality. The rainfall is essentially in the winter months between November and April. This is reflected in the curve of soil moisture for El Ardal in the Spanish province of Murcia, shown in figure 14.2. Soil moisture is the most important indicator because the water in the soil is available both for plant (and crop) growth and for release into the underlying rocks where it collects as groundwater. The graph shows the weight of water

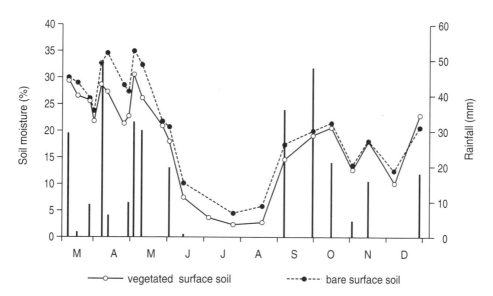

Figure 14.2 Curve of soil moisture under bare and vegetated soil, El Ardal field site, south-east Spain

Source: Data from López-Bermúdez (1993).

per weight of soil as a percentage. The Winter months show high soil moisture until mid-May and early June when there is a dramatic fall to the very low values that persist until the first Autumn rainfalls, usually in October or November. This dramatic change in soil moisture has important consequences for all the physical and biological processes, especially vegetative activities. As the soil moisture peaks in Spring, vegetative growth is strongest. As it falls dramatically, the plants try to cut down water loss by shedding leaves and facultatively deciduous trees and shrubs have a decided advantage because they can get through the Summer without dying.

The third important feature of the water regime is the very high variability from year to year. This can be as much as 625mm one year and 425mm the next. This inter-annual variability is shown in figure 14.3. Not only can a very wet year be followed by a very dry one, but the diagram shows 'runs' of relatively wet and relatively dry years. Runs of excessively wet years charge up water supplies, lead to wet soil, heavy rainfall and flooding. Runs of excessively dry years lead to struggling vegetation, poor crop yields, water level drawdown and drought.

Finally, the rainfall can come in heavy downpours in very short periods. Occasionally 90 per cent of the rain comes in 1 per cent of the time, or a whole year's rain may fall in one or two days. The season when the rain falls is critically important. If it comes in the Spring, when the ground usually has a good green cover, it may cause little damage, but if it comes at the end of Summer and falls on hard-baked bare soils, runoff and erosion can be catastrophic.

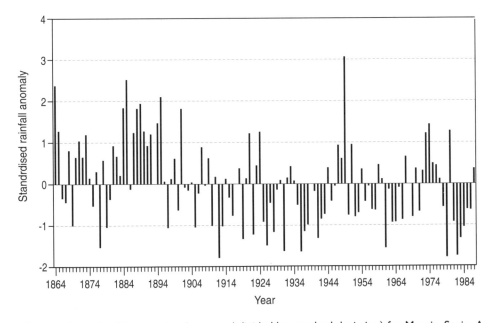

Figure 14.3 Rainfall anomalies (year total divided by standard deviation) for Murcia, Spain. A value of +3 means three standard deviations above the norm
Source: Thornes (1991).

In south-east Spain, intense cyclones normally come from the east (the sea) and are forced up over the mountains. This sets off violent atmospheric instabilities and intense downpours that produce violent floods. As noted earlier, an extremely devastating flood occurred in 1973 (figure 14.4), when a deep cyclone, this time from the west, followed the coast from Málaga to Alicante, shedding up to 250mm of rain in 15 hours on mountain basins. Runoff was swift and heavy, destroying houses, crops, irrigation channels, railways, roads and bridges, and causing landslides and mudflows over a large area (Thornes 1976).

Water and agriculture

The great inter-annual variability in mean annual rainfall means high uncertainty for vegetation and agriculture. Plants transpire water and the processes produce the carbon that comprises stems, leaves and other plant tissues. Roughly speaking, 2 grams of carbon are produced for 1mm of rainfall. This produces a 'boom and bust' cycle in plants. In years of good rainfall, the plants produce a lot of biomass. When a dry year comes, all that biomass has to be sustained by water and the plants are under great stress. In Mediterranean Portugal the yield of wheat (tonnes/hectare) is closely linked to rainfall with a feast-famine type of economy. Because of government policy, farmers were encouraged to plant wheat very extensively, but this proved to be a high-risk strategy when the inter-annual variation was so large.

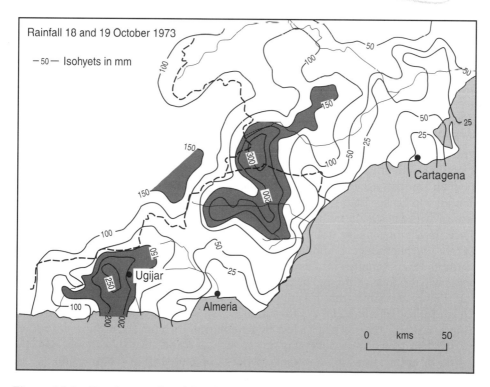

Figure 14.4 Distribution of rainfall in the flood of October 1973 in south-east Spain

Dry farming, without the benefits of irrigation, is generally practised with water-efficient crops such as trees (olives and almonds) and cereals (barley and wheat). The cereals are planted after the Autumn rains and the land is mulched by placing straw on the surface or a tilth of fine soils is produced by harrowing. These procedures limit the loss of water from bare soil evaporation.

But, of course, there is more to drought than rainfall. Some soils can store water more efficiently and for longer than others, so the impact of drought depends on the soil type or other storage capacity. Sometimes large river flows are diverted onto fields (flood irrigation) to soak the soil for plant growth, but this may be a very dangerous practice because it can lead to very severe erosion of terrace systems and deposition of mud and rocks on otherwise very fertile terrace surfaces. Irrigation agriculture, if carefully designed and managed, gives a high resistance to drought, especially where an underground supply is plentiful. Many resourceful methods of irrigation have been used over the millennia in Mediterranean countries. Some represent technology of great antiquity such as the gallery tunnels of brick built in the gravel beds of rivers to collect any percolating flood water. Another is the simple system of lifting water in a continuous bucket-belt often driven by a donkey.

Most recently, large construction works (reservoir dams) have been used to capture water for irrigation and domestic use, or water has been transferred between water-rich basins to water-poor regions. An interesting recent example is the Tajo-Segura Trasvase, over 300km long, which takes water from the head of the Tajo River system in central Spain to the Segura River basin in the dry south-eastern province of Murcia, where it is heavily used for high-value irrigated crops such as oranges and tomatoes. Another example comes from Cyprus, where a major restructuring of the water system took place following the Turkish occupation of the northern part of the island, to ensure the sustainability of agriculture in the southern part. The Southern Conveyor, a large canal, transfers water from the western rivers, draining the Troodos mountains, to the dry plainlands of the east.

One of the largest projected basin transfer schemes in the Mediterranean is the 'great man-made river' of Libya (Margat 1992) which will transfer groundwater from the south of Libya to the coastal north for domestic use, agricultural development and industry. The groundwater in the south is 'fossil' water from periods when rainfall was higher than it is today, so the term 'water mining' is appropriate. The project will deliver 5.5 million m^3/day to the region of Benghazi and the Gulf of Syrte. In the first phase, completed in 1991, two areas of extraction at Sariv and Tazebo each produce 1 million m^3/day. This water is transported through two conduits, each 4 metres in diameter and totalling 1,600km in length. The second phase will tap four additional fields of water in the Djebel Fezzan area and the third and final phase will extract 1.5 million m^3/day from the aquifer south of Koufra.

Careful use of water can ensure sufficiency. Typically, when water is conveyed from reservoirs, there is a large loss by percolation through the bottom of unlined canals. This 'conveyance loss' may be as much as 40 per cent by volume. Also the timing of applications is important. Water supplied early in the morning can be evaporated easily. Modern farmers resolve the problem by using drip or trickle irrigation controlled by computer-driven valves which supply the water when it is most needed by the plant and when it is likely to be most efficiently used. Nutrients are also applied in this manner, giving a very high level of control and efficiency in plant growth.

Water resources – the future crisis

A wide-reaching assessment of water resources in Mediterranean countries was carried out by Margat (1992) under the auspices of the Blue Plan (Mediterranean Action Plan), the programme of economic evaluation of the present and future prospects for the Mediterranean region, organised and sponsored by UNEP. Although now a little dated, it represents the major available data set and resource assessment for water. In the early 1980s (figure 14.5) the highest proportion of water in all countries, except the former Yugoslavia, was for irrigation, whereas industry took only a small proportion. The water demand per person is a most useful indicator of the seriousness of the water resources crisis. Spain had the highest per capita demand (1400 m³) compared with Algeria (about 150 m³) for all uses. Available

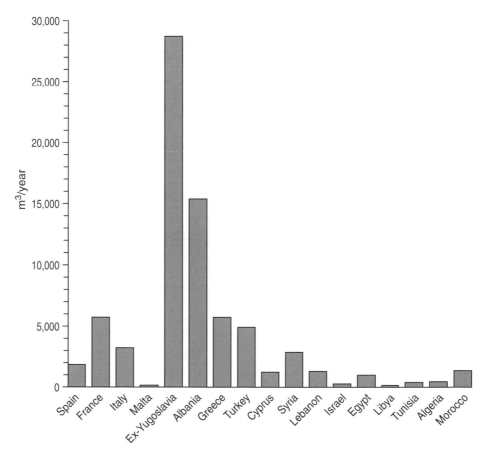

Figure 14.5 Total renewable water resources divided by national population
Source: Based on data from Margat (1992, table 30)

water resources per person were very high in the former Yugoslavia and extremely low in the North African states of Libya, Tunisia, Algeria and Morocco (Margat 1992). Malta, Israel and Libya appear to have the most serious crises ahead, because they have large populations projected for 2025 and the highest population per cubic metre of water currently used per year.

There are two major approaches to resolving these crises. One is to create more and larger structures, such as the Cypriot water system, but the number of possible sites for reservoirs that can provide multi-functional capability (water resources, electric power, flood control) is reducing year by year. The second is to encourage the conservative use of water, for example by pricing policy. Another major issue is the waste of fresh water by sea water intrusion, especially in coastal areas and small islands, where coastal aquifers are heavily pumped. Similarly, if river water is stored in reservoirs, it reduces the fresh water recharge to the aquifers, allowing salt water intrusion to take place. Collin and Barroccu (1998) describe the case of Capoterra, just west of Cagliari in southern Sardinia, where the coastal aquifer is jeopardised by salt water in areas exploited to satisfy the ever-increasing demands from agriculture, industry and domestic use for groundwater. Groundwater is withdrawn from the aquifer through about 300 wells scattered over an area of 60 km^2.

Around Tomelloso in Castilla la Mancha, Spain, groundwater has been falling about one metre per year. Recent declines in groundwater can be attributed to (i) increases in public water consumption for agriculture – the total area under irrigation increased from 159,100 hectares in 1970 to 354,452 hectares in 1989;[4] (ii) reduced inputs into the groundwater system through declining rainfall trends; (iii) landuse changes since 1970 from dryland species (Castilla la Mancha has traditionally used dryland farming techniques) to new crops and varieties with higher yields. In some areas groundwater has fallen by more than 20 metres since 1970. One consequence has been that the important Tablas de Damiel wetlands have decreased in area from 20 to just 1 square kilometre.

Flooding

Flooding occurs when rivers overflow their banks. This is usually caused by heavy rain producing overland flow. This results from changes in the capacity of the soil to take in water because the soil has a lowered infiltration capacity. A good vegetation cover can increase infiltration and the infiltration capacity tends to increase following vegetation regeneration on abandoned land. Changing the density and type of vegetation cover changes the amount and rate of overland flow. Moreover, thinner soils can hold less water, so that land degradation will also increase the likelihood of flooding. The development of calcareous crusts on soils also increases runoff. Tillage has the opposite effect, especially when accompanied by vegetation growth. As the demand for irrigable agricultural land close to water has increased, river floodplains have been encroached upon more and more for agriculture and this can be dangerous. For example in Spain the behaviour of *ramblas*, wide flat-floored gravel-filled valleys, is quite unpredictable. This is because intense storms are of small spatial extent, so that only a small area of a large catchment might be affected at any one time. The tributaries draining the sub-catchment provide a flood wave to the main channel at different times, so that there may be

Table 14.1   The most important floods in Spain

1651	Murcia	Serious damage	1000 died
1779	Coastal part of Jucar (Valencia)	Important damage	no data
1802	Lorca (Murcia)	Destruction of the city	700 died
1874	Catalonia	700 houses destroyed	600 died
1879	Murcia	Great damage	800 died
1891	Consuegra (Toledo)	Partial destruction of municipality	359 died
1957	Valencia	300 buildings destroyed 10,000 million pesetas damages (1957 values)	86 died
1962	Catalonia	5,000 houses destroyed Losses of 2,700 million pesetas	1000 died
1972	Valdepeñas (Ciudad Real)	Important damage	22 died
1973	South-east Spain (Málaga, Granada, Almería, Murcia)	Great damage over a large area	38 died
1982	Levante (Valencia)	Great damage Losses of 300,000 million pesetas	38 died
1983	Basque Country and Cantabria	Losses of 500,000 million pesetas	40 died
1989	Málaga and the south-east	Losses of 200,000 million pesetas	42 died
1997	Alicante and the south-east	Losses of 10,000 million pesetas	5 died

Source: El País, 26 November 1997.

several waves of water moving down the trunk stream. All these characteristics were seen in the 1973 flood in south-east Spain, which cannot be matched in living memory, though historical records tell of other great floods such as the one in the sixteenth century in Lorca (Gil Olcina 1976).

A report from the Spanish national newspaper *El País* (26 November 1997) lists the major flood disasters, damage and losses of life for the largest floods in history (table 14.1). The newspaper reports that, between 1971 and 1996, the consortium of insurance companies in Spain paid out 439 billion pesetas for flood damage. This enormous amount went mainly to the Mediterranean areas of the country, stretching from Catalonia to Almería. Indeed a project to examine urban areas threatened by flood found that 197 of the 592 reaches identified were in the province of Murcia, mainly in the Segura River basin.

Political aspects of water resources development
Water problems are not only national, many are trans-boundary, for example between Spain and Portugal, or between Turkey, Syria and Iraq, or Israel and the West Bank. Spain now uses about 40 per cent of all its renewable water resources (twice the EU average), though water resources in the Iberian Peninsula are not as seriously under pressure as those in the Middle East or North Africa, where water

demand exceeds 90 per cent of the renewable supply (Wolf 1995; World Bank 1995). According to Llamas (1997), problems in the management of Portuguese-Spanish rivers were an important issue in the political reforms of all political parties in the 1995 general elections in Portugal, but were never mentioned in the 1996 Spanish general elections. This reflects the fact that Portugal's resources are 'downstream' from Spain. Nevertheless, Llamas concludes that the current water problems between the two countries can probably be resolved between the two countries without the help of international arbitration.

Table 14.2 Sectoral use of water in the Iberian Peninsula (km³/year)

	Portugal	Spain
Urban water supply	0.5–1	4
Industry	1–3	2
Agriculture	5–8	25
Total	7–12	31
Irrigation requirement	0.95	0.75

Source: Llamas (1997).

The per capita annual water availability in Spain (3000m³/person/year) is rather high compared with other European countries (Barraque 1995). In Portugal the water availability is more than twice as high. If all Spanish rivers flowing into Portugal went dry, Llamas calculates that Portugal's internal per capita water resources would be the same as Spain's. But, of course, in both countries, the available resources are not where they are most needed, nor is the sectoral use the same in both countries, as table 14.2 shows. The figures are in cubic kilometres per year. Agriculture appears as the dominant water use, though the amounts needed for irrigation are small, but higher in Spain than in Portugal.

Llamas is less optimistic about the conflicts between different Spanish regions. He calls 'hydromyths' the notion that hydrological salvation can only be achieved by constructing great hydraulic works to transfer water from the wet northern and western areas of Spain to the dry central and eastern areas. This approach is the gist of the latest Spanish Hydrological Plan (Llamas 1995). Llamas (1997) concludes that 'water wars' in the Iberian Peninsula 'have not been caused, in the main, by water scarcity or serious quality degradation, but by mismanagement'.

Spain's first operational water law was established in 1879 and revised under the new *Ley de Aguas* of 1985. The new law put groundwater as a public rather than a private property and required that a water plan be developed for each watershed to allocate water rights, to establish water quality and put the emphasis on basin water management, like the schemes in other European countries. Llamas (1997) points out that, 11 years after the enactment of the new water law, no basin water plan has been approved (as required) by the government. So progress is slow, despite the supposed urgency and crisis character of the water resources problem.

One of the most important water transfer schemes within Spain, the Tajo-Segura canal already referred to, has not been without its problems. Originally it was supposed to transfer 1km³/year from the Tajo to the Segura, but this has been reduced over time to 0.15 km³. In the difficult year of 1995, under pressure from the south-east, the transfer was stepped up again, reducing the amount in the Tajo at Aranjuez by 50 per cent and causing ecological problems and damage to the recreational quality of the river, which is critical to the tourist industry.

The Spanish Water Plan of 1993 proposed a programme to build about 200 more dams in Spain and to transfer almost 4km³/year from water-rich basins to water-deficient ones. However, this plan was widely opposed and has not been approved. As in the United Kingdom, the provision of more storage capacity by dam construction is highly debated, not least as a result of pressure from ecological groups. It is often argued, with good reason, that it is more economic to make more efficient use of existing water through new technology (such as irrigation) or by manipulating the cost of water. Most users regard water as a free good and, since water consumption information is poor, it is hard to determine the true cost. By pumping groundwater, river flows into reservoirs are reduced and so hydropower generation may be threatened. Llamas (1997) quotes the case of the Jucar River which drains the eastern part of Castilla la Mancha, where the hydro-electric company, Iberdola, has calculated that each cubic metre of water extracted results in a loss of one kilowatt-hour of electricity generation.

Llamas' appraisal of water exploitation and scarcity is well-informed without being unnecessarily doom-laden and without eliciting the term 'crisis' at every possible moment. Such a balanced appraisal also needs to be adopted in relation to water quality where, again, there is a determination to view crises and to exaggerate the issues.

The land degradation crisis

Land degradation, defined earlier in this chapter, is mainly caused by loss of soil through soil erosion – a complex amalgam of many processes about which many books and papers have been written. In semi-arid environments, soil erosion by overland flow is the most important and this includes sheet wash and rill and gully erosion. The collapse of underground natural piping is also a key process. The relative dominance of these processes depends on rainfall intensity, vegetation cover and soil type. The process and principles are well-covered in standard texts such as Morgan (1980) and Kirkby and Morgan (1986) and will not be dealt with further here. Gully erosion often produces dramatic badlands, especially on easily-eroded lithologies such as marls, which are widespread throughout the Mediterranean as a result of the former greater extent of the Mediterranean Sea in Tertiary times. These badlands are often cited as evidence of an intense erosion crisis and are frequently used to illustrate the dramatic impact of erosion. There is a need for caution here because detailed fieldwork suggests that some of these badlands may be as much as 4,000 or more years old and could even be relatively low in sediment production today. They certainly should not be regarded as contemporary products of current climatic and/or socio-economic conditions (Wise *et al.* 1982).

Langbein and Schumm (1958) demonstrated a strong curve of erosion (figure 14.6), indicating that there is a peak erosion rate in areas with an 'effective' annual precipitation of 200–300mm. 'Effective' precipitation is standardised for temperature. With higher annual temperatures, the 'effective' rainfall is less than the actual. At lower 'effective' rainfalls than in the Mediterranean peak (200–300mm), the runoff reduces and so too does the erosion until, in deserts, sediment yield from runoff erosion is very low. If the effective rainfall is higher than the Mediterranean peak, this encourages more vegetation cover so, again, sediment yields fall. Data collected by Kosmas *et al.* (1997) from experimental plots throughout the Mediterranean appear to confirm this curve quite well for runoff and sediment yield (figure 14.7). In general it is confirmed that, as the percentage of vegetation cover increases, the erosion rate compared with bare soil decreases dramatically so that, with approximately 30 per cent cover, the erosion rates fall to 70 per cent of the bare soil value. Further increase in percentage vegetation cover decreases soil erosion rates only very slowly. These curves have to be qualified by rainfall intensity: with higher intensity the curve shifts towards the upper right, so greater cover is needed to effect the same reduction in erosion (Francis and Thornes 1990). Figure 14.8

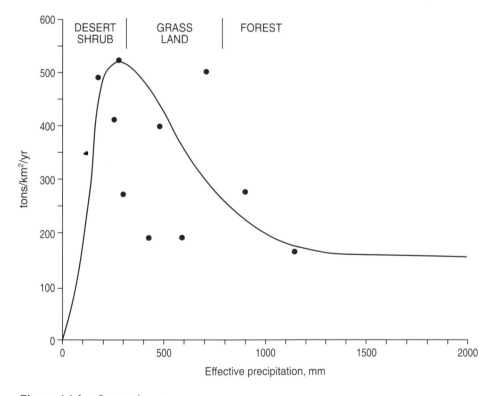

Figure 14.6   Curve of erosion

demonstrates this modification. In addition, the soil depth modifies this curve. Deeper soils can store more water so runoff rates are lower (Poesen and van Waesemael 1995).

What are the controls of land degradation in Mediterranean environments? We have shown that rainfall intensity and vegetation cover are the two main controls on erosion. Rainfall intensity is highest in the driest areas of the Mediterranean and vegetation cover is lowest, so the FAO map of erosion, based on both these criteria and on data from the CORINE European database, indicates that the Iberian, Italian and Greek peninsulas are the most eroded areas of Europe (Commission of

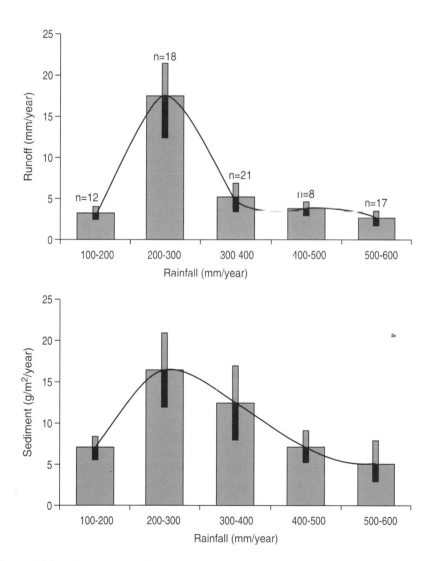

Figure 14.7 Curves of runoff and sediment yields plotted against rainfall for MEDALUS sites
Source: Kosmas *et al.* (1997).

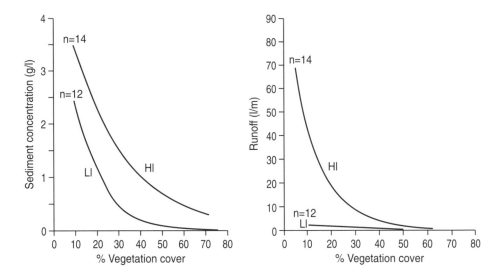

Figure 14.8 Relation of sediment concentration and runoff to vegetation cover under high-intensity (HI) and low-intensity (LI) rainfall conditions
Source: Francis and Thornes (1990).

the European Communities 1992)). This picture is compounded by soil erodibility and, as mentioned above, the Pliocene marls (a mixture of silt and calcium) are the most erodible soils. Measured rates of erosion in the Mediterranean are nevertheless small by international standards and, except in badlands, do not support a 'crisis' view, except that the affected areas form a large proportion of the available arable land. These data have to be treated with some suspicion because they are difficult to measure and easy to exaggerate. A major future problem for the European Union will be to estimate the real consequences of land degradation in terms of the cost to farmers if the commitments under the International Convention on Combating Desertification are to be met through agricultural subsidies or engineering works to mitigate the problem. Another highly contentious problem will be how the money should be spent. There are, generally, two approaches to this problem. The first is to attempt to solve the problem by engineering structures and this almost invariably results in terrace construction by excavation. Terrace structures, with stone fronts, are traditional in the Mediterranean but, to be effective, they need to be maintained. With the loss of population from rural areas, this is often one of the first things to be neglected. The second approach is to reduce runoff of water and sediment by planting an exotic or semi-natural vegetation cover. Afforestation is often seen as the panacea to erosion problems and, if carried out properly, can be very effective; however, it is neither quick nor cheap. In these environments, the establishment of trees requires irrigation and careful nursing of immature plants (Francis and Thornes 1990). One of the major factors in past degradation has been overgrazing. Careful animal husbandry, the exclusion of sheep and goats in particular, may be sufficient to allow the regeneration of the natural shrub vegetation. Francis and

Thornes (1990) showed that a shrub cover of the *matorral* or *maquis* species *Anthyllis cytisoides* may provide as much protection as a young forest of Aleppo Pine (*Pinus halepensis*) and has other advantages for restoration.

The pollution crisis

In common with the rest of Western Europe, heavy and often excessive fertiliser applications commenced with the pressures on agricultural production in the 1950s. The essential differences in the Mediterranean arise from the physical conditions: runoff is high and fertilisers, herbicides and pesticides can be readily 'mobilised' by overland flow and from there carried into rivers and other water courses and eventually to the Mediterranean Sea. Environmental awareness and consciousness came late to the Mediterranean, so legal arrangements were, and still are, rather less strict.

Moreover the flux into the Mediterranean Sea has been estimated to be nearly three times greater from the northern shore countries than from those on the southern shore, reflecting mainly the economic contrasts in the environments. France, Italy and Egypt lead strongly in terms of discharges of recycled water to rivers; in Italy and Egypt this recycled water is mainly from irrigation; in France it is cooling water from power stations. Considering the whole Mediterranean Sea, the River Po is believed to make the single biggest input of contaminating water. Concentrations of copper, nickel, zinc and cadmium are high in the northern Adriatic, compared with the southern Adriatic and the rest of the Mediterranean. Conditions are worsened by the Adriatic shelf which means that the contaminants are not lost to ocean deeps as they are elsewhere. As a consequence, copper and nickel concentrations are four times higher than in Mediterranean open waters, but levels fall off very rapidly from the mouth of the river (Lipiatou 1997).

Individual rivers show the usual expected variations and again it is from the River Segura (south-east Spain), one of the most scientifically studied smaller rivers in Europe, that most information is available, though it is certainly not the worst case in the Iberian Peninsula. The River Nervión, draining the industrialised mountain valleys of the Basque lands around Bilbao (and flowing into the Atlantic), and the River Vinalopo, draining the district of leather industry near Alicante, have their own special problems.

The River Segura drains an area of intensive irrigated fruit culture. The accompanying fruit-processing industry makes heavy demands on water and return water is large in quantity and often poor in quality. According to data in Egea Ibáñez (1995) the following volumes of water are needed to produce one tonne of fruit (including irrigation and processing needs): apricots 20,000 litres, pears 13,000 litres, peaches 12,000 litres, grapefruit 4,000 litres, tomatoes (whole) 2,200 litres. The result is significant pollution of the river from near its source at Cieza to its outlet to the sea at Guardamar.

The quality of river water in the tributaries of the River Segura has gone down significantly and consistently since 1991. Taking an index of 100 as high-quality water and values less than 60 as unacceptable, the index has had the following annual values over the years since 1991: 58, 52, 47 and 45. This is for the River Argos, but conditions in another tributary, the River Mula, are no better, with its large number of jam-making factories. Only a periodic flooding of great intensity pushes the stagnant pools into the sea.

There are two other problems in these semi-arid environments. The first is the pollution of the aquifers under the river beds that are used for drinking water and irrigation. Another problem is the build-up of fertilisers, herbicides and pesticides in soils with little vertical downward movement or lateral movement of water. The 'chemical time-bombs' have been building up since the mid-1950s and only severe erosion carries away this dangerous material if it can absorb soil particles, but of course the soil is then re-deposited elsewhere. This phenomenon has only been poorly studied so far. It is reasonable to suppose that there is a real crisis around the corner for which remedial action could and should now be taken by regional authorities.

Concluding remarks

In this review, the main purpose has been to dispel the concept of the Mediterranean as a 'Garden of Eden'. The Mediterranean Basin is a tough and very uncertain environment in which there is a constant battle of humankind against the natural forces. At the same time an attempt has been made in this chapter to defuse the notion of a crisis, actual or imminent, by pointing out that remedial measures are at hand and are to be sought in the Mediterranean environment itself. The situation is most critical with regard to water, not least because with this resource the main problems in the future are going to be in the North African and Middle Eastern sectors of the Mediterranean, rather than the northern shore countries. The view of desertification as a 'myth' is not sustained however, and a transnational effort for its mitigation is appropriate, starting with the establishment of the requisite political and technical machinery, which is regrettably not yet in place.

Notes
1   Much of Southern Italy also received 'Objective 1' status.
2   UNEP (United Nations Environment Programme), *United Nations Convention to combat desertification in those countries experiencing serious drought and/or desertification, particularly in Africa*. New York: United Nations, 1994.
3   The steady decrease in mean rainfall noted earlier probably means that the deficits shown in figure 14.1 are now somewhat more severe.
4   According to the *Censo Agrario de España*, Madrid: Instituto Nacional de Estadística, 1990.

References
Bakker, M. *et al.* (1998) *MEDALUS III Meeting Lesvos 24–28th April 1998, Field Trip Guide*. Thatcham: MEDALUS.
Barraque, B., ed. (1995) *Les Politiques de l'Eau en Europe*. Paris: La Découverte.
Collin, J. and Barroccu, G. (1998) The increasing demand for water, in Mairota, P., Thornes, J. B. and Geeson, N. (eds) *Atlas of Mediterranean Environments in Europe: the Desertification Context*. Chichester: Wiley, pp. 98–101.

Commission of the European Communities (1992) *CORINE Soil Erosion Risk and Important Land Resources in the Southern Regions of the European Community.* Luxembourg: Office for Official Publications of the European Communities, EUR 13233.

Delano-Smith, C. (1979) *Western Mediterranean Europe: a Historical Geography.* London: Academic Press.

Egea Ibáñez, J. (1995) La evolución del uso del agua y la agricultura del futuro, in Senent Alonso, M. and Cabezas Calvo Rubio, F (eds) *Agua y Futuro en la Región de Murcia.* Murcia: Asamblea Regional de Murcia, pp. 209–20.

Francis, C. F. and Thornes, J. B. (1990) Runoff hydrographs from three Mediterranean cover types, in Thornes, J. B. (ed.) *Vegetation and Erosion.* Chichester: Wiley, pp. 363–85.

Gil Olcina, J. (1976) *Inundaciones en la Ciudad y Termino de Alicante.* Alicante: Instituto Universitario de Geografía.

Gilman, A. and Thornes, J. B. (1972) *Land Use and Prehistory in Southeast Spain.* London: Allen and Unwin.

Hadley Centre (1992) *The Hadley Centre Transient Climate Change Experiment.* London: The UK Meteorological Office.

Kirkby, M. J. and Morgan, R. P. C., eds (1986) *Soil Erosion.* Chichester: Wiley.

Kosmas, K. *et al.* (1997) The effect of land use on runoff and soil erosion rates under Mediterranean conditions, *Catena*, 29(1), pp. 45 59.

Langbein, W. B. and Schumm, S. A. (1958) Yield of sediment in relation to mean annual precipitation, *American Geophysical Union Transactions*, 39, pp. 1076–84.

Leontidou, L. (1990) *The Mediterranean City in Transition. Social Change and Urban Development.* Cambridge: Cambridge University Press.

Lipiatou, E. (1997) *Interdisciplinary Research in the Mediterranean Sea. A Synthesis of Scientific Results from the Mediterranean.* Brussels: Commission of the European Communities, Targeted project (MTP) Phase I 1993–96, EUR 17797.

Llamas, M. R. (1995) La crisis del agua, mito o realidad *Atti dei Convegni Lincei*, 114, pp. 107–15.

Llamas, M. R. (1997) Transboundary water resources in the Iberian Peninsula, in Gleditsch, P. *et al.* (eds) *Conflicts and the Environment.* Amsterdam: Kluwer Academic Publishers, pp. 335–53.

López Bermúdez, F. (1993) Field site integration, data base management desertification response units, in *First Annual Report, MEDALUS II Project.* Brussels: Commission of the European Community DGXII, pp. 24–6.

Mairota, P., Thornes, J. B. and Geeson, N. (1998) *Atlas of Mediterranean Environments in Europe: the Desertification Context.* Chichester: Wiley.

Margat, J. (1992) *L'Eau dans le Bassin Méditerranéen.* Paris: Economica.

Morgan, R. P. C. (1980) *Soil Erosion and Conservation.* London: Longman Scientific and Technical.

Palutikoff, J., Goodess, C. M. and Guo, X. (1994) Climate change, potential evapotranspiration and moisture availability in the Mediterranean Basin, *International Journal of Climatology*, 14, pp. 853–69.

Poesen, J. and van Waesemael, B. (1995) Effects of rock fragments on the structure of tilled topsoil during rain, in Derbyshire, E., Dikstra, T. and Smalley, I. J. (eds) *Genesis and Properties of Collapsible Soils.* Dordrecht: Kluwer Academic Publishers, NATO Advanced Studies Institute Series, Series C: Mathematical and Physical Sciences, 486, pp. 333–43.

Thornes, J. B. (1976) *Semi-Arid Erosional Systems.* London: LSE Department of Geography, Special Publication No.7.

Thornes, J. B. (1991) Environmental change and hydrology, in Giraldez, J. V. (ed.) *III Simposio sobre el Agua en Andalucía, Vol. 2.* Madrid: Instituto Tecnológico GeoMinero de España, pp. 555–70.

Thran, P. and Broekhuizen, S., eds (1965) *Agro-Climatic Atlas of Europe.* Wageningen: Centre for Agricultural Publications.

Wise, S. J., Thornes, J. B. and Gilman, A. (1982) How old are the badlands? A case study from south-east Spain, in Yair, A. and Bryan, R. (eds) *Piping and Badland Erosion.* Norwich: Geo Abstracts, pp. 29–56.

Wolf, A. T. (1995) International dispute resolution: the Middle Eastern Multilateral Working Group in Water Resources, *Water International*, 20(3), pp. 141–50.

Woodward, I. (1994) Vegetation modelling, in *Third Interim Report for MEDALUS II.* Brussels: Commission of European Community DGXII Section 2.02, pp. 61–5.

World Bank (1995) *From Scarcity to Security: Averting a Water Crisis in the Middle East and North Africa.* Washington DC: World Bank.

Guide to further reading

The literature on the geography of the Mediterranean as a macro-region is surprisingly limited; until very recently, it seemed as though geographers had lost interest in the region. The last few years, however, have seen a flurry of activity, including some significant new books published during the final stages of the production of the present volume. In the guidance notes which follow, emphasis is placed on books rather than articles, although a list of the most relevant journals is appended. Priority is given to works by geographers, although some important books by other scholars such as historians and anthropologists are included where they have special importance and to reflect the fact that some of the best work on the Mediterranean region has been produced from an interdisciplinary perspective. Most of the literature referred to is in English, although there are a few key references in other languages.

Amongst classic texts on the Mediterranean, pride of place must still go to the two-volume epic by Fernand Braudel, also available in an abridged single-volume edition:

Braudel, F. (1972, 1973) *The Mediterranean and the Mediterranean World in the Age of Philip II*, London: Collins, 2 vols (shorter illustrated version published by Book Club Associates, 1992).

Two more recent books which may eventually come to enjoy a similar status as *tours de force* are:

Horden, P. and Purcell, N. (2000) *The Corrupting Sea: A Study of Mediterranean History*, Oxford: Blackwell.

Matvejvic, P. (1999) *Mediterranean. A Cultural Landscape*, Berkeley: University of California Press.

Curiously, none of the authors of the above three studies is a geographer, but their books, all of which have a broad sweep, encompass geography, economic and social history, ecology, cultural studies and anthropology in a rich interdisciplinary tapestry which can only inspire the enthusiastic student of Mediterranean geography. Amongst books of a more conventional geographical mien, the following are the most wide-ranging and recent:

Conti, S. and Segre, A., eds (1998) *Mediterranean Geographies*, Rome: Società Geografica Italiana (Geo-Italy vol. 3).

King, R., Proudfoot, L. and Smith, B., eds (1997) *The Mediterranean: Environment and Society*, London: Arnold.

From an earlier era, when regional description was an acceptable paradigm, several texts can be mentioned as elementary (and now dated) introductions:

Beckinsale, M. and R. (1975) *Southern Europe, the Mediterranean and Alpine Lands*, London: University of London Press.

Branigan, J. J. and Jarrett, H. R. (1975) *The Mediterranean Lands*, London: MacDonald and Evans.

Robinson, H. (1970) *The Mediterranean Lands*, London: University Tutorial Press.

Semple, E. C. (1932) *The Geography of the Mediterranean Region: Its Relation to Ancient History*, London: Christophers.

Siegfried, A. (1949) *The Mediterranean*, London: Cape.

Walker, D. S. (1965) *The Mediterranean Lands*, London: Methuen.

Two geographies which adopt a more sophisticated, cultural landscape approach (although confined to the Western Basin) are:

Delano Smith, C. (1979) *Western Mediterranean Europe: A Historical Geography of Italy, Spain and Southern France since the Neolithic*, London: Academic Press.

Houston, J. M. (1964) *The Western Mediterranean World: An Introduction to its Regional Landscapes*, London: Longman.

Important general works in other languages include:

Bellicini, L., ed. (1995) *Mediterraneo. Città, Territorio, Economie alle Soglie del XXI Secolo*, Rome: Credito Fondiario.

Birot, P. (1964) *La Méditerranée et le Moyen Orient*, Paris: Presses Universitaires de France (earlier 2-volume edition by P. Birot and J. Dresch, 1953, 1956).

Isnard, H. (1973) *Pays et Paysages Méditerranéens*, Paris: Presses Universitaires de France.

Ribeiro, O. (1983) *Il Mediterraneo. Ambiente e Tradizione*, Milan: Mursia (a much earlier version of this book was published in Portuguese: *Mediterrâneo. Ambiente e Tradição*, Lisbon: Fundação Calouste Gilbenkian, 1968).

So much for general texts on the geography of the Mediterranean which encompass all, or at least a major part, of the region. On environmental issues, the past decade or so has seen a rush of books, some of them of the coffee-table genre, but all worth dipping into in view of the ecological sensitivity of the region and the massive environmental pressures that future growth will bring. Hence, in addition to useful sequences of chapters in King *et al.* (1997), chs 2–4, 15–17 and in Conti and Segre (1998), chs 2–3, 12–14 (full references given above), see:

Attenborough, D. (1987) *The First Eden: the Mediterranean World and Man*, London: Collins/BBC Books.

Brandt, C. J. and Thornes, J., eds (1996) *Mediterranean Desertification and Land Use*, Chichester: Wiley.

Conacher, A. J. and Sala, M., eds (1998) *Land Degradation in Mediterranean Environments of the World*, Chichester: Wiley.

Grenon, M. and Batisse, M. (1989) *Futures for the Mediterranean Basin: the Blue Plan*, Oxford: Oxford University Press.

Haas, P. M. (1990) *Saving the Mediterranean: the Politics of International Environmental Cooperation*, New York: Columbia University Press.

Jeftic, L., Milliman, J. D. and Sestini, G., eds (1992) *Climatic Change in the Mediterranean, Vol. 1*, London: Arnold.

Jeftic, L., Keckes, S. and Pernetta, J., eds (1996) *Climatic Change in the Mediterranean, Vol. 2*, London: Arnold.

Mairota, P., Thornes, J. B. and Geeson, N., eds (1998) *Atlas of Mediterranean Environments in Europe: the Desertification Context*, Chichester: Wiley.

Manzi, E. and Schmidt di Friedberg, M., eds (1999) *Landscape and Sustainability, Global Change, Mediterranean Historic Centres*, Milan: Guerini (Geo & Clio vol. 4).

Margalef, R., ed. (1985) *Western Mediterranean*, Oxford: Pergamon Key Environments Series.

Pastor, X., ed. (1991) *The Mediterranean*, London: Collins and Brown for Greenpeace.

Thirgood, J. V. (1981) *Man and the Mediterranean Forest*, London: Academic Press.

Geographers' interests in the economic development of the Mediterranean fall into two approaches: an earlier period – late 1970s to mid-1980s – when the Mediterranean (or, more specifically, Southern Europe) was viewed as a periphery or semi-periphery of the North European core; and the more recent upsurge of interest in the Mediterranean as an economic divide between Europe and North Africa, culminating in the Euro-Mediterranean partnership launched by the EU in 1995. For geographical literature from the former period see:

Hadjimichalis, C. (1987) *Uneven Development and Regionalism: State, Territory and Class in Southern Europe*, London: Croom Helm.

Hudson, R. and Lewis, J., eds (1985) *Uneven Development in Southern Europe: Studies in Accumulation, Class, Migration and the State*, London: Methuen.

Seers, D., Schaffer, B. and Kiljunen, M. L., eds (1979) *Underdeveloped Europe: Studies in Core–Periphery Relations*, Hassocks: The Harvester Press.

Williams, A. M. (1984) *Southern Europe Transformed: Political and Economic Change in Greece, Italy, Portugal and Spain*, London: Harper and Row.

And on the more recent period:

Gillespie, R., ed. (1997) *The Euro–Mediterranean Partnership: Political and Economic Perspectives*, London: Frank Cass.

Joffé, G., ed. (1999) *Perspectives on Development: the Euro–Mediterranean Partnership*, London: Frank Cass.

Ludlow, P., ed. (1994) *Europe and the Mediterranean*, London: Brassey's.

Pierros, F., Meunier, J. and Abrams, S. (1999) *Bridges and Barriers: The European Union's Mediterranean Policy, 1961–1998*, Aldershot: Ashgate.

Reiffers, J.-L., ed. (1997) *La Méditerranée aux Portes de l'An 2000*, Paris: Economica.

Van Oudenaren, J., ed. (1996) *Employment, Economic Development and Migration in Southern Europe and the Maghreb*, Santa Monica: RAND.

In addition to economic relations, most of the above group of books also deal with questions of migration, security and geopolitics. Specifically on trans-Mediterranean migration see:

King, R. and Black, R., eds (1997) *Southern Europe and the New Immigrations*, Brighton: Sussex Academic Press.

King, R., Lazaridis, G. and Tsardanidis, C., eds (2000) *Eldorado or Fortress? Migration in Southern Europe*, London: Macmillan.

Tourism is another form of human spatial mobility – short-term, mostly originating from Northern Europe, but with a tendency to become longer-term as more and more northerners are attracted to live in the 'sunny south'. The literature on tourism and travel in the Mediterranean spans a range of approaches: travelogues, historical studies of the Grand Tour and the nineteenth century, and more conventional academic studies of tourism. Of the many dozens of travel books on the region the following are a brief sample, offering diverse representations of the Mediterranean at different points in time:

More, J. (1956) *The Mediterranean*, London: Batsford.

Newby, E. (1984) *On the Shores of the Mediterranean*, London: Harvill Press.

Theroux, P. (1995) *The Pillars of Hercules: A Grand Tour of the Mediterranean*, London: Hamish Hamilton.

The historical geography of European tourism to the Mediterranean is covered in contrasting ways by:

Aldrich, R. (1993) *The Seduction of the Mediterranean: Writing, Art and Homosexual Fantasy*, London: Routledge.

Pemble, J. (1988) *The Mediterranean Passion: Victorians and Edwardians in the South*, Oxford: Clarendon.

Reynolds-Ball, E. (1914) *Mediterranean Winter Resorts*, London: Kegan, Paul, Trench, Trüber and Co.

For more conventional geographical studies of contemporary tourism, with special reference to mass tourism, Spain and the increasingly popular phenomenon of international retirement to the Mediterranean, see:

Barke, M., Towner, J. and Newton, M. T., eds (1996) *Tourism in Spain: Critical Issues*, Wallingford: CAB International.

Jenner, P. and Smith, C. (1993) *Tourism in the Mediterranean*, London: Economist Intelligence Unit.

King, R., Warnes, A. and Williams, A. M. (2000) *Sunset Lives: British Retirement Migration in the Mediterranean*, Oxford: Berg.

Williams, A. M. and Shaw, G., eds (1998) *Tourism and Economic Development: European Experiences*, Chichester: Wiley, chs 1–6.

On urbanism, urbanisation and other aspects of Mediterranean cities, see:

Cortesi, G., ed. (1995) *Urban Change and the Environment: the Case of the North-Western Mediterranean*, Milan: Guerini (Geo & Clio vol. 3).

Leontidou, L. (1990) *The Mediterranean City in Transition: Social Change and Urban Development*, Cambridge: Cambridge University Press.

And on Mediterranean mountains and islands:

Carli, M. R., ed. (1994) *Economic and Population Trends in the Mediterranean Islands*, Naples: Edizioni Scientifiche Italiane (Collana Atti Seminari 5).

Kolodny, E. (1974) *La Population des Iles de la Grèce*, Aix-en-Provence: Edisud.

McNeil, J. (1992) *Mountains of the Mediterranean: an Environmental History*, Cambridge: Cambridge University Press.

Finally, a note on some journals. The only periodical devoted to the study of Mediterranean geography is the French journal *Méditerranée*, which has been published since 1960. Other journals, which mix some geographical material with anthropology, history, politics, economics etc., include *Journal of Mediterranean Studies*, *Journal of North African Studies*, *Mediterranean Historical Review*, *Mediterranean Politics*, *Peuples Méditerranéens* and *South European Society and Politics*.

Index

Note: Information in notes is shown in the form 58n7, i.e. note 7 on page 58